# GABRIEL HARVEY

His Life, Marginalia
and Library

# GABRIEL HARVEY

## His Life, Marginalia
## and Library

BY

VIRGINIA F. STERN

OXFORD
AT THE CLARENDON PRESS
1979

*Oxford University Press, Walton Street, Oxford* OX2 6DP

OXFORD LONDON GLASGOW
NEW YORK TORONTO MELBOURNE WELLINGTON
KUALA LUMPUR SINGAPORE HONG KONG TOKYO
DELHI BOMBAY CALCUTTA MADRAS KARACHI
NAIROBI DAR ES SALAAM CAPE TOWN

*Published in the United States by*
*Oxford University Press, New York*

**British Library Cataloguing in Publication Data**
Stern, Virginia F.
  Gabriel Harvey
  1. Harvey, Gabriel – Biography  2. Poets, English –
Early modern, 1500–1700 – Biography
821'.3  PR2288  78-40241
ISBN 0-19-812091-5

*Printed in Great Britain*
*at the University Press, Oxford*
*by Eric Buckley*
*Printer to the University*

TO
ELIZABETH STORY DONNO

# PREFACE

MY interest in Gabriel Harvey began some ten years ago when on a visit to the British Museum I first handled some of his fascinating marginalia volumes. But when I sought to learn more about this man with the beautiful penmanship and amazing annotating habits, I found insufficient information available to satisfy my curiosity. The only biographical sketch of value was the groundbreaking but all too brief one of G. C. Moore Smith prefacing the 1913 edition of selected marginalia transcriptions.[1] Harold S. Wilson had in 1948 published a helpful article on Harvey's methods of annotation,[2] but, this too was incomplete and I was curious to find out more.

Harvey seemed to me unique as a marginalist. Other Renaissance men (as, for example, the printer Henri Estienne, Ben Jonson, and John Milton) sometimes annotated their books more or less extensively, but none did so with the abundance, variety, the artistry, and the consistency of Harvey. And as I examined more and more of his annotated volumes, I became impressed with what must have been the size and scope of his library. I realized, too, that there was need for a new biography of Harvey to supplement the sketch of Moore Smith and to incorporate material not then available or not included by him. Thus interest in Harvey's library and in the man who had so assiduously collected and annotated the volumes prompted me to investigate further.

As John Lievsay has so aptly said, 'Nowadays no piece of literary research can be anything other than, directly or indirectly, a piece of intricate collaboration.'[3] Especially is this true of a survey of Harvey's library and marginalia. Based on widely scattered sources and touching on multifarious disciplines this enterprise has been dependent upon the help of many scholars.

In addition to my obvious debt to such bibliographers as W. Carew Hazlitt,[4] G. C. Moore Smith, and Harold S. Wilson, who laid the foundations for work on Harvey, and to the helpful staffs of the many libraries

[1] *Gabriel Harvey's Marginalia*, collected and edited by G. C. Moore Smith, Stratford-upon-Avon, Shakespeare Head Press, 1913.

[2] 'Gabriel Harvey's Method of Annotating His Books', *HLB*, ii (1948), 344–61.

[3] *Stefano Guazzo and the English Renaissance (1575–1675)*, Chapel Hill, Univ. of N. Carolina Press, 1961, p. lx.

[4] Chapter on 'Gabriel Harvey' in part xiii of Bernard Quaritch's *Contributions toward a Dictionary of Book Collectors*, London, Bernard Quaritch, Ltd., 1899.

which gave me access to their marginalia, boundless thanks are due to William Nelson and Elizabeth Story Donno whose enthusiasm and steady guidance increased my courage and ability to tackle so large and amorphous a subject. Paul Oskar Kristeller's keen erudition was of invaluable assistance at various points in my progress, and I am indebted for help in unravelling other of Harvey's crypticisms to S. F. Johnson, James A. Coulter, and Leonard Trawick. I wish also to express my gratitude to those who gave me bits of information or suggestions which opened up avenues leading to fruitful exploration: Jane Apple, William Elton, William Urry, Daniel Javitch, Susan Senneff, and in addition to my husband DeWitt who was handily available to help solve some of my perplexities as to historical facts or Latin and Greek translation. I cannot conclude without mentioning my deep indebtedness to *Renaissance Quarterly* for publishing in 1972 the bibliographical basis of my further study.[5]

VIRGINIA F. STERN

1979

---

[5] Virginia F. Stern, 'The *Bibliotheca* of Gabriel Harvey', in *RQ*, xxv. 1 (Spring 1972), 1–62.

# CONTENTS

# LIST OF PLATES

# NOTE ON TRANSCRIPTION

IN transcribing Harvey's hand, I have made certain alterations from the original in order to provide greater ease of reading: I have changed his use of capital 'A' for the article to the customary lower-case 'a', 'yᵉ' to 'the', have expanded contractions and most abbreviations (exceptions are: '&', 'etc.', 'Mʳ', and 'Dʳ'), and have observed the modern conventions in the use of 'i' and 'j' and of 'u' and 'v', and have substituted 'F' for 'ff' where applicable and 's' for long 's'. The same procedures have been followed in my use of all printed texts of the period (except for Latin texts where I have retained the 'i' and 'j' as found). I have kept the original punctuation except where otherwise noted. Italics within Harvey's marginalia represent his underlinings of words or phrases.

# PART I

# BIOGRAPHY

━━

'Axiophilus . . . will remember to leave
sum memorials behinde him.'★

★ Marginalia on folio 422ᵛ of Harvey's Chaucer

# 1. Saffron Walden boyhood

ALONG the river Cam in the north-west corner of Essex about fifteen miles from the University of Cambridge lies the little town of Saffron Walden, which in the sixteenth century was a market town of considerable importance situated in a rich agricultural district. From documents and maps of the period one learns that it was customarily called 'Walden'. The adjective 'saffron' seems not to have been added until the mid sixteenth century, and until nearly the end of the century one finds only isolated cases of its use. Possibly Robert Greene, in using it in his 1592 *Quip for an upstart Courtier,*[1] popularized the double name. Thomas Nashe reiterated it in *Strange Newes* (1592) and made it unforgettable in *Have with you to Saffron-walden* (1596). It was, of course, an apt name: Walden was famous for its extensive culture of the purple saffron crocus, a plant used as dye, condiment, perfume, and medicine.

The town still retains a number of fine architectural relics which date from its eminent past. The beautiful Parish Church of St. Mary is a handsome example of Perpendicular Gothic and nestled alongside is its graveyard with the tombstones of many early Walden residents. Not far from the church, at Myddleton Place, can be seen some of the original Elizabethan dwellings with their distinctive half-timbered dark beam and white plaster construction and overhanging eaves.

The town centre is marked by a broad Common having at its eastern end a large earthen Maze (about 300 feet in circumference)[2] which has proved an attraction since Walden's earliest days. At the western side of the Common on what is known as 'Common Hill' stood the nearly adjacent mansions of Sir Thomas Smyth (later to become Principal Secretary to Queen Elizabeth) and of Mʳ John Harvey, father of Gabriel. Less than three miles from the town centre still stands a palatial structure, the shell of the early seventeenth-century rebuilding of the great manor of Audley End, seat during Gabriel Harvey's lifetime of Thomas Howard, fourth Duke of Norfolk and of his heirs.

At the other end of the spectrum in a modest section of Saffron Walden on Gold Street are located the still functioning but rebuilt King Edward VI

---

[1] On sig. C3ᵛ of the first issue of the first edition.
[2] The Maze, which is circular, consists of a network of very narrow dirt paths cut into the turf. The paths are said to be nearly a mile in length.

Almshouses in which the town's indigent were cared for through the beneficence of its more prosperous citizens and the jurisdiction of Walden's governing Corporation. The sixteenth-century Corporation was comprised of a Treasurer, two Chamberlains, and twenty-four Assistants. The Treasurer and Chamberlains were elected annually from among the Assistants who, freemen of the town, were to be 'of most honest behayvor, wyse, sobre, & discreate & meete to be Cownsellers to a common wealthe'. The members of this governing board were among the wealthier residents, most of them yeomen, farming land around Walden and living in the town, their prosperity deriving from the fertility of the countryside. Many of them combined other activities with their husbandry and carried on trades at their premises in town.[3] John Harvey was among the more active members of the Corporation throughout much of his lifetime and for at least one year (October 1572 to October 1573) held the chief office of the town, that of Treasurer.[4]

Gabriel Harvey's family were substantial citizens of Walden for at least two generations before his birth. There is extant the will, dated 1556, of his grandmother Christian Harvye, widow of Walden. The bequests left to her various relatives and friends and to the 'pore householders of Golstrete' suggest that she was fairly affluent. Her 'landes both free and copie'[5] she bequeathed to her daughter Elizabeth Stockbredge (wife of Richard Stockbredge) but to her son John Harvey she left only her feather-bed and sundry 'chattels' which were to be divided with his sister. His mother was evidently on close terms with John, as she appointed him an executor. It seems reasonable to infer that she left him no lands because he already had plenty of property of his own and had less need of them than his sister. In fact, there is evidence that he had inherited at least one large parcel of land from his father[6] and there are extant mid sixteenth-century documents concerning the buying and transfer of land in which John Harvey was involved. To John's wife, Ales, his mother left her best gown and petticoat. Perhaps, like her son Gabriel, Ales was fond of fine attire.[7]

John Harvey seems to have been a strong-minded man determined to rise in the world and to see that his sons did likewise. He enrolled Gabriel,

---

[3] See S. A. Culliford, *William Strachey, 1572–1621*, Charlottesville, Univ. Press of Va., 1965, pp. 18–19. Culliford derives quoted specifications for members from Ca. Pat. Rolls, Edward VI, 18 Feb. 1549.

[4] Saffron Walden Town Hall MS. 313, 'Guild of the Holy Trinity Accounts 1545–1651', p. 82.

[5] Copyhold land was that derived from the manorial lord (here the Howards of Audley End). Such transactions were copied down and recorded in the manorial court-roll.

[6] C2 Eliz. Cc 22/34 refers to lands of 'John Harvy' derived by 'copy of court roll' from Thomas, Lord Howard (at P.R.O.).

[7] See transcript of Christian Harvye's will in Appendix A, p. 256.

Richard, and John the younger at Cambridge as pensioners. The fourth and youngest brother Thomas has left few traces, but according to Sonnet xx in *Foure Letters* (1592),[8] John Harvey's four sons cost him at least a thousand pounds.

Gabriel's father, like other members of the Corporation, was a yeoman farmer, with considerable farmland probably on the outskirts of Walden, but he was also a master rope-maker and it is possible that a good share of his income derived from this trade. It is not known whether he had his rope-making shop behind his house or at another location in the town.

To make rope he would have needed a so-called 'rope-walk' of some 1,000 to 1,400 feet in length covered on top and perhaps also enclosed at the sides. Here the worker would have walked back and forth as he twisted the strands, one portion of which had been fastened at the end of the walk. The procedure was for a young assistant to hold or wind around himself a bundle of hemp and then to hackle or straighten out the fibres by drawing the material in a steel-tooth comb. The master or an experienced assistant would then twist several groups or strands of fibres in one direction and afterwards twist the completed strands together in the opposite direction. The process was one of twisting fibres into strands and strands into ropes as the worker proceeded backwards along the length of the rope-walk. The finished rope, when properly twisted, was extremely strong and could be sold for nautical rigging, cargo handling, farm use, bell ropes for church or school, or even for less pleasant use by the hangman. In the Churchwarden's accounts of Bishops Stortford Parish Church is the following item: '1579. Paid to Harvie of Walden for a rope for the bell ij$^s$ iij$^d$.'[9]

Rope-making was a skilled occupation of which John Harvey and his family originally were very proud, for they commemorated it in a carved stone frieze surmounting one of the large fireplaces in their home. In the *Essex Review*, 7 (1898), 22 ff. in an article entitled 'The Harveys of Saffron Walden', Algernon R. Goddard tells of the final demolition of the Harvey mansion in 1855 and the finding of this frieze. He relates that when the 'grand old house' was pulled down and the materials sold

a new and interesting piece of evidence was brought to light. On the removal of some wainscoting two dignified Elizabethan mantels were uncovered. They were of soft clunch stone and no doubt owe their preservation to the fact of their recessment at the rear of the great chimney. . . . Also [preserved] . . . is

---

[8] This English pamphlet by Harvey includes an attack and answer to Robert Greene who had denigrated John Harvey and his four sons in *Quip for an upstart Courtier* (1592), written shortly before Greene's death. See discussion of this on pp. 91–2 below.

[9] *Essex Review*, 7 (1898), 22.

a small oak window from the dwelling house—apparently suggesting an earlier date for the bulk of the house than the chimney or mantel, which were probably added by Gabriel Harvey's father.

Both mantels are characteristic examples of domestic work of the period. The frieze of one is particularly interesting. . . . [It] is sculptured in high relief and is divided into three compartments by means of two trees. The largest of these is occupied by representation of a rope-walk, with three men busily engaged at their calling. At the head of it sits the master—as we may see by his superior dress—probably Father Harvey himself, in effigy. In the left-hand division is a pack-ox with a load. On the right is a hive with a swarm of huge bees all about it. Various other details relative to the work of the farm, which was attached to the place, are introduced.

At the time of demolition the carved relics were removed to the Saffron Walden Museum where one can still examine them.

Below the relief and spaced so as to apply to the whole is incised the partly obliterated inscription NOSTRI PLACENTE[S] [S]UNT LABOR[ES] which can be translated 'Our labours are pleasing'. Above the left-hand panel depicting a heavily laden pack-ox is the inscription ALIIS NON NOBIS, indicating that its work achieves pleasure only for others. The inscription over the middle panel portraying the rope-makers reads NEC ALIIS NEC NOBIS. At least one of the products of the halter-makers brings no joy either to others or to themselves. Over the right-hand panel showing bees engaged in the process of making honey the message is ALIIS ET NOBIS, the fruits of their toil are pleasurable to all.

One wonders whether Gabriel Harvey was the author of this rather witty device. If so, he is himself in part responsible for the association of rope-maker and hangman which Greene and Nashe were later to harp on so devastatingly.[10]

The approximate location and nature of one of John Harvey's houses is known but not the full extent of his property. There is, however, something of a description of his holdings in a deposition by James Crofte, a Notary Public, who in 1593 had drawn up his will; here Crofte recapitulated its provisions. Among them was John Harvey's bequest to his wife of his 'mansion house wherein I nowe dwell in Walden'. He describes this piece of property as 'All that my messuage or tenement wherein I nowe dwell scituat in Walden aforesaid withall the houses yardes gardens edifices & buildinges therunto adjoyning or belonginge withall and singuler ther appurtenaunces wholie as I now occupie the

---

[10] See below, p. 92, for discussion of Robert Greene's allusions in *Quip for an upstart Courtier* (1592) in which he closely connects rope-maker and hangman and suggests that the rope-maker is like a witch, since both do their work backwards.

same'. Upon his wife's death the above was to go to Gabriel provided that within four years of his father's death he had paid to his sister Mary sixty pounds; otherwise this property was to go to Mary.

John Harvey also bequeathed to his wife all his other 'landes tenementes and hereditamentes whatsoever as well free as in stowarie scituat' whether within the town and parish of Walden or elsewhere. After his wife Ales's decease there is left to Gabriel, his heirs, and assigns another dwelling of John Harvey's (now in the tenure or occupation of Tobias and William Malyn or their assigns) in Walden on Hill Street with all its houses, barns, stables, yards, gardens, buildings, and other appurtenances, provided that within three years of Ales's death twenty pounds a year are paid to Gabriel's brothers, Richard and Thomas (John the younger had died in 1592). If Gabriel defaults in these payments, then this property is to revert to Richard and Thomas.

Movable goods, chattels, household stuff, corn, grain, and utensils are left to Ales to dispose of as she sees fit. All other property not otherwise bequeathed is left to Gabriel outright after his mother's death. She is made sole executrix of her husband's will.[11]

The will contains no bequest to Gabriel's sister Alice but, since she had been married to Richard Lyon in 1570, it is possible that some sort of settlement was made at that time.

Gabriel, the eldest of John and Ales Harvey's children, was probably born in 1550.[12] The parish register of baptisms, marriages, and burials does not commence until 17 November 1558, but in a letter of 26 April 1573, written from Pembroke Hall, Gabriel himself remarks: 'If Mai proove no better with me, then March and April have dun, I must needs sai, and mai sai truly, it wilbe the worst spring, yea the wurst and rouhist winter for me, that hapnid this xxii. years.' (Sloane MS. 93, fol. 19ʳ.) That his birthday was in July is suggested by two facts: his occasional noting in his marginalia when the sun is in the constellation of Leo,[13] and his being given by his relative Dr. Henry Harvey in July of 1568 a handsome folio volume of Livy, obviously a fitting birthday present.

His sister Alice and younger sister Marie (apparently it was she who was sometimes called 'Marcie') were born before 3 December 1556, for they, like Gabriel, are mentioned in the will of Christian Harvye which bears this date. 'To Gabriell Harvey the sonne of John Harvye my sonne'

11 Chancery Town Depositions, 24, 346/31 as quoted by Irving Ribner in 'Gabriel Harvey in Chancery—1608', *RES*, 2, no. 6 (Apr. 1951), 142–7.

12 Moore Smith, Harold S. Wilson, and Mark Eccles all concur on a birth date of 1550–1.

13 For example, the note next to Gabriel's signature at the end of Cicero's *Epistolae* reads: 'Mense Julio, sole in Leonis corde flagranti. 1582.' A catalogue and brief description of each of Harvey's books can be found on page 198 of Part III.

his grandmother bequeathed a great brass pot with six shillings eight pence.[14] Of the other children who lived to adulthood Richard was baptized 15 April 1560, John on 13 February 1563/4, and Thomas on 5 September 1567.[15]

Gabriel and his brothers attended the local grammar school, located on the north side of the churchyard, which had a tradition of rigorous but excellent training, with a number of its students going on to the universities or Inns of Court. It had been founded in 1525 with its curriculum modelled on that of Eton. Latin, grammar, and 'higher learning' were taught to the scholars (natives of Walden, Newport, Widdington, and Little Chesterford) who started their school day at 6 a.m., and concluded at 5 p.m. That Harvey approved of its strict and rigorous schooling is suggested by a manuscript note in his copy of J. Foorth, *Synopsis Politica* (1582): 'Good bringing upp, we call breaking, as well in children, schollars, and Servants, as young coultes &c. which cannot be withowt sum mixture of severity.' (fol. 6ʳ.)

The Master of the Grammar School was to be in holy orders and 'a profound gramarion'. John Disborowe (M.A. of Trinity College, Cambridge), who became Head in 1564/5, supervised Gabriel's last year and a half. Although the name of the previous Master of Walden Grammar School is not recorded, one can speculate that it might have been one of the Dove family of Saffron Walden, for in the marginalia of a copy of Ramus' *Oikonomia*, acquired by Harvey in 1574, is the following observation: 'Dooves saying, what neede I wright owt, or studdy, that I knowe alreddy. If I knowe it once; all is on, whether it be spoken in Latin, Greeke, Inglish, &c.' (fol. 7ʳ.)[16]

With his inimitable raillery Nashe writes of Harvey's early schooling:

He ran through *Didimus* or *Diomedes* 6000. books of the Arte of Grammer, besides learned to write a faire capitall Romane hand. . . . Many a copy-holder

[14] This is equivalent to one-half mark. Other monetary bequests in the will are in fractions or multiples of a mark (e.g. ⅔).

[15] Moore Smith in his introduction to *Gabriel Harvey's Marginalia* (p. 5) includes in the family a Mary, baptized 15 May 1567, and explains with difficulty why Thomas's and her baptisms were four months apart. But, although this Mary was a daughter of 'John Harvey', it is undoubtedly another John Harvey of Walden (probably a cousin), one who had a son Richarde, baptized 6 Nov. 1564, and a son John who was buried 20 July 1570. The parish register, from which the above information is derived, makes evident that there were at least two contemporary John Harvey families in Walden. The names of Gabriel's brothers and sisters are confirmed in the will of John Harvey the elder as recounted by James Crofte (see p. 131, below), in the will of Richard Harvey (p. 132), and in other legal documents connected with the family.

[16] Thomas Dove, a younger member of the family, was a fellow student of Gabriel's at Cambridge, first at Christ's (matric. 1566) and later at Pembroke. He was Vicar of Walden from 1580 to 1607.

or magistrall scribe . . . comes short of the like gift. . . . But in his Grammer yeares he was a very gracelesse litigious youth, and one that would pick quarrels with old *Gulielmus Lillies Sintaxis* and *Prosodia*, everie howre of the daye. A desperate stabber with pen-knives, and whom he could not over-come in disputation, he would be sure to break his head with his pen and ink-horne.

If Nashe is to be believed, Gabriel from earliest childhood was completely absorbed in learning and earnest in the pursuit of it. He apparently was analytical of all that he heard or read—even of William Lily's time-honoured grammar. Constantly involving himself in verbal contention and disputation, he must have quickly become aware that words could serve to obscure truth or to make it clear and could be used as a weapon of attack.

Nashe continues with his lightly satirical portrait and intimates that Gabriel was a bit of a prodigy, writing ballads and simple verses from the age of nine onwards:

Scarce nine years of age he attaind too, when, by engrossing al ballets that came to anie Market or Faire there-abouts, he aspired to bee as desperate a ballet-maker as the best of them; the first frutes of his Poetrie beeing a pittiful Dittie in lamentation of the death of a Fellow that, at Queen *Maries* coronation, came downward, with his head on a rope, from the Spyre of *Powles* steeple, and brake his necke. Afterward he exercised to write certaine graces in ryme dogrell, and verses uppon everie Month, manie of which are yet extant in Primers and Almanackes.[17]

Gabriel lived in an area frequented by busy and colourful fairs which were customarily of two or three weeks' duration. Local ones were held in early spring and summer on Walden Common and the famous Sturbridge Fair took place at the beginning of September not far from Cambridge.

[17] *Works of Thomas Nashe*, ed. R. B. McKerrow, New York, 1966 (reprint of 1910 edn.), iii. 60.

# 2. Christ's College Cambridge

FROM a relatively protected boyhood in Walden, this gifted, aspiring, and intense young man was now to proceed to the wider arena of Cambridge where he would spend the greater part of the next eighteen years. The University was to be a fostering mother in making available the essential nourishment and training of higher learning and the opportunity to become well acquainted with its great minds and forceful personalities. A number of these men were to inspire and interact with Harvey and he in turn was to give of himself in encouraging and training younger students.

But Cambridge life had its harshly abrasive side and gave the yeoman rope-maker's eldest son many a rude slap in the face. Gabriel's progress was to be recurrently impeded by petty jealousies, derision, and cruel injustices, and he would painfully learn that his Renaissance world was not one of unlimited potentiality in which merit and industry automatically led to recognition and reward. These eighteen years were to transform a trusting, handsome, and eager young student to an embittered, proud, touchy, and sharp-tongued scholar ever on his guard, prepared to manœuvre and defend himself by the subtlety of his wits. Nevertheless, to his closest friends and to those members of his family to whom he was devoted, Harvey, although wounded, seems somehow to have remained very nearly the loyal, likeable, self-sacrificing, open person of his youth.

On 28 June 1566 (probably just before his sixteenth birthday) Gabriel matriculated at Christ's College, Cambridge, as a pensioner: as a 'pensioner', housing, commons, and other expenses would have been paid for by his father. Harvey's tutor at Christ's was William Lewin who was then a Fellow. He had received his M.A. in 1565 and had been selected as tutor to Burghley's daughter, the Lady Anne Cecil. By 1576 Lewin had obtained his LL.D. and become Judge of the Prerogative Court of Canterbury; in 1582 he was to receive a Doctor of Civil Law degree from Oxford. Lewin while M.A. and a civil law student had obtained, perhaps through Burghley's influence, a dispensation from the Archbishop of Canterbury to hold a benefice without spiritual charge.[1]

That Harvey became his erstwhile tutor's close and esteemed friend is

---

[1] Harvey also followed the procedure of LL.D. at Cambridge and D.C.L. at Oxford but was never able to secure a similar dispensation and thus never obtained the financial security of a benefice.

indicated in a Latin letter from Lewin written to him on 15 December
1576. In it Lewin assures Harvey that he cherishes and esteems him more
than anyone else at Cambridge with the exception of his father-in-law
Thomas Byng (Regius Professor of Civil Law and Master of Clare Hall),
whom he venerates above all others. Reminding Harvey of his devotion
to him, he adds a bit of practical advice which translates: 'You seek that
you may live as a free man. Moreover, you will accomplish this if you
will devote yourself to some lucrative profession as soon as possible.'[2] He
seems aware that if Harvey is to achieve his ambitious goals, he must
provide for his financial needs.

Two months later (11 February) Lewin evidences his estimation of
Harvey in a more public manner in another letter printed as a preface to
his 1577 *Ciceronianus*, which is, of course, dedicated to Lewin. Here in
glowing terms Lewin praises Harvey's remarkable ability and achievement, although he admits that he does not altogether share some of
Harvey's more liberal views when, for example, it comes to preferring
Ramus to Sturm. Lewin closes by ranking Harvey among Cambridge's
foremost professors.

It was during Harvey's undergraduate years at Christ's that he also
became acquainted with Sir Walter Mildmay, brother-in-law of Sir
Francis Walsingham and friend of Lord Burghley. A cultivated man, his
field of expertise was financial matters, and in 1566 Elizabeth appointed
him Chancellor of the Exchequer. Mildmay, who had been educated
at Christ's, remained closely affiliated with it; in 1568 he agreed to
give £20 annually to establish six scholarships, a Greek Lectureship, and
a Preachership. It seems likely that Gabriel was awarded one of these
scholarships, for he refers to Mildmay as his benefactor.[3] It is also possible
that he took advantage of this opportunity to pursue the study of Greek,
though he might have learned the rudiments at grammar school. By
1573 Harvey was to be proficient enough to be appointed Greek Lecturer
for Pembroke.

Mildmay is commemorated as a patron of studies and instruction in
Harvey's dedication to him of *Smithus; vel musarum lachrymae* (1578),[4]
an elegy to Sir Thomas Smyth in Latin verse. Its structure may well have
served as a model for Spenser's 1591 *Teares of the Muses*.

---

[2] 'Tu quaere, liber ut vivas; Hoc autem efficies, si alicui quaestuosae Arti, teipsum
quamprimum consecraveris.' (Baker MS. 36, fols. 110–11.)

[3] See copy of Latin letter to Mildmay in Baker MS. 36, fols. 111–12 at Univ. Lib.,
Cambridge.

[4]         An mage Mildmaio sis deditus, optimus olim
            Qui studiorum esset, doctrinarumque Patronus?
                                    (sig. *ij[v].)

Mildmay was also a patron to Hugh Broughton from whom Harvey in 1580 received a gift of Lhuyd's *Breviary of Britayne*.[5] A young man of about the same age as Harvey, Broughton became a divine, and an outstanding Hebrew scholar. Of Welsh descent, he evidently wished to call to Harvey's attention this little book written by his countryman.

Though Harvey received his Baccalaureate degree from Christ's in 1569/70 and was ninth in the Ordo Senioritatis, he was not elected a Fellow. Distressed at the time, he sought a fellowship elsewhere and on 3 November 1570 was elected, through the recommendation of Sir Thomas Smyth, to one at Pembroke Hall.

[5] Title-page of this 1573 tract is inscribed by Harvey, 'Ex dono M$^{ri}$ Browghton, Christensis'.

# 3. Pembroke Hall

HARVEY would have had the opportunity to become intimate with Smyth between April 1566 and March 1571, when he was living almost continuously in Essex. Before and after this and during a very brief trip to France in 1567, Smyth was out of England on government service; but for most of the five years after Sir Philip Hoby succeeded him as ambassador, Smyth was living either at his country estate at Theydon Mount or at his town residence in the central square of Walden close to the Harveys' home. By 1573 the elder statesman had certainly become 'intellectual father' to the gifted young scholar. Harvey's letters to Sir Thomas[1] refer to the advice he has given him, his guidance in studies, and to his orienting Harvey toward a life of service to the state. He visited him at his country home at Theydon Mount,[2] studied with him, sought his counsel, and corresponded regularly. In a 1573 letter Harvey writes of 'the special frendship that I alwais hetherto sins mi first cumming to Cambridg have found at your hands (as suerly I do, and must neds remember it often, having continually had so ful trial thereof)'. He refers to Smyth's having aided him in attaining his fellowship at Pembroke 'not past thre yers ago', and he discusses whether or not he should take up the study of civil law:

I know wel both your wisdum to be sutch, that you can easily discern what is best for me, and I assure mi self your gud affection to be sutch, that you wil gladly counsel me for the best. . . . I am now almost determinid, notwithstanding al douts, that miht hinder or slack mi purpose, to be shortly novus Justinianeus, as mi other busines wil suffer me: if your wisdum shal think good or not mislike of the matter. Inded ons I suppose verrely Christ Collidg fellowship, whitch I had over great a fansi to, miht have drawn me in to the ministeri, as it hath dun a great mani mo. And as I remember, I tould your wurship so mutch at that time.

But suerly in this, and in sum other respects, I thank God as mutch, that I was not chosen fellow there, as I have caus to do, that I was chosen in au other place.

---

[1] Copies in Harvey's hand of his letters to Sir Thomas are to be found in Sloane MS. 93, fols. 91$^v$–92$^r$, 96$^r$–98$^r$ (Harvey's 'Letter-Book').

[2] In a letter of 21 Mar. 1573 to John Young, Master of Pembroke, Harvey writes: 'I am occasionid to go abroad about Whitsuntide (and in deed it is to Sir Thomas Smyths hous in the cuntri, as I was certifiid).' (Sloane MS. 93, fol. 11.)

He then alludes to his study of Cicero's *Topica*, of the German philologist Hegendorff's writings on law logic, and of the first book of the *Institutes*. He adds that, as soon as he has leisure from his other studies, he intends to take pains in the direction of study of law, 'and therefore am thus imbouldnid to use you as a Crassus, or a Scaevola in the matter, uppon hose advise, as uppon an oracle, I am to set down mi staf, both what wai I were best to turn me to, and how to cum most conveniently at mi jurnies end.'[3]

In a subsequent letter from Pembroke Hall (not specifically dated), he writes:

most harti and esspecial thanks for your continual goodnes towards me, thes shalbe to let you understand that first mi busines and then mie helth hath bene sutch, sins mi last being with you, that according to your wil and mi desier I could not wel cum unto you. I have bene il at ease this thre or fower wekes, and am yet as il almost as ever I was, or els I had trublid your wurship long ere this. But as soone as I shal recoover mi helth, and begin to stur abroad again, I purpose, God willing, to crave your favorable and gentle furtheraunc. Suerly I took wunderfull pleasure and great profit bi your last frendli or rather fatherli taulk. I thank your worship most hartely for it; and I thought it sins a singular benefit and blessing of God, that I had sutch a patron, or rather a father, to resort unto. And as for the points you stud uppon, I did so mark them at the praesent, that I am suer I shall never forget ani of them as long as I live.

He then recapitulates Smyth's advice as to how he should go about the study of civil law.[4]

Two younger members of Sir Thomas's family were also friends of Harvey's. The first of these was Smyth's illegitimate but cherished only son, M. Thomas Smyth, later Colonel of the Ardes in Ireland.[5] He and Gabriel read the 'First Decade' of Livy's Roman history together, according to marginalia in Harvey's folio Livy (sig. Z3).

The second was Sir Thomas's secretary and supposedly favorite nephew John Wood. On the title-page of Harvey's copy of Smyth's *De Recta & Emendata Linguae Anglicae Scriptione* (1567) is the inscription: 'ex dono Ioannis Woddi, aulici, et amici mei singularis. Anno 1569. Acceptissima semper munera sunt, Auctor quae preciosa facit. G. Harveius. Ioannis

---

[3] Sloane MS. 93, fols. 87–9. Crassus signifies a wealthy patron and Scaevola a famed civil lawyer.

[4] Sloane MS. 93, fols. 91ᵛ–92ʳ. In the punctuation of the above and in subsequent letters from Sloane MS. 93, I have reduced Harvey's myriad use of commas.

[5] On sig. C3ʳ of *Foure Letters* (1592) Harvey writes of 'my Cosen, M. Thomas Smith'. Nashe refers to Harvey's telling a boasting tale 'of the Funerall of his kinsman, *Sir Thomas Smith*, (which word *kinsman* I wonderd he causd not to be set in great capitall letters,) and how in those obsequies he was a chiefe Mourner' (*Works*, iii. 58).

Woddi liber, ex ipso Auctoris dono.'[6] In addition there is extant Harvey's copy of his letter (undated) to Wood asking him for news of life at Court,[7] and finally, in *Smithus; vel musarum lachrymae* (1578), there are two sets of verses by Gabriel addressed to Wood and an Epilogue to him by Richard Harvey.

As previously mentioned, Harvey was in November 1570 through the influence of Sir Thomas elected to a Fellowship at Pembroke. Soon after he must have become acquainted with Edmund Spenser, a young Londoner of modest circumstances who, after attending the recently founded Merchant Taylors' School, entered Pembroke as sizar[8] on 20 May 1569. Although no documentary evidence of the earliest years of Spenser's and Harvey's ripening friendship has been found, marginalia of Harvey's, dated December 1578, in his copy of *Howleglas*[9] and the Latin verse letter (October 1579) of Spenser's[10] are among the many indications that their *entente* had become so close by the late 1570s that it was probably of some years' standing.

Spenser was three years Harvey's junior in academic rank—in 1570 Spenser was a sophister whereas Harvey was in his first year of graduate study and was probably about two years senior in age.[11] Their extant letters suggest that the influence of the older youth on the younger was to spur him to deeper and more disciplined study, to disclose the magical vistas that learning could make visible, and to urge him towards service to the state as the most worthy of aims. Spenser's influence on Harvey (as the Spenser–Harvey correspondence, Harvey's marginalia, and his printed comments on Spenser's work suggest) must have been to stress the necessity for compromise between ideal values and the expediencies of the everyday world, and to encourage tolerance for human foibles, a feeling for the melody of language and for the sentiment of life, and an appreciation of the need for occasional whimsical playfulness. Perhaps much of Spenser's influence did not take effect until years later.

---

[6] 'from a gift of John Wood, courtier, and my special friend. In the year 1569. Those gifts which the author makes precious are always the most welcome ones. G. Harvey's. A book from a personal gift by the author'.

[7] Sloane MS. 93, fol. 101ᵛ.

[8] A sizar (i.e. a student unable to pay his own way) received an allowance from the college and in return performed certain menial duties.

[9] See annotations in Murner's *Howleglas*, as described p. 49, below, showing Spenser's efforts to get Harvey to open his eyes to lighter and more playful literature.

[10] In the Earl of Leicester's service, Spenser, believing himself about to set sail for France, writes Harvey this long and affectionate letter of temporary farewell, including considerable thoughtful counsel.

[11] According to Alexander Judson, *Life of Spenser*, Baltimore, 1945, p. 1, Spenser was probably born in 1552.

Two young men of inquiring minds, opposites in temperament, they were much alike in altruistic ideals and practical aims: the ideals involved in earnestly working to create a better world, the pragmatic aims concerned with manœuvring for posts at Court, from which each in his own way could eventually exert some politico-moral influence on the English people. Such posts, if obtained, would further serve to augment personal status and finances, for each man was throughout his lifetime to be keenly aware of financial need.

As we have seen, Harvey wrote to Sir Thomas Smyth[12] to the effect that he now was glad to have been denied a Fellowship at Christ's since this had resulted in his going to Pembroke instead. He referred to the likelihood that, if he had remained at Christ's, he would have been drawn into the ministry as had happened with so many who had fellowships there. But, although he did not elaborate, he mentioned that there were other respects besides this in which he was glad that he instead became a Fellow of Pembroke. Undoubtedly, he was grateful because of the firm friendships he had formed with Spenser and with Humphrey Tyndall (who later became President of Queens' and Dean of Ely) but also because of the aura of élitism at Pembroke, and because of its tradition of broad learning and scholarship (although Harvey was soon to find that the tradition was not an unbroken one). Above all, Harvey was grateful for Pembroke's superb, although non-resident, Master, John Young, described by Archbishop Parker in 1570/1 as possessing a princely nature and a princely pronunciation.[13]

As a young Pembroke Fellow, Gabriel Harvey applied himself with enthusiasm to the studies required for his Master of Arts degree, giving evidence of his aptitude for scholarship and intellectual exploration.[14] He had become so absorbed in the University's academic offerings that he probably was totally unaware that he had a malevolent enemy within the walls of his own college. Much like the beautiful and splendid butterfly of Spenser's poem *Muiopotmos* (published in his 1591 *Complaints. Containing sundrie small Poemes of the Worlds Vanitie*) Harvey was handsome,[15] confident of his own excellence, and inclined to be incautious. Apparently he was oblivious of the antagonist lurking within his paradise waiting for the opportunity to entrap him when in March 1573/4 he

[12] Sloane MS. 93, fols. 88ᵛ–89ʳ, as quoted on p. 13, above.

[13] Letter to Burghley, 2 Feb., cited in Alexander Judson, 'A Biographical Sketch of John Young', *Indiana Univ. Studies*, 21 (1934), 8.

[14] Marginalia in Livy, Aristotle, and Wilson attest to this.

[15] Nashe writes that Harvey as a young man was handsome: 'He is ... distractly enamourd of his own beautie.' (*Works*, iii. 67.) 'A smudge peice of a handsome fellow it hath beene in his dayes, but now [1596] he is olde and past his best'. (Ibid. 94.)

was ready to seek grace for the granting of his M.A. degree. The web-spinning spider was Thomas Neville who, like Harvey, had been elected to a Pembroke Fellowship in 1570, but who had already received his M.A. in 1572.

Neville, the younger and apparently less able brother in a noble and important family, was proud, disdainful, and fond of personal ornamenta-tion[16] and, since he was not as dedicated a scholar as his brother Alexander, evidently determined to be a leader in other ways. No specific information has been found as to what, if anything, Harvey had done to warrant Neville's enmity. In a letter of Harvey's of 21 March 1573 to John Young he refers to

certain jars that had faln out betwixt M. Osburn[17] and me, and M. Nevil and me long ago: thai them selvs, not I being the beginners and continuers of them .... it was likeli that thai went about of private grudgis to make them commun causis. . . . And to sai troth, I suppose verily a litle breach betwixt thes twoo and me was the tru and onli caus of al these sturs.[18]

Reading between the lines of Harvey's letters to his father, to Young, and to Tyndall,[19] and relating this to what can be inferred about Harvey himself from the content of his later correspondence with Spenser, one can surmise that it may well have been Harvey's earnest dedication to excellence and his occasional aloofness from the mundane world which irked Neville. No doubt, also, he was jealous and annoyed that one from a middle-class family should dare to consider himself a peer of noble-men's sons.

As for Harvey himself, although he had a number of close personal friends, he seems to have lacked the art of being convivial in a relaxed way among groups of his contemporaries. The want of gentlemanly ease, modest grace, and light-heartedness could earlier have counted against his being offered a fellowship at Christ's. Harvey seems to have been well aware of this deficiency in himself, for his marginalia show that he studied Castiglione and Guazzo with considerable care, evidently intent on learning the art of courtierly manners and *sprezzatura*.[20]

Harvey may, of course, have been a chance victim with whom Neville at first idly toyed and then discovered that he could readily bait him

[16] Sloane MS. 93, fol. 4r.

[17] Richard Osburn seems to have been a frequent companion of Neville's at Pembroke.

[18] Sloane MS. 93, fol. 9r.

[19] Sloane MS. 93: fols. 23–5 to his father, fols. 2–11, 14–22, 27–34, 85–6 to Dr. Young, and fols. 12–13 to Humphrey Tyndall.

[20] In 1572 he acquired Hoby's translation of Castiglione's *Courtier* and annotated it copiously. He bought and studied Guazzo's *Civil Conversation* in the early 1580s. He owned copies of both books in Italian and in English as well as a Latin copy of the *Courtier*.

because of his naïveté, defencelessness, and earnestness. Neville also quickly learned that he could rally certain disgruntled Fellows around him to enjoy the sport and thus he became the leader and strategist of a sadistic project.[21]

Although Neville is today remembered as one of the important Masters of Trinity College (1593–1615), he may have been less than exemplary in his youth. There is also the possibility that he was a man of two faces for, whereas Cambridge has always looked upon him as a great bene-factor for his bountiful gifts to Trinity College, Canterbury Cathedral (where from 1597 until his death in 1615 he was Dean) has considered him a purloiner of its treasures, treasures which found their way to Cambridge.[22] Indications that his high opinion of himself was not necessarily shared by those who knew him may be gleaned from the inscription on his sepulchre at Canterbury, a monument which, although no longer intact, is illustrated with large engravings and described in detail in the Revd. John Dart's *History and Antiquities of Canterbury*.[23]

On the east side of the great cathedral's south aisle stood a handsome marble monument divided by two arches: set into the niche formed by the left-hand arch was the sculptured full-length figure of Thomas Neville viewed in profile in his robes as Dean of Canterbury, kneeling at his reading desk; under the right-hand arch was the profile effigy of his older brother Alexander Neville, Esquire (1544–1614), in armour, also in a de-vout posture. Below each of the marble figures was a Latin inscription. That for Thomas Neville has been translated as follows:

Illustrious by Birth, remarkable for Piety, of extraordinary Genius and un-common Learning; of most engaging Temper and Behaviour, and a worthy and approv'd Divine. In his early youth (being at Cambridge, in Pembroke Hall where he continued fifteen Years) he was embellish'd with all the improve-ments which adorn the early Years of Life, and in that University which he adorn'd and enrich'd to his utmost Power, by his Studies and Industry. . . .

The epitaph lists his various academic and ecclesiastical posts, concluding:

Lastly, Dean of this Church, most Moderate, Faithful, and Beneficent; which Church he Govern'd          Years, with strict Justice, extraordinary Modesty and singular Integrity. This Monument therefore of his Virtue and Honour, he, in spight of Death, erected for himself. He died Anno          Aged

---

[21] The various steps and strategies in the Fellows' non-placeting of Harvey's grace are recounted in his letters to his father and to John Young (Sloane MS. 93, fols. 2 ff.).

[22] William Urry (former Archivist of Canterbury) in *The Marlowes of Canterbury* to be published shortly. See his chapter entitled 'Some Other Famous Elizabethans'.

[23] London, 1726. Pp. 40–5 describe the monuments of the Neville Chapel which was destroyed about the end of the eighteenth century. The figure of the Dean is, however, still extant but has been placed on the south side of the Choir aisle.

the Day of        And in this Chapel, which (while he liv'd) he Embellish'd
for himself and his Family; He (not without being much Lamented by all that
knew him) was buried under this Tomb, and expects the coming of our Lord
Jesus Christ, Favour and eternal Glory.[24]

In the same chapel there is a monument to Neville's parents and his
uncle which bears a Latin inscription stating that it and the double
monument to Thomas and Alexander Neville were erected August 3,
1599 by Thomas Neville, Dean of Canterbury. But, although he had the
foresight to provide (along with the other more modestly inscribed family
tombs) a handsome sepulchral tribute to himself, he evidently did not
foresee that no one at Canterbury would care enough when he succumbed
in 1615 to fill in on his monument the date of his death, his age, or the
number of years he had served as Dean of Canterbury.

An ideal opportunity for the baiting of Gabriel Harvey was afforded by
the Master's non-residence at Pembroke and the consequent jurisdiction
of the Fellows. Harvey himself has left a detailed account of the attempt
to block the award of his M.A. degree. There is extant a notebook containing copies in his hand of the series of letters describing this episode
in all its intricacies. On its first leaf the notebook bears two mottos, the
first in Greek and the second in Latin:

MY SUFFERINGS HAVE BEEN MY LESSONS
It is better to be wronged than to do wrong.[25]

On 21 March 1573 (here the year is begun on 1 January, not on Lady
Day) Harvey wrote an epistle of some eight thousand words addressed to
John Young (presumably then in London, where he was Prebendary of
Westminster), appealing to him for aid and giving a straightforward
yet poignant description of the course of events. The account recites
Neville's and a few other Fellows' non-placeting of Harvey's grace, his
asking them reasons for this unmerited action, and their countering with
charges and ruses by which to procrastinate and gradually manœuvre
more and more of the Fellows into the anti-Harvey camp.

According to this letter, Neville's first accusation was 'that I was not
familiar like a fellow, and that I did disdain everi mans cumpani', to
which Harvey answers:

What thai cale sociable I know not: this I am suer, I never avoidid cumpani:

[24] Ibid., p. 42.

[25] τὰ παθήματα μαθήματα. This is a paraphrase of a quotation from Herodotus, i. 207, and
was transmitted by Chrysippus as Stoic doctrine: 'Satius est ἀδικεῖσθαι quam ἀδικεῖν'
derives from Plato's Gorgias. The inscriptions are found on fol. 1ᵗ of Sloane MS. 93; the
letters follow on fols. 2–34.

I have bene merri in cumpani: I have bene ful hardly drawn out of cumpani. And as for disdaining of others, I wuld thai culd have found in their harts to have made that account of me that I have made of them, and bene as willing to accept of mi cumpani as I have bene reddi to seek theirs. And of troth this dare I avouch bouldly to your wurship, whitch sum others can testifi to be most tru, that I have not shoun mi self so surli towards mi inferiors, as M. Nevil hath shown him self disdainful towards his oequals, and superiors too. For mi self, whitch in deed am an inch beneath him, as he ons made his vaunt; he can not deni it, he hath confest so mutch to me him self, that I passing bi him, and moving mi cap, and speking unto him, he hath looked awri an other wai, neither afording me a word, nor a cap: purposing, as I take it, to make of his inch a good long el, and to show a lusti contempt of so silli a frend. . . . But it is almost impossible that he shuld se him self, which is so mutch gevin to look uppon others. Nether do I speak this so mutch to accuse him, whome I use to speak wel of, and mutch better then he hath deservid, as it proovith now; as to excuse mi self whome he hath so unjustly and falsly chargid with arroganci and disdainfulnes.

The second of Neville's accusations was 'that I culd hardly find in mi hart to commend of ani man; and that I have misliked those which bi commun consent and agrement of al, have bene veri wel thout of for there lerning'.

Harvey, so he reports, answered:

I thout it mi duti, to speak wel of those, that deservd wel; and that he might sundri times have harde me commend mani a on. . . . He ouht to give me, and others as gud leave to use our judgments in that behalf, as I, and others had givn him. Stil he harpid uppon that string, that I could not aford ani a gud wurd.[26] And I made him aunser, that this were great arroganci, and extreme folli. Hereuppon we had farther descanting of them both. . . . In which discours amongst other things I hapned to kast out thus mutch, that although thei were both veri il, yit of the two, it were better for a man to be thout arrogant, then foolish. Whereuppon aris a wunderful accusation made bi M. Nevil in a gud great audienc, I mi self not being there, that sutch an on semid flatly to allow of arroganci, that he praeferd it before folli, that he wuld not steak to sai bouldly he had rather be arrogant, then foolish. And this was a great big matter, and a hie point forsooth, for them to beat there heds, and whet there tungs about. So that your wurship mai here plainly se, as in a glas, how reddi thai ar to take the vantage, and how able thai ar to make sumwhat of nothing.[27]

Neville's third charge was 'that I made but smal, and liht account of mi

---

[26] Harvey's writings and marginalia are filled with lists of men and works he admires. Such lists of commendations are sometimes continued *ad nauseam*. One wonders if Harvey at such times is making certain that no one can ever again accuse him of not speaking well of deserving men or their works.

[27] The above passages are excerpted from Sloane MS. 93, fols. 3ʳ–4ᵛ.

fellowship'.[28] In point of fact, Harvey stated that he would not have been able nor willing to pay twenty pounds for it:

I tould him I culd have made better shift without a fellowship, then I culd have made shift to pai so mutch for a fellowship. I said also, that I made as great amount of the bennefit, as ethir he, or anie man els did, or coold: marri so, that I wuld have bene loth to have bouht it so dear, as he spake of.

Harvey's last statement implies that he values a Pembroke fellowship too much to be willing to cheapen its value by making it financially purchasable.

Neville's fourth point was 'that I was a great and continual patron of paradoxis and a main defender of straung opinions, and that communly against Aristotle too'.

In 1568 Harvey purchased a copy of Aristotle's *Organon*; in 1572 he was given a copy of Aristotle's *Rhetoric*. Both Greek texts were copiously annotated and thus indicate that they were closely studied. Harvey's response to Neville's charge shows his intellectual nonconformity to be the product of an inquiring and analytical mind which refuses to accept authorities *in toto* merely because they have been traditionally venerated. Probably about this time he became a strong proponent of Peter Ramus's method and tenets. Though Harvey continued to show respect for much of Aristotle's thought, he agreed with Ramus that the Aristotelian classifications of 'logic' and 'rhetoric' could be clarified and made more functional. Harvey's discrimination is again shown in his later *Ciceronianus*, the opening lecture delivered to his Cambridge rhetoric classes in the Easter term of 1576 and printed the following year.[29] An advocate of Cicero's style, he here points out that the classical rhetorician's example is now becoming much abused through the imitation of its surface features without adequate understanding of its spirit.

---

[28] Sloane MS. 93, fol. 5ʳ.

[29] Marginalia in Harvey's copy of Peter Ramus (Pierre de la Ramée), *Ciceronianus* show that it was carefully read and annotated by Harvey in 1568–9. In his own *Ciceronianus* (1577) he relates that the discovery of Ramus's work had opened his eyes and he was quick to recognize the Frenchman's contribution to the teaching of liberal arts.

About 1570 Harvey purchased and read the *Academia* of Audomarus Talaeus a close associate and disciple of Ramus in his programme of teaching reform. After Ramus's tragic death in the 1572 Massacre of St. Bartholomew's Day, Harvey wrote his elegiac tribute, *Ode Natalitia . . . In memoriam P. Rami, optimi, et clarissimi viri*, published in 1575. Warren B. Austin in *MLN*, 61, 242–7 identifies *Ode Natalitia* as apparently Harvey's first printed work. The Latin verse consists of two eclogues: the first, an allegory of a youth who, using Ramus's method as guide, progresses smoothly along the road to Liberal Arts knowledge; the second begins with a description of the French humanist's contributions, cut short by his untimely death, and concludes with an appeal for perpetuation of his teachings. For discussion of Ramus's tenets and of his influence on English thought, see Wilbur Howell, *Logic and Rhetoric in England: 1500–1700*, New York, 1961, pp. 146–281.

Commenting, then, on Neville's final accusation, Harvey assures Young that in most respects he is a staunch admirer of Aristotle but there are five points on which he differs, for (evidently tending toward a modern scientific outlook) he believes that the earth is not eternal, the sky is not of the fifth nature (i.e. it is composed of a substance not in any way different from the four elements), the sky is not animated, nothing is physically infinite in force, and virtue is not within our power.[30]

In addition, he explains that what Neville and his cohorts deride as 'singulariti in philosophi' is Harvey's choice of challenging problems and controversial questions as subjects for philosophic disputation in the 'chappel' (apparently the college auditorium) in preference to the customary topics of 'de nobilitate, de amore, de gloria, de liberalitate, and a few the like'. But these themes, Harvey observes, have been 'thurrouly canvassid long ago: and everi on that can do ani thing is able to write hole volumes of them, and make glorious shows with them', so that they are now more suitable for students' declamations than for 'masters problems to dispute uppon'. Perhaps Neville is afraid, comments Harvey sardonically, that he will prove to be 'sum noble heretick like Arrius and Pelagius: and so disturb and disquiet the Church as I do now the Chappel'. Harvey assures Young that no one need have any fear of this.

Harvey continues sarcastically, reporting that the above are the 'forcible and waiti reasons M. Nevil hath usid against me; besides certain glauncis at two or thre od things mo': criticisms such as that Harvey sits in chapel reading his oration rather than kneeling (Harvey answers that the procedure is optional and as many men do one as the other), and that he wears a hat 'at problem' (a classroom exercise where the students were given some difficult question to solve).[31]

Having talked with Neville, Harvey, as he reports, next tried to communicate with Osburn, requesting that he not deter Harvey from his grace. He volunteered that had he offended Osburn or Neville in any way, he would do his utmost to redress and amend it. Osburn, it appears,

---

[30] Harvey lists his propositions against Aristotle as follows: 'Mundus non est aeternus: Cœlum non est quintæ naturæ: Cœlum non est animatum: Nihil est $\phi\upsilon\sigma\iota\kappa\tilde{\omega}\varsigma$ infinitum potentia; Virtus non est in nostra potestate.'

[31] To this Harvey replies that he has not done so for the past year; before that he wore a hat because he was just recovering from a severe illness and did not want to get his head chilled. At the time no one criticized him; furthermore M. Neville himself wore a hat on frosty evenings. Harvey explains that he extended the practice a bit after his illness since he became chilled rather easily. At that time M. Nuce, a senior Fellow who was the Master's deputy, reminded him that it was not customary for bachelors to be covered and that it was offensive to certain of the company. Harvey responded that he had not previously known of this custom but now that M. Nuce had so informed him, he would no longer wear a hat. And, according to Harvey's statement, he never wore a hat thereafter.

seemed mollified, but would make no promise without first speaking to Neville.

Harvey, now hopeful, then sought out Neville to see if he could find him any more tractable but, as he reports:

M. Nevil was M. Nevil stil, the veri self same man that I left him, nethir to be allurid bi promissis, nor persuadid bi wurds, nor wun bi intreati, nor moovid bi greef. . . . He brake out into this jolli and brave jybe: We have you at the suords point: we wil hould ye thare whilst we have ye thare, les if you get ons within the half swurd you chaunc to give us the lamskin. So that now he was altogither . . . setting al at nauht that I culd sai unto him. And yit I am suer I wooid him so, as mani a gud man never did his wife: and I am suer he rejectid me so, as few honist masters have dun there servants.

Now Neville got hold of a new argument of 'singulariti': that Harvey was busy studying at Christmas-time when others were busy amusing themselves, that Harvey was hottest at his book 'when the rest were hardist at there cards'. To this Harvey replied:

I had veri urgent busines this last Christmas, more then everi man knew of, or els I had bene man like enouh to have dun as other did. He tould me again, na, I had dun so everi Christmas sins I cam to the house. Wel there was no shift of it, M. Nevil wil and must needs have his own wurds: and thus he and his fellows seek to pick quarrels with me, and to make huge mountains of smal low molhills. . . . M. Nevil is flinti, and is M. Nevil stil: I must needes abide the brunt of his displeasure: what is a gentleman but his pleasure? . . . After this I tooke mi time to intreate M. Osburn anu: and I found him not so quiet and calme as before. . . . He revilid me to mi teeth: and tould me veri flatli that he was bewitchid before, yea and that it was mi flatteri and serviliti (for so it pleasd him now to term it) that bewitchid him.

Having answered quietly that Osburn was dealing very roughly and 'uncurtuusly' with him to interpret so ill what was meant so well, he continues, 'Now I parsaived plainly that nothing culd content them. Before I was arrogant and now I am servile', to which Osburn retorted that Harvey went 'ab extremo ad extremum sine medio'.

Meanwhile Harvey sought out the Master's deputy, Nuce. But Neville and Osburn had reached him first and Nuce asserted that he could do nothing until their grievances were thoroughly examined. However, he did admit 'it was likeli that thai went about of private grudgis to make them commun causis'.

Nuce thought it best that the whole matter be heard by the Fellows and judged by them. In the morning he therefore moved Harvey's grace anew among the assembled Fellows, whereupon Osburn addressed the

supplant, 'Sʳ Harvie[32] you have bene veri surli and too surli heretofore:
and therefore if I shuld addere autoritatem, I think I shuld ignem igni
addere'.

Neville now asked leave to be excused because of urgent business and
he requested Osburn to proceed. Osburn proceeded and charged Harvey
with contentiousness because he 'had bene at wurds with him, with
M. Nevil, and I cannot tel with home besides'. Harvey here comments,
'To sai troth, I suppose verily, a little breach betwixt thes twoo and me
was the tru and onli caus of al thes sturs'.

Harvey then goes on to quote aptly from Virgil's *Aeneid* (ii. 97–100).
Freely translated these lines read: 'From then on, Ulysses always terrified
me with new charges; from that time on, he threw out double-meaning
words to the crowd, and conscious of his guilt, he sought means of attacking me. Nor did he rest until, using Calchas as his tool . . .'[33] These words,
although in themselves apt, are uttered by Sinon, a pretended deserter
from the Greeks, who spins a yarn of lies to gain the sympathy of his
Trojan captors by recounting his past misfortunes. Ultimately, he wins
their confidence sufficiently to persuade them to allow the wooden
horse within the gates of their city. One would think that Harvey might
have found an appropriate quotation from a more reputable character
than Sinon!

After this little literary digression, Harvey (continuing his account to
Young) writes that he countered Osburn's charge with the accusation
that the plaintiffs themselves played the major part in contentiousness.
The meeting broke up when the bell rang for Congregation. Harvey
bemoans:

And now tales run up and down the toun that Sʳ Harvi of Pembrook Hale
hath dun thus and thus, abusid thes men, and thes men, behavid him self after
this manner and this manner and thereuppon is staid in the hous: in so mutch
that thai that knew me not, can not choose but think veri il of me, and thai that
knew me too (unles it be sum of mi nearist acquaintanc) mai now begin to dout
of me. . . . Matters are made wurs and wurs in the telling: and now forsooth
mi not being sociable, arguith great arroganci; mi reprehending of others,
arguith great arroganci: . . . and to be short, everi thing, mi going, mi speaking, mi reading, mi behaviur arguith great and intolerable arroganci.

He comments bitterly, 'I suppose verily never bare was so baitid at the
stake with bandogs and masties as sum of them and namely gud M.
Nevil, a gud benefactur of mine, hath baitid and tuslid and chasid me'.

[32] 'Syr' is the title given to all Bachelors and thus a reminder to Harvey that he is not
an M.A.

[33] Trans. by Kevin Guinagh in Holt, Rinehart, and Winston edn. of *Aeneid* (1966),
p. 32.

Apparently now at his wits' ends, Harvey beseeches Young for help so that he may somehow get his grace in the house:

There is now no other wai, thai have handelid the matter so clarkly, and armid them selves so strongly. So gracius ar thai that have mi grace in there hands: and so ungracius am I, that am to seek mi grace at sutch mens hands. Whereas almost al the toun are gracid yea and admitted too alreddi. . . . But thes men, for ani thing I can see, wil never relent of them selvs, but bi your worships means, and commandiment. . . . For Gods sake, wai mi caus, what it is, and considder mi case how it standith: and as soon as you mai conveniently rid me out of sum of this greef.

In closing, Harvey apologizes, 'I know I have bene too too taedius unto you. Wherein I have most ernestly to request your wurship to pardun me: as also in mi over flat, and homeli kind of writing'.[34]

Young promptly wrote to Nuce and the other Pembroke Fellows, urging the passing of Harvey's grace without further delay. But this, too, brought no result except the Fellows' anger at being dealt with in such a high-handed manner. Harvey, becoming ever more desperate, as subsequent letters reveal, consulted with his friend, a Senior Fellow, Humphrey Tyndall. Tyndall seems to have advised Harvey to retire to Saffron Walden until the atmosphere cleared, while he himself volunteered to ride to London to persuade Young to come to Cambridge in person.[35] This was done and was effective, for Young, a forceful personality with shrewdness and a sense of timing, was able to push Harvey's grace through in short order. No sooner was this accomplished than the Senior Proctor, Walter Allen of Christ's College, gave Harvey first place in the University's Ordo Senioritatis. Highest rank in the Ordo seems to have been customarily assigned to those of noblest birth but there are some exceptions when top ranking was given to one of outstanding accomplishment (e.g. in 1532/3 to John Caius of Gonville Hall).

Another lesser honour redounded upon Harvey at this time: appointment as Greek lecturer. During Young's brief visit to Cambridge he, quite of his own accord, without solicitation from Harvey, arranged to have the young man installed in this post at Pembroke on 10 October 1573. But no sooner had the Master departed for London than the Fellows found means to retaliate.

Harvey describes the episode in his letter to Young of 1 November 1573 (several weeks after the happenings of which he writes):

Uppon Mundai, the same dai, that your wurship took your jurni towards Lundon, even immediatly before I should go to read the grek lecture (for the

34 Sloane MS. 93, fol. 11.
35 Sloane MS. 93, fols. 12 and 23 (letters to Tyndall and to Harvey's father).

whitch, as for mani things mo I account mi self infinitely bound unto you,
and the rather, bycaus it was frely offrid of you, not ambitiusly souht of me)
being in ded fully purposid, and providid to read, mi father sent for me of the
sudden, to go praesently to him to the Griffin. Whereuppon the bel being
tould to the lecture, as I had willid the butler before, I cam by, and by into the
hall, and tould the schollars mi busines was sutch, that I could not in ani wise
read that dai, willing them to provide them selvs of books against the next
dai, and telling them what book I intendid to read unto them.[36]

Gabriel's father was apparently not a person to be kept waiting.
Marginalia on sig. † 6ᵛ in Joannis Ramus's *Oikonomia* (1570) suggest that
he was readily irascible and a dominating person.[37] Not only had John
Harvey ruled his family, but from the beginning of October 1572 to October 1573 he had been Treasurer of Walden,[38] an office roughly equivalent
to that of a present-day Mayor. Shortly after his municipal duties were
completed and his term of Treasurer had come to an end, he made the
trip to Cambridgeshire and summoned his eldest son to meet him at
'the Griffin', probably a hostel in or near Cambridge.

As soon as Gabriel had departed from Pembroke for the meeting with
his father, Osburn took advantage of Harvey's absence to stir up irritation
against him, and with Neville's help it was rapidly fomented. By the
next morning (when Harvey had returned and both Fellows and Scholars
were assembled after prayers), Nuce affirmed that Dr. Young would be
content to have the Fellows hold a new election for the Greek lecturer
so that a more fitting man might be chosen. Since many of the Fellows
were annoyed that the Master had 'forcid mens voices', they were glad
of this seeming opportunity to reverse Young's actions. However, when
approval for a new election was sought by Nuce from the assembled
group of Fellows, a few refused and one preferred to wait for Young's
answer. Thereupon the Junior Proctor (Launcelot Browne) asked Nuce
in a rage if he went about to mock him, for it now appeared that Browne
was desirous of becoming Greek lecturer himself. There were heated
words, and in the hearing of all the Scholars Browne called Harvey
'Syr', asserting that Harvey should recognize that he was still only a
Bachelor and that Browne intended to pronounce him so openly in the
schools. At this point Harvey requested that the Scholars should be

---

[36] Sloane MS. 93, fol. 27. This letter runs from fols. 27ʳ to 34ᵛ.

[37] Gabriel here describes a flare-up of his father's: 'My father began to chyde and square
with me at the Table: I praesently, & doing my duty, ryse from the bowrd, saying only:
I pray you good Father, pray for me and I will pray for you'.

[38] 'Guild of the Holy Trinity Accounts 1545–1651', p. 82, has an accounting record from
3 Oct. 1572 to 3 Oct. 1573 headed by the names of 'John Harvey Treasurer' and of 'John
Jackson and William Malyn Chamberlains'.

dismissed so that no more of the dispute would be conducted in their hearing. Browne refused. Hot words were directed against both Harvey and Young: against Harvey because he had not, so the Fellows said, been selected by proper procedure; against Young because he had been too authoritarian.

Harvey's long letter to John Young recounts the petty bickering between the two factions and the humiliations to Harvey which ensued. His letter also makes plain his reverence for learning and his awareness that the majority of the student body did not share his values, that 'if sum miht have ther wils, thai shuld have nether greek, nor latin, nor ani thing els red unto them: but shuld run at randon, to what soever thai lusted'. Yet he has hopes that Pembroke may once again become a centre of learned men as it had been in an earlier day.[39] Harvey concludes his poignant letter to Pembroke's Master with the following passage:

For the bestowing of the lecture, do in it as you shal think best for the behoof of the Collidg. For mi part I am the more desirus of it, I must needs confes, bicaus of the stipend, whitch notwithstanding is not great;[40] and yit suerly I wuld refuse no pains, to do the schollars good, and to help forward lerning in the meanist, yea if there were no stipend at al. I know, and confes, I am able to do litle: but that, whitch I am able to do, bi mi private studdi, I wilbe reddi to do, to the Collidg proffit; and everi dai more, and more, as I shalbe better, and better hable. Only I wuld desier men to think the better, not the wors of me for so doing: as suerly I shuld do of them, if it miht like them to take ani pains that wai.

The fracas over the Greek lectureship was eventually settled by Dr. Young: the behind-the-scenes instigators, Neville and Osburn, were reprimanded; Browne was made aware of his unreasonableness; and in due course Harvey proceeded with his lectures. His *De Discenda Graeca Lingua* is a product of his teaching of Greek at Pembroke. Its two orations cover nine printed pages at the end of Jean Crespin's Greek–Latin dictionary.

This time, as in the previous 1573 attack by 'the spider' Neville, Harvey, despite being seriously threatened, escaped the ultimate fate of Spenser's butterfly Muiopotmos. Unfortunately, there were to be other spiders in Harvey's life, notably Greene and Nashe—the last the most

---

[39] He refers to 'the late ornaments of Cambridg, and the glori of Pembrook hal, bisshop Ridli, bisshop Grindal, M. Bradford, Doctor Car, M. Girlingtun, Doctor Hutton, and sum other that I could name'.

[40] A tithe of forty shillings had been given to the College before 1568 by Edmund Grindal (later Archbishop of Canterbury) to pay a fee to the Greek lecturer.

clever and deadly enemy of all, for Nashe, by irreparably destroying Harvey's hopes for a brilliant career, would in the 1590s leave him critically wounded, a pitiful 'spectacle of care'.

But, in 1573 Harvey managed to escape from the spider's web and was free to fly again, for a brief time in 1578 soaring close to the upper reaches of the atmosphere. After his reinstatement as Greek lecturer, Harvey seems quickly to have won the esteem of his colleagues, for he was chosen Bursar (or Junior Treasurer) before the end of 1573 and Senior Treasurer in 1575.[41] Other honours and a busy academic life followed. On 23 April 1574, while he was teaching Greek and continuing his tutorial duties,[42] he became University Praelector of Rhetoric.[43] He maintained this professorship until at least midsummer of 1576.[44] In this post he lectured a minimum of four days a week to all the first-year students and any other University members who wished to attend. His lectures seem to have drawn unusually large audiences. In *Rhetor* (sig. Ciiiᵛ) he refers to an attendance of almost four hundred; he also alludes (sig. Ai) to the fact that his lectures are embarrassingly popular, for far more distinguished professors, such as Thomas Byng (Regius Professor of Civil Law) and Bartholomew Dodington (Regius Professor of Greek) have on occasion been forced to address the walls and empty benches. Surely a rather indiscreet observation on Harvey's part!

The opening orations of spring 1574 and 1575 were printed as *Rhetor* by Henry Bynneman in November 1577. *Ciceronianus*, the inaugural lecture of Easter term 1576, was a slightly earlier publishing venture (June 1577) also printed by Bynneman, but it seems preferable to discuss them in the order in which they were delivered rather than that in which they were later printed.

*Rhetor, Vel duorum dierum Oratio, De Natura, Arte, & Exercitatione Rhetorica* (dedicated to Bartholomew Clerke of King's College, Cambridge, orator and jurist) sets forth three means to effective oratory: natural inclination, theory, and practice. In the first day's oration Harvey deals with *natura* and *ars* and in the second with *exercitatio*. Agricola,

---

[41] 'Aulae Pembrochianae Socii ab anno MDLXII ad MDCCX', tom. ii, p. 7.

[42] The beginning of 'Oratio Secunda' of *De Discenda Graeca Lingua* refers to his then having taught Greek at Cambridge almost two years. A letter of approximately Apr. 1574 to Sir Thomas Smyth (Sloane MS. 93, fol. 96ʳ) refers to his Rhetoric Lecture 'together with the reading to mi pupils'.

[43] John Venn, ed., *Grace Book Delta (1542–89)*, Cambridge Univ. Press, 1911, p. 274.

[44] *Ciceronianus* was the inaugural lecture of Easter term 1576, a term which continued until July. Marginalia on sig. M8ʳ of Quintilian are signed 'Gabrielis Harvejus, Rhetoricus Professor Cantabrig. 1573, 1574, 1575'. In 1573 he had been lecturing as Rhetoric Professor at the request of his predecessor, Robert Church (see Sloane MS. 93, fol. 96ʳ). His formal appointment to the office was 23 Apr. 1574.

Sturm, and Erasmus are mentioned, but Harvey's preferred authorities are the 'moderns', Ramus and Talaeus. Like Ramus, Harvey states that oratory uses both logic and rhetoric since it must have both matter and manner. Like Ramus, he dissents from traditional studies of logic and rhetoric because they include invention and disposition in both categories whereas, together with memory, they belong to the study of logic while style and delivery properly appertain to the study of rhetoric. In the following passage Harvey explains the way in which the five traditional parts of Ciceronian rhetoric should be redistributed:

For of that fivefold division, which has almost alone prevailed among our ancestors, how many now do not see that invention, disposition, and memory are not the property of speech but of thought, not of tongue but of mind, not of eloquence but of wisdom, not of rhetoric but of dialectic? Therefore two sole and as it were native parts remain as proper and germane to this art . . . —style and delivery, the former bright in the splendors of tropes and the involutions of schemes; the latter agreeable in the modulation of voice and the appropriateness of gesture. . . .[45]

In the second *Rhetor* oration Harvey deals primarily with the topic of practice or exercise and in this context presents an engaging allegorical character 'Exercitatio' whose mistress is 'Eloquentia'. Harold S. Wilson finds 'Exercitatio' reminiscent of the didactic eagle of Chaucer's *House of Fame*.[46] Thus Harvey indirectly instructs his students quite delightfully with a solemn playfulness. Here, as in the first oration and in *Ciceronianus*, despite his basic seriousness, he shows that he is well aware of the value of fun and humour.[47] In his copy of Quintilian he notes (on sig. V6[r]) that jesting is an important part of an orator's equipment.

In this second *Rhetor* lecture Harvey discusses at length the need for 'analysis' (scrutinizing someone else's composition to study its use of logic and rhetoric) and 'genesis' (creating a composition of one's own in which are applied what one has learned of invention, disposition, style, and delivery).

Like the *Rhetor*, the *Ciceronianus* is a work strongly influenced by

---

[45] The translation and some of the analysis of this work is taken from W. S. Howell, pp. 248–9. For Ramus's teachings, see Walter J. Ong, *Ramus: method, and the decay of dialogue: from the art of discourse to the art of reason*, Cambridge, Mass., 1958.

[46] 'Harvey's Orations on Rhetoric', *ELH*, 12, 3 (Sept. 1945), 178.

[47] At the beginning of *Rhetor* there is an amusing contrast between those who are serious auditors (*auditores*) of Harvey's lectures and those who are merely casual spectators (*spectatores*). The most humorous section of *Ciceronianus* is the Erasmian portrait of the ridiculously Ciceronian devotee Nosoponus.

Ramus's teachings, but it is a briefer, broader, and probably more important work. W. S. Howell writes of it:

Harvey's *Ciceronianus* is interesting not only as a plain indication of the methods of rhetorical analysis in process of being demonstrated to Cambridge undergraduates of the fifteen-seventies but also as an expression of a Cambridge man's idea of the change that was occurring in the intellectual climate of England. In one sense, this change meant the renunciation of a counterfeit Ciceronianism and the adoption of a true one. In another sense, it meant the end of a literary school devoted only to style, and the beginning of a school devoted to subject matter as well. In still another sense, it meant a realignment of English learning towards France and Germany as opposed to Italy. And finally it meant an endorsement of Cambridge as the brightest future star in the firmament of European as well as strictly English scholarship.[48]

In this work one becomes aware of certain positive qualities of Harvey's. He was no doubt an excellent teacher, for he presents his erudite subject matter in an interesting, well-organized way. His discourse itself is an exemplary model of the form of a Ciceronian oration. Yet he does not forget that he is speaking before an audience composed primarily of first-year students between thirteen and seventeen years of age. He conveys to them his enthusiasm for his subject and assures them that there are within their ranks potential future Ciceros. He makes clear his liberal-mindedness, unwilling to be trammelled by time-honoured notions because they have traditionally been accepted. He examines these notions anew and reaches independent conclusions. One senses that he must have inspired his students, opening their eyes in many ways to new perspectives.[49]

The sub-title of *Ciceronianus* is *Oratio post reditum, habita Cantabrigiae ad suos Auditores*, an obvious allusion to Cicero's *Oratio post reditum in Senatu*. In his discourse Harvey refers to his return to Cambridge after a sojourn of something less than twenty weeks at his home in Walden. This long absence from Cambridge preceding the 1576 Easter term has not been explained with certainty,[50] but Harvey himself gives some suggestive background.

He begins the *Ciceronianus* by explaining that he is returning to Cambridge somewhat as Cicero returned to Rome after a visit to his Tusculan

---

[48] *Logic and Rhetoric in England*, pp. 250–1.

[49] For a translation of *Ciceronianus* and a discussion of its content and background, see the excellent edition by Clarence A. Forbes and Harold S. Wilson, *Gabriel Harvey's 'Ciceronianus'*, Univ. of Nebraska Studies, Nov. 1945.

[50] Moore Smith believes it was because of plague at Cambridge, but Harold S. Wilson is doubtful since there is no record of a University adjournment at this time. Wilson presents several possible alternative explanations but reaches no firm conclusion.

villa for recollection and refreshment. There he had rusticated for some time with friends and kinsmen and had also spent enjoyable hours in studying, reading, and writing—not weighty treatises but pleasant works for private relaxation and recreation. At Harvey's 'Tusculan villa' he has similarly spent his leisure. His home has been for him a sort of suburban school of rhetoric and philosophy where in a relaxed way he could dine on the fruits of classical authors as well as on those of certain interesting later writers. Most recently he has been studying Macrobius' *Saturnalia*.

The two previous years must certainly have been extremely busy ones for Harvey with his Rhetoric and his Greek lectures, his tutoring, and other University duties, as well as the writing of what was probably his first printed work, the elegy to Ramus, *Ode Natalitia* (published in 1575).[51] Harvey might well have been temporarily exhausted and eager for a change of scene and studies. We know, too, that he had political aspirations, so he may have felt that this was an opportunity to renew his talks with Sir Thomas Smyth and perhaps engage in discussions with those who might visit his home at Theydon Mount. By 1578-9 Harvey was well acquainted with Leicester, Philip Sidney, and Daniel Rogers. It is possible that he may have met them as early as 1576[52] in Essex or in excursions from there.

Harvey may also have welcomed the chance to remove himself from the Cambridge area because of the plague which had been prevalent in the previous term (see p. 30 n. 50, above). Perhaps he preferred not to teach in the Easter term and obtained a temporary leave of absence. In *Ciceronianus* he states that his friend Jones (the new Praelector of Philosophy) had remarked that it was not surprising that his predecessor Duffield had deserted philosophy, for Harvey's beloved profession of eloquence had been forsaken by him.[53] Harvey is quick to deny this,[54] but, as the following evidence suggests, it seems likely that for a while he considered just such a course.

In his copy of Quintilian,[55] as previously noted, Harvey signed his name and appended to it the title 'Rhetoricus Professor, Cantabrigiae, 1573, 1574, 1575' as though 1575 were for him the termination of an episode. In the letter prefacing *Ciceronianus* William Lewin refers to his

[51] Gascoigne's *Posies* (1575) includes a ten-line commendatory poem signed 'G.H.' which is undoubtedly Harvey's, but *Ode Natalitia* (1575) seems to be the first complete published work of Harvey's own. See note 29 above.

[52] See *Gratulationes Valdinenses, Liber II*, sig. fiiʳ.

[53] *Ciceronianus* (Forbes–Wilson edn.), pp. 58–9.

[54] Ibid. But p. 28 n. 44, above, suggests that Harvey thought of his Praelectorship as concluding in 1575.

[55] On sig. M8ʳ.

esteemed friend Harvey who 'had he remained in his illustrious function as praelector, would have won unbelievable profit and glory both for himself and his entire university'.[56] This was written in February 1576/7 (almost a year after Harvey's 'Tusculan villa' sojourn), by which time he, apparently of his own choice, had definitely relinquished the professorship. It seems reasonable that he had considered the move a year earlier. One can speculate that word reached Harvey before classes were to recommence for the Easter term that no satisfactory replacement had been found for the Rhetoric Praelectorship and he was asked to continue for at least another term.

[56] Translation is by Clarence A. Forbes, *Ciceronianus*, ed. cit., p. 43.

# 4. Essex interim: reaching for the world outside Cambridge

In October 1577 Harvey's friend George Gascoigne died. Shortly thereafter Harvey composed a verse tribute to him, for in Sloane MS. 93 (usually referred to as Harvey's 'Letter-book'[1]) following the series of Pembroke letters discussed on pages 17–27 above, there is in Harvey's hand a scribbled inscription reading:

A neue Pamflett conteininge a fewe delicate Poeticall Devises of Mr. G. H., extemporally written by him in Essex, at the ernest request of a certain gentleman a worshipfull frende of his, and made as it were under the gentlemans owne person, immediatly uppon the reporte of the deathe of M. Georg Gascoigne Esquier, and since not perusid bye the autir. Published by a familiar frende of his, that copyed them owte praesently after they were first compiled with the same frends praeface of dutifull commendation and certayne other gallante appurtenances worth the readinge.

Directly following this inscription four pages have been cut from the manuscript. Preceding the excerpted pages are some brief Latin verses in praise of Gascoigne and following the missing section are a number of English quatrains scrawled all over the page in one of Harvey's least legible hands. They tell of the various eminent countrymen Gascoigne will meet upon his entry to heaven: Chaucer, Gower, Lydgate, Surrey, More, and the Withipolls (Daniel and Batt).[2] A stanza which mentions Skoggin and Skelton[3] has been scored through. It is unlikely that any of the remaining verses were part of the 'neue Pamflett'. Probably it was

---

[1] It was so titled by Edward John Long Scott who transcribed the manuscript and had it published by the Camden Society, London, in 1884 as the *Letter-book of Gabriel Harvey A.D. 1573–1580*.

[2] Bartholomew Wythipoll ('Batt'), third oldest of eleven brothers, died in 1573, Daniel, the fifth brother, was born about 1540 and died before 1577. Both were friends of Gascoigne's and belonged to an Essex family with which Harvey seems to have been well acquainted. In his marginalia in Hopperus (sig. Aa3r), he refers to a younger brother Peter who was a contemporary of his at Trinity Hall (see p. 69, below). There are also references in his writings to 'Old Maister Withipoll' (Edmund) indicating that Harvey admired and knew the boys' father well. See G. C. Moore Smith, 'The Family of Wythipoll', in *Walthamstow Antiquarian Society Official Publication*, no. 34 (1936), p. 59.

[3] 'And cause thou art a merry mate / Lo Schoggin where he lawghes aloane / And Skelton that same madbrayned knave / Looke how he knawes a deade horse boane.' The last line is a reference to Skelton's *Colin Cloute*, line 478.

circulated in manuscript but there is no record of its having been published in print.

Harvey's so-called 'Letter-book' contains in its mid section a number of other original writings dating from 1573 to about 1579. Beginning on folio 58 is the rough draft of a series of amusing, sardonic verses entitled 'The Schollars Loove: or a Reconcilement of Contraryes'.[4] It is, Harvey notes, 'the very first peece of Inglish Ryme that ever the autor committed to wrytinge: and was in a rage devised and deliverid pro and contra according to the quality of his first and last humour'. He further describes it as 'an amorous odious sonnet, intituled the Students Loove or Hatrid, or both or nether, or what shall please the looving or hating reader, ether in sport or ernest to make of such contrary passions as ar here discoursid'. This outburst, unless it is purely fictional, seems to have been occasioned by his having been jilted or at least 'treacherously' treated by a young lady whom in the poem he calls 'Ellena'. He says of her:

> Parisses Helena not like my Ellena.
> A primrose with a witnesse, a heavenly demigoddesse
> Descended from the sky, in aethereal bravery,
> To ravish every eie that regardith her bewtie.
> I dare bowldely sownde it, as sum eies have fownde it.
> Her very excrementes nothing but incrementes
> Of pothecarye restoratives, to comforte yungemens lives.
> No shame at all to tell such a tale.
> Her spitt is Hypocrasse, the rest I lett passe,
> The dunge of a muskecatt not like unto that.
> A compownde of marchepane, a very Diane
> Now may I well bragg of my piggesnye,
> The sunne did never her like espye.
> Or she is matchelesse, or I am senselesse.
> The oddist wight, in my eiesight,
> That ever man begatt of woman.

Aware that his Skeltonic verse leaves something to be desired, he writes:

> Looke for no measure in my ryme,
> Passions, you know, observe no tyme.

In another section of what is apparently the same long poem, he comments on Venus:

> A likely tale,
> And smelles of ale:
> That Venus cam of water.
> Likelier it is,

4 It is dated September 1573.

And yet amisse,
That whott wine was the matter.
If I can gesse the Reason whye,
Me thinkes, is mente the Contrarye

Fyer was father,
Fyer was moother,
Fyer was nurse and all:
Fyer was the matter,
Fyer was the manner,
Fyer was the cause finall.

Can cowldnes heate?
Can water burne?
Doth sea ingender flame?
You gabb fonde poetts, or in bowrde,
You blason Neptunes name.

A strange effecte
If it were true,
That water showlde inflame.
Ile sooner howlde,
The fyer is cowlde,
And pleade it with lesse shame

.     .     .     .     .

Death and life
In the same knife,
Sea and lande
In the same hande,
Fier and water,
In one platter?[5]

His awkward, longer, convoluted verses in this collection are also an interplay of conflicting passions. One suspects that Harvey was a highly emotional person who was constantly struggling to keep intense feelings under control. Nashe in *Have with you to Saffron-walden* (*Works*, iii. 81) writes, 'Gabriell was alwayes in love'. He evidently was also capable of much bitterness.

About Christmas-time of 1574 Gabriel seems to have played a role in preventing the seduction of his young sister Mercy (or Marcie) by the seventeen-year-old nobleman Philip Howard, Earl of Surrey and later of Arundel.[6] Philip was the elder son of Thomas Howard, Duke of

---

[5] Verses quoted are from fols. 58ᵇ and 59ᵇ of Sloane MS. 93.
[6] G. C. Moore Smith in *Notes and Queries*, 8 Apr. 1911, pp. 261–3, demonstrates convincingly that the young lord of this incident was Philip Howard (born 28 June 1557).

Norfolk, who had been executed in 1572 for suspected treason with Mary, Queen of Scots. Philip, who together with his young wife Anne was living at Audley End about three miles from Walden, had, according to Harvey's account,[7] caught a glimpse of Mercy in the fields of her father's farm a mile from town,[8] was attracted to her, and through messages carried by his servant attempts to arrange an assignation. Mercy, flattered by the nobleman's attentions (for he sent her various gifts), is torn between his enticement and her determination to preserve her virtue. There is an exchange of letters in which she reminds him that her 'bringing upp hath bene allwaies so homelie and milkmaidelike', that he is a great lord with a wife, and she signs herself 'Pore M'. Seemingly afraid to reject his insistence on a rendezvous, each time she cleverly but precariously manages to elude his grasp. He, however, is undeterred in his pursuit. Accusing her of cruelty, he writes continually of his unhappiness.

As Gabriel relates, on a trip home from Cambridge he happened to intercept one of Philip's letters and, having appraised the situation, he handles it in a tactful but effective manner by writing deferentially to the young lord to inform him that someone is evidently masquerading in his name in order to attempt a seduction of Gabriel's sister. He suggests that Philip investigate who this charlatan may be and put a stop to his actions so that 'Milord's' name will not suffer.

In his marginalia Harvey alludes to this incident in several places. For example, in his copy of Erasmus's *Parabolae* he brackets and underlines the following passage:

qui viro *malo* addit *opes* & *gloriam*; is *febricitanti* ministrat *vinum*; *bilioso mel*; *coeliacis opsonia*; quae morbum animi, hoc est, *stultitiam augeant.*[9]   (sig. C8ᵛ.)

---

[7] Sloane MS. 93, fols. 71ᵛ–84ʳ.

[8] There is extant a 1670 survey of a large tract of land of about 139 acres then known as 'Harvies Farm' located in Wimbish (on the outskirts of Walden near Audley End). It included hopland, groves, a mansion house, etc. There is a reference to a small portion of it having been purchased from one Ralph Collins. The document is now in the Town Hall safe at Saffron Walden within a volume titled 'Book of Almshouse Properties: 1540–1713'. The volume is unfoliated but the account referred to is found on five pages headed 'Harvies Land' and having the date 1670 in the left margin. It describes 'The Farme called Harvies lying in Wimbish now in the occupacion of Richard Bands'. Although I have not traced the provenance of this tract it seems possible that this may have been the farm of Gabriel's father where hops were grown and malt brewed, and that a small portion of it at sometime came into the possession of a descendant of Gabriel's brother-in-law, Phillip Collyn. Gabriel's grandmother's will mentions 'a great kettel with a pan to cole in worte'. (See note 1 in Appendix A, p. 256, which indicates that brewing was carried on by the family).

[9] 'He who adds wealth and glory to a wicked man is serving wine to one ill of a fever, honey to one who is bilious, viands to those afflicted with pain in the bowels, an illness to this mind, thus it is, they increase folly.' Italics represent Harvey's underlinings.

Just below this passage in the left margin he comments:

iᵃ85 ¹⁰

You know who used to write, Unhappy Philip. At bonus titulus, Bonum omen: et animosa Alacritas, splendidum prognosticon Victoriae.¹¹

On page 85 (sig. f3ʳ), which is Harvey's cross-reference, we find the following text bracketed and underlined by him:

⎧Huius servus interrogatus quid ageret dominus; cum adsint, inquit bona,
⎨
⎩quaerit mala.¹²

Next to these lines of text Harvey annotates: '= unhappie Philip.' The volume was acquired by Harvey in 1566, read by him at some time thereafter, and re-read in September of 1577.¹³ One presumes that these annotations were made at the time of re-reading.

Nashe, always ready to point out likely embarrassments to Harvey, alludes to his 'bawdy sister' in *Have with you to Saffron-walden*, 1596 (*Works*. iii. 129):

I will not present into the Arches or Commissaries Court¹⁴ what *prinkum prankums* Gentlemen (his nere neighbors) have whispred to me of his Sister, and how shee is as good a fellow as ever turned belly to belly; for which she is not to be blam'd, but I rather pitie her and thinke she cannot doo withall, having no other dowrie to marie her. . . . Had it not been for his baudy sister, I should have forgot to have answerd for the *baudie* rymes he threapes upon me.

In Sloane MS. 93 (fols. 92ᵛ–93ʳ), a few folios after the account of the Christmas 1574 incident there is the copy of a letter from Pembroke Hall dated '29 March' (year unspecified)¹⁵ to Lady (Philippa) Smyth, the wife of Sir Thomas Smyth, asking whether she could find a place in her household for 'a pore sister of mine' who 'for sundri good causis' is 'marvelous desirous to do you service'. Harvey hopes that her Ladyship will accept her as she is eager for such a position. He assures Lady Smyth that she will have 'a diligent, and trusti, and tractable maiden of hir, besides sutch service as she is able to do in sowing, and the like qualities

---

¹⁰ 'iᵃ' together with a page number is used by Harvey as a cross-reference symbol to signify '*infra*' (i.e. 'see below)'.

¹¹ '. . . But a good title is a good sign; and bold eagerness is a splendid omen of victory.'

¹² 'When this man's servant was asked what his master was going to do, [he replied] "when they arrive, he says good things, he seeks evil ones".'

¹³ On sig. l4ᵛ Harvey writes: 'Relegi mense Septembri 1577.'

¹⁴ Harvey was a lawyer in the Court of Arches.

¹⁵ The contents of the letter indicate that it was written while Sir Thomas was still alive.

requisite in a maid'. Perhaps Gabriel's solicitation to Lady Smyth is an attempt to keep Mercy out of further trouble.[16]

On 12 August 1577 after long and painful illness Sir Thomas Smyth died at his home in Essex and with his death Harvey lost a close friend, wise counsellor, and benefactor. His funeral must have been a sad occasion for the young man, but Nashe ridicules his mournfulness as he does almost everything else about Harvey:

[Harvey] tells a foolish twittle twattle boasting tale (amidst his impudent brazen fac'd defamation of Doctor *Perne*,) of the Funerall of his kinsman, *Sir Thomas Smith*, (which word *kinsman* I wondered he causd not to be set in great capitall letters,) and how in those Obsequies he was a chiefe Mourner.[17]

Harvey himself refers to the funeral at which the Cambridge Vice-Chancellor Andrew Perne preached and where he and Harvey exchanged some sharp words after the ceremonies. The subject was some 'rare manuscript bookes' of Smyth's given to Harvey by Lady Smyth and her co-executors. Perne, who like Harvey was a bibliophile and had an extensive library, apparently coveted these manuscripts and had hoped to acquire them. When he found that they had already been bestowed upon Harvey, he accused him of foxiness, probably unjustifiably, for Harvey was the logical person to receive them from the estate since he had been Sir Thomas's intellectual protégé and a close friend of his son Thomas and of his nephew John Wood. But the Vice-Chancellor resented being thwarted and apparently was annoyed at Harvey's retort to him.[18]

---

[16] It has been suggested by Janet Biller in the biographical introduction to her excellent critical edition of Harvey's *Foure Letters*, 1592 (Columbia Univ. diss., 1969), p. xiii, n. 1, that Harvey's tale about the attempted seduction of his sister is a fiction—'a novella in miniature', but I fail to agree. For the following reasons I believe it to be a true account (although perhaps an overly dramatized one) of a factual incident: (1) Harvey seems never to have been a particularly imaginative writer, (2) there are corroborative notes in Harvey's marginalia and perhaps in his letter to Lady Smyth, (3) Harvey's purpose in writing this account was, I believe, to have an accurate record of the events of this incident in case it were ever needed in his or his sister's defence, but not otherwise to be used for publication. Besides, he would hardly have dared connect Lord Surrey with such a story if it were completely fanciful. Harvey gives so many clues that the identification of the seducer would have been unmistakable to his contemporaries, even though Harvey does not give his full name.

[17] *Works*, iii. 58.

[18] In *Pierces Supererogation* (1593) (sig. Ddi), Harvey gives the following description of the incident: '[Perne] once in a scoldes pollicie called me Foxe between jest, and earnest: (it was at the funerall of the honorable Sir Thomas Smith, where he preached, and where it pleased my Lady Smith, and the co-executours to bestow certaine rare manuscript bookes upon me, which he desired): I aunswered him betweene earnest, & jest, I might haply be a Cubb, as I might be used, but was over young to be a Fox, especially in his presence. He smiled, and replyed after his manner, with a Chameleons gape, and a very emphaticall nodd of the head.'

Ever after Perne seems to have harboured a deep grudge against him which was manifested in devious ways including opposing Harvey's candidacy for the University Oratorship in 1579[19] and manœuvring to prevent his election as Master of Trinity Hall in 1585.[20]

On the day after the funeral Harvey began to compose a group of Latin elegies eulogizing Smith. Five months later (January 1578) they were published by Bynneman as *Smithus, vel Lachrymae Musarum*. The work consists of a series of verse laments uttered by each of the Muses in turn, a format to be used again in 1591 by Spenser in *The Teares of the Muses* as part of his *Complaints . . . of the Worlds Vanitie*.

Harvey seems now to have set his sights on the wider horizons of The Court, for he had by this time become acquainted with such luminaries as the Earl of Leicester, Philip Sidney, Edward Dyer, and Daniel Rogers.[21] In early 1578 there apparently were plans for Harvey's going abroad, in attendance, to a conference of Protestant princes to be held on 7 June at Schmalkalden in Germany.[22] At the urging of Duke Casimir whom Sidney had recently visited, Elizabeth had chosen four deputies to represent England at the Conference: Laurence Humphrey, John Still, John Hammon, and Daniel Rogers. Rogers and Still were friends of Harvey's and may have secured some appointment for him in connection with the mission.[23] Possibly because of Elizabeth's reluctance to involve England in foreign commitments, the project fell through and the deputies never left English shores.

On 26 July 1578 in the course of one of her frequent 'Progresses', the Queen stopped at the great estate of Audley End in Essex and during her few days' stay was presented with gifts and entertained by Cambridge dignitaries and some of its outstanding scholars. Harvey was there to

---

[19] See p. 53, below.

[20] See p. 77, below.

[21] Harvey's *Gratulationes Valdinenses*, 1578 has on sig. Fiiʳ a poem described as having been presented to Leicester in 1576.

On fol. 53ʳ of Sloane MS. 93 is the copy of a letter from Harvey to Spenser in which he writes: 'The twooe worthy gentlemen, Mʳ Sidney and Mʳ Dyer, have me, I thanke them, in sum use of familiaritye.' In *Three Proper and wittie, familiar Letters*, 1580, sig. F3ʳ Harvey refers to 'my good friend *M. Daniel Rogers*, whose curtesies are also registred in my Marble booke'. Harvey also alludes to Rogers in some of his early marginalia. Rogers who was born about 1538 and died in 1591 was a courtier and foreign agent for Elizabeth. See J. A. Van Dorsten, *Poets, Patrons, and Professors: Sir Philip Sidney, Daniel Rogers, and the Leiden Humanists*, Leyden and London, 1962, pp. 9–75.

[22] Richard Harvey in his dedication to Bishop John Aylmer of the 1583 *Astrological Discourse* refers to Aylmer's 'singular curtesie toward my brother Gabriel when he should have travailed to Smalcaldie'.

[23] See Moore Smith, *Marginalia*, p. 21, n. 1 which discusses the aborted plans to send English deputies to the Schmalkalden conference.

take part in two philosophical disputations in Latin by University Masters of Art on the topics: 'Clementia magis in Principe laudanda quam severitas' and 'Astra non imponunt necessitatem'. The positive of each topic was presented by Master Fleming of King's College while Harvey headed the opposition assisted by Master Palmer of St. John's and Master Hawkings of Peterhouse. The negative of these questions argued by Harvey's team won the decision of the judge, Thomas Byng of Clare Hall. The disputation, which is said to have lasted over three hours, was moderated by Master Fletcher of King's College, informally assisted by the Lord Treasurer Burghley who showed great interest in the proceedings, as did the Queen. Official gifts of gloves elaborately embroidered with appropriate mottoes and coats of arms were made by the University to Elizabeth and her chief nobles.[24]

Harvey presented as gifts of his own four manuscripts of Latin verse written on large folio-sized sheets in his ornamental Italian hand. One of these manuscripts is preserved in the British Library (Lansdowne MS. 120, fols. 179–87) and is entitled 'Gabrielis Harveii χαῖρε, vel Gratulatio Valdinensis, ad Honoratissimum, clarissimumque virum, Dominum Burgleium, magnum Angliae Thesaurarium, summumque Academiae nostrae Cantabrigiensis Cancellarium; Audleianis aedibus una cum Regia ipsa Maiestate, reliquis Nobilibus honorificentissime exceptum'. It consists of elegiac verses of Harvey's supplemented by an epigram by Abraham Hartwell and an ode by Janus Dousa,[25] and concluded by two epigrams of Harvey's, 'in effigiem Democriti' and 'in effigiem Heracliti', on the lamentable depravity of the age.

In September of 1578 Henry Bynneman printed a volume composed of the four manuscripts plus certain additions. This volume, the *Gratulationes Valdinenses* is comprised of four books of Latin verse: Book I addressed to Elizabeth, Book II to Leicester, Book III to Burghley, and Book IV to Oxford, Hatton, and Sidney. The printed edition was presented to the Queen by Harvey on 15 September[26] at Hadham Hall, Hertfordshire, the estate of his friend Arthur Capel.[27] In the September

---

[24] The proceedings at Audley End are described in John Nichols's *Progresses and Public Processions of Queen Elizabeth*, London, 1823, ii. 222 f. and in Richard, Lord Braybrooke's *History of Audley End*, London, 1836, pp. 74–7. Both accounts seem to be based on Cole MSS.

[25] Abraham Hartwell (born *c.* 1542), an eminent scholar and Fellow of King's College, Cambridge, became known for his Latin poetry; Janus Dousa, governor of Leyden, was a poet and scholar and member of the Sidney circle. For further background on him, see Van Dorsten, pp. 1–9 and *passim*.

[26] Nichols, op. cit., ii. 222, records the stop at Capel's that day.

[27] Two letters of Harvey's to Arthur Capel are found in Sloane MS. 93 (fols. 90ᵛ and 102ʳ). They bespeak close friendship.

Eclogue of Spenser's *Shepheardes Calender* (1579) E. K. writes as a gloss to Spenser's mentions of 'Hobbinoll':

Mayster Gabriel Harvey: of whose speciall commendation, aswell in Poetrye as Rhetorike and other choyce learning, we have lately had a sufficient tryall in diverse his workes, but specially in his Musarum Lachrymae, and in his late Gratulationum Valdinensium which boke in the progresse at Audley in Essex, he dedicated in writing to her Majestie. Afterward presenting the same in print unto her Highnesse at the worshipfull Maister Capells in Hertfordshire.

A comparison of the manuscript presented to Burghley at Audley End with Book III, its printed version, reveals that the latter was amplified by the addition of two epigrams: one by Walter Haddon and one by Pietro Bizari.[28] Books I, II, and IV were evidently similarly amplified by verse additions. Although the original manuscripts of these three books are no longer extant, one can conjecture as to which sections were added in the printed edition. For instance, in Book I to Elizabeth there are two fairly lengthy passages that refer to happenings at Audley End during the Royal visit at the end of July: they treat of the Queen's permitting Harvey to kiss her hand ('Epilogus, de Regiae Manus osculatione') and her remark that Harvey looked like an Italian ('Pars Epilogi secunda: De vultu Itali'). In Book II to Leicester there are curious satirical verses relating to Machiavelli and his apotheosis which suggest topical allusions. Thomas Hugh Jameson has submitted that 'the satires in this revised book are not authentic satires on Machiavelli, but warnings against Alençon [as suitor to Elizabeth], his tools, and his Italianate connections, the de' Medici'. Although Machiavelli 'about whom there was this foolish hue and cry, was dead; . . . the people who were the real exemplars of his supposed methods were not only very much alive, but were seeking to gain entrance to England. Their ambassadors [Quissy and Bacqueville] were even then at the English Court!'[29]

In the printed Book II (addressed to Leicester) are certain *nova carmina*: a bitter burlesque representing Machiavelli speaking in person (which sounds like a forerunner of Marlowe's prologue to the *Jew of Malta*) and a short dramatic poem, 'The Florentine Mercury or Machiavelli's Apotheosis', in which Mercury pondering the writings of Machiavelli (which Cosimo de' Medici has shown him), notes the similarity in their methods

---

[28] Walter Haddon, a friend of Leicester's, was Regius Professor of Civil Law at Cambridge and was one of the foremost English Latinists. Pietro Bizari (1530?–1586?), one of Leicester's protégés, was an Italian historian and poet who because of his Protestantism had settled in England.

[29] Thomas Hugh Jameson, 'The "Machiavellianism" of Gabriel Harvey', *PMLA*, 56 (1941), 647, 649.

and the near fusion of their identities.[30] The poem is followed by a sardonic hymn in which all the Medici join in praise of their fellow townsman. The hymn concludes:

He who was Machiavelli, is Mercury, a god. Him then you robbers, merchants, hard tyrants, vagabond lawyers, smooth magistrates, perjurers, two-faced Januses, traitors, savage and league-breaking leaders . . . all worship! . . . O happy house of the Medici . . . who are led by such a genius, such a star![31]

Since in Book II Harvey urges his patron Leicester to marry Elizabeth, Jameson's view that Harvey's 'added' verses are directed against the proposed Alençon marriage seems a tenable one. Irony, although less explicit, is also found in the assumed additions to Book I, especially in the 'De Vultu Itali' epilogue. At Audley End, Leicester had apparently introduced Harvey to the Queen as an attractive and able young man now in her service, one whom he was planning to send to Italy and France. Elizabeth graciously gave the young man her hand to kiss and evidently remarked to Leicester that he had made a good choice in Harvey, for 'even now he has the face, the look of an Italian!'[32] Harvey uses the Queen's comment as the point of departure for a seemingly rapturous verse passage, 'Epilogus, de Regiae Manus osculatione; deque eo, quod vultum Itali habere, ab excellentissima Principe diceretur. Pars prima, de Osculo. Pars Epilogi secunda: De vultu Itali'. In the second part he exaggeratedly praises everything Italian; but the very excessiveness of the lines and the frequently ambiguous praise indicate irony. For instance:

Certainly, I venerate Italians. They are a polished people and singly, too, most wise; you'd think most of them are sons of Mercury. Eloquence swims on their lips, their minds are like a labyrinth, blocked with manifold turns. No one of them but wins honor by some merit! A magnificent, spirited animal that holds everything inferior to itself, Kings, crowns, scepters, whatever had dominion over the wide world!—This is the special merit of the Italians, like so many flies to throng courts and Kings' palaces, to gorge themselves on golden fare, both drink and eat regal tid-bits, be present on all occasions, at no time and from no place be excluded. . . .[33]

[30] Although Harvey's marginalia show him to be an admirer of Machiavelli, he is here using 'Machiavellianism' in the popular sense of 'something cleverly treacherous'.

[31] Translation by Thomas Hugh Jameson from his edition of 'The Gratulationes Valdinenses of Gabriel Harvey', unpublished Yale University dissertation (1938), p. 652. The Latin passages which he has translated are on sigs. F1ᵛ–F2ʳ of Harvey's Gratulationes Valdinenses, Book II.

[32] Harvey denies Italianate qualities but he admits that he may be somewhat Italian in appearance for he has black hair and a rather dark complexion (sig. Ciiiiᵛ).

[33] This is Jameson's translation (dissertation, op. cit., p. 650) of Harvey's Latin lines on sig. D1ʳ.

As Jameson notes, if by 'Italians' Harvey means representatives of Catherine de' Medici and her son, this is 'bowld Courtly speaking'.[34] Harvey suggests that Elizabeth herself is in part responsible for allowing 'Italianate' influence to seep into England. He intimates that she fails to discriminate between admirable Italo-Roman cultural influences and 'Machiavellian' devious treachery. One should be accepted, the other guarded against.

Given Harvey's balance of praise and criticism and his excessive subtlety his message undoubtedly escaped the general reader. It is quite likely, however, that his argument was recognized by Leicester and perhaps the Queen and Burghley—and if it was, it is not surprising that Harvey never moved higher in Court circles. Leicester kept him dangling in his service for a while, during which time Harvey studied Italian and French,[35] becoming very fluent in the former though acquiring little more than a reading knowledge of the latter. But, so far as can be ascertained, Leicester never did entrust him with the foreign assignment that Harvey anticipated.

Book II was damaging to him for yet another reason. By urging Leicester's suitability as a spouse for Elizabeth, he put his patron in an embarrassing situation: although Harvey was completely unaware of it, Leicester was at this time secretly married to the Countess of Essex.

In Book IV, as noted, are a group of encomia: to Oxford, Hatton, and Philip Sidney, the last being recognized by Harvey as a model of conduct. Harvey seems one of the first to point to him as an exemplar of the perfect courtier, for the Sidney tributes[36] are followed by two verse commentaries: *Castilio, sive Aulicus* and *De Aulica*. Both are in the 'courtesy book' tradition, based primarily on Castiglione's *Il Cortegiano*[37] whose precepts Harvey states in his own words in a concise and memorable way. He even agrees with its emphasis on skill in arms as being more important for the courtier than skill in letters. In accord with the Humanists' programme he

---

[34] Advocated by Harvey in the marginalia of his copy of Ramus's *Oikonomia* (in the preliminary index pages).

[35] In Florio's *First Fruites* (1578) on sig. Eei[v] Harvey refers to John Florio and John Eliot as 'mie new London Companions for Italian & French. Two of the best for both.' From this time on Harvey bought and studied a number of Italian grammars and texts, also some in French and some in Spanish.

[36] They are entitled 'Ad Nobilissimum, humanissimumque Iuvenem, Philippum SIDNEIUM, mihi multis nominibus longe charissimum' and 'Elegia ad eundem, paulo ante discessum'. According to the State Papers, Foreign (1578–9), no. 149, dated 7 August, Leicester in a letter to Sir Francis Walsingham refers to the Queen's comments when his nephew Philip was about to take his leave and receive his dispatch for Duke Casimir. Probably Sidney had planned to leave Audley End before the proceedings were over.

[37] Harvey owned copies of *The Courtier* in Italian, English, and Latin. The English copy (now at the Newberry Library) contains profuse annotations by Harvey. See p. 205, below.

stresses imitation as a method of learning especially important in the study of languages, and he requires for his courtier a knowledge of law. Like Castiglione, he insists on nobleness of birth in the lady of the Court; he is less explicit (perhaps for personal reasons) about the courtier. Harvey's *aulicus* must have skill in poetry and he must study music, painting, dancing, hunting, and various physical exercises.[38] Although there seems to be no evidence that Harvey himself studied music or painting, the would-be courtier did develop reasonable skill in all the other required areas. Harvey (like Spenser)[39] valued courtesy and defined it as the civilized ideal of making oneself adaptable to one's fellow men.

Nashe with his pen of humorous ridicule sketches a picture of Harvey's deportment at Audley End.[40] Reading between the lines, one gathers that Harvey was intent upon using this golden opportunity to make an impression and curry favour with the great lords and ladies of the Court. Garbed in his suit of velvet and scintillating with witty repartee, he undoubtedly in his eagerness had forgotten Castiglione's important lessons in *sprezzatura*, the nonchalance that was becoming to a gentleman. According to Nashe's version:

There did this our *Talatamtana* or *Doctour Hum*[41] thrust himselfe into the thickest rankes of the Noblemen and Gallants, and whatsoever they were arguing of, he would not misse to catch hold of, or strike in at the one end, and take the theame out of their mouths, or it should goe hard. In selfe same order was hee at his pretie toyes and amorous glaunces and purposes with the Damsells, & putting baudy riddles unto them . . . . making love to those soft skind soules & sweet Nymphes of *Helicon*, betwixt a kinde of carelesse rude ruffianisme and curious finicall complement; both which hee more exprest by his countenance than anie good jests that hee uttered.[42]

After Harvey was allowed to kiss the Queen's hand and she commented that 'he lookt something like an Italian', he, according to Nashe, became ecstatic and started speaking with a marked Italian accent.

Harvey now followed the train of the 'delicatest favorites and minions' who had 'withdrawne a mile or two off, to one Master *Bradburies*'

---

[38] Detailed analysis of Harvey's *Aulicus* and *De Aulica* is found in George L. Barnett's 'Gabriel Harvey's *Castilio, sive Aulicus* and *De Aulica*', in *SP*, 42 (1945), 146–63.

[39] See especially *Faerie Queene*, VI.

[40] Nashe was only eleven years old in 1578. He must have drawn on hearsay or on *Pedantius* for this description.

[41] These seem to be topical allusions but have not been traced, although 'Talatamtana' may refer to the Clare Hall play described on p. 71, below. It must have been given about 1583 but Nashe in *Have with you to Saffron-walden* (1596) is writing in retrospect and often inserts anachronistic allusions.

[42] *Works*, iii. 75–6.

where Margaret, Countess of Derby[43] was staying, and after supper, like the others, he fell to dancing. As chance would have it, his partner was (in Nashe's version):

the foulest ugly gentlewoman or fury that might be, (then wayting on the foresaid Countesse,). . . . A turne or two hee mincingly pac't with her about the roome, & solemnly kist her at the parting: Since which kisse of that squinteyd *Lamia* or *Gorgon*, as if she had been another *Circe* to transforme him, he hath not one houre beene his owne man. For whilst his lips smoakt with the steame of her scortching breath . . . he ran headlong violently to his study as if he had been born with a whirl-winde, and strait knockt me up together a Poem calde his *Aedes Valdinenses*, in prayse of my L. of Leycester, of his kissing the Queenes hand, and of her speech & comparison of him, how he lookt like an Italian. . . .[44]

One wonders whether this 'gentlewoman' may perhaps be the one who later became Harvey's 'championess' against Nashe?[45]

If the humourist is to be believed, the four original manuscripts were not prepared in advance but were put together during the few days that the Court was at Audley End[46] and then presented to the Queen and her chief nobles before they left. There is plentiful evidence in Harvey's letters and marginalia that he prided himself on his ability to work amazingly fast.

In the printed *Gratulationes Valdinenses* which was presented to the Queen in September, additional verses were added, some newly composed by him and some previously written by various other scholars in honour of an earlier visit (1564) of Queen Elizabeth to the Cambridge area. All of the latter would have been available in printed collections in 1578 except for the poem by the Vidame of Chartres which had not yet been published.[47] Thus the volume *in toto* is an anthology of praise and celebration of Elizabeth and her chief nobles,[48] primarily by Harvey, but supplemented by numerous epigrams written by native and foreign scholars. All or nearly all seem to have been protégés or friends of

[43] She was the wife of Henry Stanley, fourth Earl of Derby. See McKerrow's note in *Works of Thomas Nashe*, iv. 339.

[44] *Works*, iii. 77–8.

[45] See *Pierces Supererogation*, 1593 (sigs. Dd3–4 and *passim*) and *New Letter of Notable Contents*, 1593 (sigs. B4r–C3v and D3v).

[46] Nichols, op. cit., ii. 111–15, reports that Harvey and the University scholars found no lodging in Walden after the disputation so walked back to Cambridge that night. When Harvey returned to Audley End he may have brought with him copies of poems which he could use to supplement his own.

[47] Jameson dissertation, op. cit., pp. xxxvii–xxxviii.

[48] Harvey apologizes that he cannot include more of them.

Leicester's, and a number were probably personal friends of Harvey's as well.[49] Within some of Harvey's verses, as previously noted, are certain satirical lines of monitory counsel.

T. H. Jameson in the introduction to his critical edition of *Gratulationes Valdinenses* includes a discussion of some of the poetic techniques that Harvey uses in this work and suggests that they are an emulation of the 'singularity' and 'hyperbole' of Aretino so much admired by Harvey. For instance, he includes several prosopopoeias: in Book I the personification of the stone and brick walls of Audley End are made to utter greetings to Elizabeth as does the town of Walden; in Book II Machiavelli is represented delivering a satirical message in person. Harvey also makes prominent use of hyperbolical amplifications. He describes complete subjection to an emotion or state: before Queen Elizabeth he is all made of sleeplessness (sig. C1$^v$), in order to hail Leicester whole-heartedly he would like to become a joyous elegy (sig. D7$^r$). Jameson believes that examples of Harvey's 'singularity' and 'hyperbole' such as these may have contributed to his contemporary reputation as an outstanding Latin poet.[50]

At about this time Harvey was apparently summoned to Court to serve as secretary to the Earl of Leicester. Although the young man's subsequent dismissal was not necessarily as abrupt as Nashe implies (*Works*, iii. 79), Harvey's sojourn at Court was probably of short duration and (if Nashe is to be believed) Harvey was told that he was 'fitter for the Universitie then for the Court'. It is possibly his reaction to such a termination of his services which occasioned the following outburst on sig. C1$^r$ of Ramus's *Oikonomia*:

Common Lerning, & the name of *a good schollar* was never so much contemned, and *abjected of princes*, pragmaticals, & common Gallants, as nowadays: in so much that it necessarily concernith, & importith the lernid, ether praesently

---

[49] The list of poets and scholars (all Protestants) contributing epigrams includes the following:
English: Walter Haddon—see p. 41 n. 28, above. Abraham Hartwell—see p. 40 n. 25, above.
Scottish: George Buchanan (1506–82)—eminent professor and tutor of James VI, a bitter enemy of Mary's. Harvey owned some of his writings against her.
French: Camille de Morel—member of the Pléiade. Jean de Ferrières, Vidame of Chartres—a man of great learning and one of the chief French Protestant noblemen.
Flemish: Charles Utenhove—born in Ghent in 1536, see p. 158 n. 27, below.
Dutch: Jan Dousa—see p. 40 n. 25, above.
Italian: Pietro Bizari—see p. 41 n. 28, above. Benedetto Varchi—Florentine historian, (1502–65). Bernardino Tomitano—physician and writer (1506–76). Cornelius Amaltheus —Latin poet (1530–1603). Vulpianus of Verona.
[50] Jameson dissertation, op. cit., pp. lxxiii–lxxv.

to hate ther bookes; or actually to onsinuate, & enforce themselves, by very special, & singular propertyes of emploiable, & necessary use, in all affaires, as well private, as publique; amounting to any commodity, ether œconomical, or politique.[51]

[51] These annotations were probably made in 1580 since marginalia were added to the volume during this year.

# 5. Trinity Hall and Oxford

SEVERAL years earlier at the advice of Sir Thomas Smyth, Harvey had determined on the study of civil law. Smyth had suggested the basic texts with which he should begin, but academic demands pressed hard on Harvey's time and for the next few years he was unable to indulge in much legal study. At the conclusion of the Rhetoric Praelectorship in 1576 he turned in earnest to his chief interest and by the summer of 1578 had become fascinated with the field of civil and canon law and its potentialities.

Unfortunately, his Pembroke Fellowship was due to expire in November of this year and it was necessary for him to seek re-election. The statutes required that in order to continue as a Fellow he would need at this time to study Divinity intensively. Harvey, however, was unwilling, even at the risk of losing his Fellowship, to discontinue the study on which he was now launched. He therefore sought a dispensation from Divinity training.[1] Arguing on Harvey's behalf, William Fulke, Master of Pembroke Hall since early May of 1578,[2] wrote on 22 August to the Pembroke Fellows:

Whereas my lorde, the Earle of Leycester hath made earnest request for the continuance of M[r] Harveyes fellowshipp for one yeare, and that the tyme of the expiringe thereof is very neere, this is to certify you, that I am not only well contente as mutch as lyeth in me, to dispense with him for one yeare longer, but also am becum an ernest suter for him unto you, that you will graunte your consente unto the same dispensation, as you will require the like Curtesye of me in any of your reasonable requestes.[3]

Both the Earl of Leicester and William Fulke thus supported Harvey's request for re-election, but the Fellows themselves were unwilling to accede to the necessary dispensation.

The dilemma was finally solved by Harvey's election to Fellowship at Trinity Hall, the 'home of civil law', where a relative of his, Dr. Henry Harvey, was Master. The document recording the election states: '18: December: 1578: Gabriell Hervy M[r] Artium famosus electus fuit et

---

[1] See p. 10 *re* William Lewin's receiving such a dispensation.
[2] On 10 May 1578 John Young had become Bishop of Rochester.
[3] Sloane MS. 93, fol. 48[r].

admissus socius huius coll: in eam societatem quam M<sup>r</sup> Hammond prius habuit et a qua per reliquationem suam ultro recessit.'[4] Once a student had become a Fellow at one college, it was not customary for him to transfer elsewhere, but on rare occasions colleges did accept Fellows from outside their own ranks if a candidate had achieved renown beyond the University's walls. Harvey evidently qualified in this respect since he was already well known for his Latin writings and perhaps had even become somewhat of a public figure. This is undoubtedly why the above statement describes him as *famosus*.

Perhaps in celebration of the new Fellowship, Harvey two days later met Edmund Spenser in London. When in the early spring John Young had left Cambridge to become Bishop of Rochester he had taken Spenser with him as secretary.[5] The reunion of the two young men on 20 December is amusingly recorded in Harvey's handwriting in his copy of Murner's *Howleglas*:

This Howleglasse, with Skoggin, Skelton, & Lazarillo, given me at London, of M<sup>r</sup> Spensar xx. Decembris, 1578, on condition [I] shoold bestowe the reading of themover, before the first of January immediatly ensuing; otherwise to forfeit unto him my Lucian in fower volumes. Wherupon I was the rather induced to trifle away so many howers, as were idely overpassed in running thorowgh the foresaid foolish Bookes: wherin methowgh[t] not all fower togither seemed comparable for subtle & crafty feates with Jon Miller whose witty shiftes, & practises ar reportid amongst Skeltons Tales.[6]

At about this time Harvey must have become concerned about the state of his finances and perhaps recalled William Lewin's advice that he should not overlook the need for the lucrative. On 24 April 1579 he wrote a long letter to the Earl of Leicester from Trinity Hall urging that he try to persuade Her Majesty to procure for him 'Doctor Byddles Praebende at Litchfeylde, the Cancelour[ship] very lately falling voyde, by his suddayne disease'. He rather plaintively alludes to 'your good Lordshippes poor schollar [and][7] servant, that only in respect of sum hinderances thorough want of sufficient hability, cannot [word illegible] undertake those Services, that otherwise have bene often purposid, and might shortely by any sutch help in sum reasonable sorte be performid'. It sounds as though the services to his Lordship which 'have bene often

---

[4] As copied by the seventeenth-century Cambridge historian Thomas Baker (Harleian MS. 7031, fol. 140).

[5] Harvey's annotations in his copy of Turler's *Traveiler* show that this book was a gift from Spenser, 'Episcopi Roffensi Secretarii. 1578.'

[6] See p. 228, below (Murner).

[7] Words which I have enclosed in brackets are either defaced or missing because of a torn corner of the page.

purposid' are the dedication of his epic poem 'Anticosmopolita'.[8] Evidently Leicester has not been enough interested in it to offer him a financial reward, so having heard that a chancellorship is now available Harvey requests that he be considered for this benefice:

if that my hability might in the least poynte be made answerable unto my purposes; I doute not but your Excellency shoulde in shorte tyme perceyve ... that on litle poore schollar would do your L. more honour in sum speciall respectes, than sum of your gallantist and courtlyest servants; whose servyces notwithstanding in ther kynde are very commendable. It is no greate matter that would suffyce for the mayntenance of on litle body: and if it might please your good L. to bestowe this small Praebend upon me (as my great hope is, it will) I were to thynke myself in an indifferent reasonable state for the praesente (without beying otherwyse chargeable unto your L.) whether I should travayle abroade, or abyde yet awhile longer at home. Especially if withall it might [word illegible] Honour to be so good and bownetyfull Lord unto me, as to moove the next Bysshop there for his Cauncelorship, which is nowe likewise vacant by the same mans death.

The letter proceeds to recommend for the bishopric Doctor [John] Still, 'My oulde Tutor, and continuall frende', who Harvey is certain would choose him as chancellor 'albeyt sum other did supply the roome, as a substitute, for a twelvemonth untill I were Doctor'.[9]

Despite his stated expectation of becoming Doctor within a year, Harvey did not actually receive his Civil and Canon Law degree until 1586. But within a short while of his transfer to Trinity Hall he was eager to proclaim his new status as a civil law student: when making plans for a second edition of his rhetoric orations Harvey changed the original title-page of *Ciceronianus* by adding '*Legista*' to his name.[10]

When Nicholas Bacon died in February 1578/9 Harvey wrote a Latin epitaph on the former Lord Keeper and Chancellor of England. It is extant in manuscript form[11] and is a conventional but well-expressed tribute. Throughout 1579 Harvey seems to have been bending every effort toward securing a niche for himself at Court. Although his epic

---

⁸ See p. 51, below.

⁹ Dudley Papers (Longleat), vol. ii, fol. 202. Transcription made by Dorothy Coates, Librarian at Longleat (1 Aug. 1957).

¹⁰ In a Bodleian volume of Harvey's works (shelfmark Bliss A110) are transcriptions by Thomas Baker from Harvey's annotations to his copy of *Ciceronianus* (apparently no longer extant). In Baker's hand on the title-page of his copy of *Ciceronianus* are the following transcriptions of Harvey's notes: 'secunda editio, paulo, quam prima emendatior The next title—of my Rhetoric orations put Legistae Gabrielis Harveii Rhetoricarum orationum Liber. In Academia—Cantabrigiens: publice—habitarum.'

¹¹ Baker MS. 36, fol. 113 (Cambridge Univ. Library). Next to his copy of the poem, Baker notes: 'G. H. faciebat. (Sed neutiquam tam foelici genio, quam, *Musarum Lachrymae*, quibus praemittitur.)' Harvey's tribute seems to have received little notice.

poem 'Anticosmopolita or Britanniae Apologia' had not met with a warm reception from Leicester,[12] he was still thinking in terms of publication, for the work was entered to Richard Day (without the name of the author) on 30 June at Stationers' Hall for the sum of sixpence. Harvey refers to it in his 1580 letter about the earthquake, at the end of which he writes of 'my *Anticosmopolita* . . . remayning still as we saye, *in statu quo*, and neither an inche more forward nor backewarde, than he was fully a twelvemonth since in the Courte, at his laste attendaunce upon my Lord [Leicester?] there'.[13] It is again mentioned in the dedication of Richard Harvey's *Astrological Discourse* (1583) where he alludes to 'my brothers Anticosmopolita'. The long epic poem was evidently never published and perhaps never completed. In a letter to Sir Robert Cecil on 8 May 1598 from Walden, Harvey mentioned the many writings he had ready or nearly ready for publication. Among them, he reported, were: 'sundrie royale Cantos, (nigh as much in quantitie, as Ariosto) in celebration of her majesties most prosperous, & in truth glorious government: sum of them devised many yeares past at the particular instance of the excellent knight, & mie inestimable deare frend, Sir Philip Sidney'.[14] These 'royale Cantos' undoubtedly are the 'Anticosmopolita or Britanniae Apologia'.

Though Harvey's epic seems never to have reached the public eye, he himself attained a kind of public fame at this time as Colin's close friend, the poet and wise rustic character Hobbinoll in Spenser's first major verse effort, the anonymously published *Shepheardes Calender*, which achieved immediate popularity after its printing in 1579.[15] The work contains an interesting prefatory epistle by 'E. K.' (Edward Kirke?) dated '10 of Aprill 1579'. Edward Kirke (1553–1613) was at Pembroke during Spenser's and Harvey's residence there and probably became a friend of both.[16] The epistle is addressed 'To the most excellent and learned Orator and Poete, Mayster Gabriell Harvey, his verie special and singular good friend, E. K. commendeth the good lyking of this his labour, and the patronage of the new Poete'. After praising the 'new Poete' and verses, E. K. closes with a postscript:

Now I trust M. Harvey, that upon sight of your speciall frends and fellow Poets

[12] For discussion of Harvey's patronage by the Earl of Leicester see Eleanor Rosenberg's *Leicester, Patron of Letters*, New York, 1955, Chapter IX: 'Leicester's "Neglected" Writers: Harvey, Spenser, Florio.'

[13] *Three Proper and wittie, familiar Letters*, 1580, sig. Dii[r].

[14] Salisbury Papers (Hatfield House), 61, 5 (8 May 1598).

[15] Hobbinoll appears in the January, April, June, and September eclogues.

[16] Spenser twice refers to 'Mystresse Kerke' (perhaps Edward's mother) in *Two other very commendable letters*, 1580, sig. G4[r] and H3[r].

doings, or els for envie of so many unworthy Quidams, whitch catch at the garlond, which to you alone is dewe,[17] you will be perswaded to pluck out of the hateful darknesse, those so many excellent English poemes of yours, which lye hid, and bring them forth to eternal light. . . .[18]

Another reference to writings of Harvey's is found in the gloss to the September Eclogue. Here E. K., after informing the reader that the rustic Colin Clout is really the author himself, discusses Colin's 'especiall good freend Hobbinoll . . . or more rightly Mayster Gabriel Harvey'

of whose speciall commendation aswell in Poetrye as Rhetorike and other choyce learning, we have lately had a sufficient tryall in diverse his workes, but specially in his Musarum Lachrymae, and his late Gratulationum Valdi-nensium. . . . Beside other his sundrye most rare and very notable writings, partely under unknown Tytles, and partly under counterfayt names, as hys Tyrannomastix, his Ode Natalitia, his Rameidos, and esspecially that parte of Philomusus, his divine Anticosmopolita, and divers other of lyke importance.

There seems to be no trace of the works referred to as 'Tyrannomastix'[19]

---

[17] E. K. evidently had great admiration for Harvey and felt that he had had a share in the development of Spenser as a poet. It is worth noting that E. K. identifies Harvey by name but not Spenser, probably because Spenser was then unknown as a writer and Harvey had already attained considerable prestige.

[18] One wonders whether 'those so many excellent English poemes' are those of which early drafts are found in the 'Letter-book' (Sloane MS. 93, fols. 35, 39, 58–70) or whether there were still others? I tentatively suggest the possibility that among the sixty-nine anonymous poems in Francis Davison's 1602 anthology, *A Poetical Rapsody Containing, Diverse Sonnets, Odes, Elegies, Madrigalls, and other Poesies, both in Rime and Measured Verse, Never yet published*, some may be Harvey's. In an address to the reader Davison refers to some of the poems having been written 'almost twentie yeers since, when Poetrie was farre from that perfection, to which it hath now attained'. One of these, an 'eglogue' written in the language of the *Shepheardes Calender*, entitled 'Perrin areed what new mischance betide' is upon the death of Sir Philip Sidney. It is worth noting that the poem entitled 'Loves Contrarieties' resembles (in condensed form) a poem of Harvey's in Sloane MS. 93 (fol. 58ᵛ–59) called 'The Schollars Love, or Reconcilement of Contraryes', although the similarities may, of course, be due to *topoi* of the period. Of the other poems to which no specific authorship is credited a number are in hexameters or other classical metres; one is designed in the shape of an altar. (This recalls Nashe's statement [*Works*, iii. 67] that Harvey wrote verses in all shapes.) In the first edition of the anthology the sixty-nine poems re-ferred to above are signed 'Anomos'. In his preface Davison refers to them as verses printed 'under the name Anomos' but his twentieth-century editor Hyder Rollins believes it was an error in spelling due to carelessness on Davison's part, for in a later edition the phrase is corrected to 'under the name of Anonymos'. By 1602 Harvey would no longer have been legally permitted to publish under his own name because of Whitgift's 1599 ban on his writings (see p. 129, below). One wonders whether it is a coincidence that the Greek word *anomos* translates as 'illegal'.

[19] Perhaps the subject of the 'Tyrant-eater' was Junius Brutus to whom Harvey alludes in these terms in his marginalia in Guicciardini (sig. E3ʳ): 'Il divino Aretino, Flagellum Principum: et esso Apollonio Tianeo, malleus Tyrannorum. Etiam hodie Iunius Brutus, Tyrrannomastix.'

or 'Rameidos' although the latter may have been the title of the 'nocturna ecloga' in praise of Ramus and his contemporaries which Harvey in his *Ode Natalitia* (dated kalends of March 1575) states he has recently written but has not included in this work as he intends to reserve it for another occasion.[20] It is not clear whether 'that parte of Philomusus' is a separate work or a description of the subsequent title 'Anticosmopolita'. In characterizing Harvey's writings as being 'partly under unknown Tytles', E. K. may be referring to works not yet published (as we know was the case with *Anticosmopolita*). That they were 'partly under counterfayt names' may mean that some were put forth under pseudonymous authorship or a 'pen name'. The *Ode Natalitia*, for instance, lists no author on its title-page but concludes with the initials 'A.P.S.' (i.e. Aulus Pembrochianus Socius). This work has been positively identified as Harvey's by Warren B. Austin,[21] first, because of its content, secondly, because of its mention by E. K., and, thirdly, because a title-page has been found which Harvey owned and on which he appended his initials. One becomes curious with regard to the other works referred to by E. K. and wonders whether or not perhaps they are extant in some as yet unrecognized form.

On 9 April (1580)[22] Harvey wrote to the Chancellor of Cambridge, William Cecil, Lord Burghley, asking for a letter to the University recommending him for the post of Public Orator. Richard Bridgewater had just resigned, having been in that office for seven years, the maximum term allowed by University law. Burghley gave his endorsement to Harvey. His letter of 14 June 1580 (Lansd. MS. 30, no. 57) thanks Burghley for it but unhappily reports that it was of no avail: Harvey's old enemy Andrew Perne had manœuvred to defeat Harvey's candidacy by arranging for the reluctant Bridgewater to stay in office an additional year.[23] By the time another year had passed and Bridgewater finally relinquished the Public Oratorship, Anthony Wingfield, nephew of the Countess of Shrewsbury, was easily elected under the auspices of the new

[20] Mentioned by Harold S. Wilson in *TLS*, 9 Mar. 1946, p. 115.

[21] 'Gabriel Harvey's "Lost" Ode on Ramus', *MLN*, lxi (Apr. 1946), 242–7.

[22] Although Harvey's letter (Lansd. MS. 28, no. 83) is dated by him 'Pridie Idus Aprilis, 1579', Moore Smith in the *Marginalia* (p. 35, n. 1) convincingly argues that Harvey has here made a mistake in the year, understandably so, since the new year had officially only begun on 25 Mar.

[23] In Harvey's *Foure Letters*, 1592 (sigs. C2ʳ and C3ᵛ) he recalls the pangs of this period: 'I was supposed not unmeet for the Oratorship of the university, which in that springe of mine age, for my Exercise, and credite I earnestly affected: but mine owne modest petition, my friendes diligent labour, our high Chauncelors most-honourable and extraordinary commendation, were all peltingly defeated, by a slye practise of the old Fox [Perne]. . . . He indeed was the man that otherwhiles flattered me exceedingly, otherwhiles overthwarted me crosly, alwaies plaied fast, and loose with me.'

Vice-Chancellor Andrew Perne.[24] This time Harvey's candidacy was at a particular disadvantage because he had just been made the butt of ridicule (probably with the help of Wingfield or his sponsors) in the Cambridge comedy *Pedantius*[25] performed on 6 February 1580/1 at Trinity College, the large college adjacent to Trinity Hall. Although the comedy was probably written primarily by Edward Forsett, as Moore Smith believes, it seems likely that Anthony Wingfield may have had a hand in promoting it, for Nashe refers to it as 'M. Winkfields Comoedie',[26] and Moore Smith himself points out that its production was in Wingfield's best interest.[27]

In late 1580, annoyance at Harvey had been provoked at Cambridge by the ill-advised publication[28] of his and Spenser's *Three Proper and wittie, familiar Letters: lately passed betwene two Universitie men: touching the Earthquake in Aprill last, and our English refourmed Versifying. With the Preface of a wellwiller to them both. Two Other very commendable Letters, of the same mens writing: both touching the foresaid Artificiall Versifying, and certain other Particulars: More lately delivered unto the Printer.* Although 'more lately delivered' to the printer and thus printed at the end of the volume, the latter were written earlier (October 1579) than the *Three Proper . . . Letters.* The volume consists of two title-pages and the following letters (which I have numbered in order to facilitate the ensuing discussion):

*Three Proper, and wittie, familiar Letters.*
Preface dated 19 June 1580 addressed 'To the Curteous Buyer, by a Welwiller of the Two Authours'.

[1] 'To my long approved and singular good frende, Master *G.H.*' dated from 'Westminster, Quarto Nonas Aprilis 1580' and signed 'Immerito' [i.e. Spenser].

[2] 'A Pleasant and pitthy familiar discourse, of the Earthquake in Aprill last.' 'Master H[arvey]'s short, but sharpe, and learned Judgement of Earthquakes.'

[3] 'A Gallant familiar Letter, containing an Answere to that of M. Immerito, with sundry proper examples, and some Precepts of our Englishe reformed Versifying.' This letter of Harvey's includes the poems 'Encomium Lauri' and 'Speculum Tuscanismi'.

---

[24] In April 1580, when Harvey originally sought the Oratorship, the Vice-Chancellor of Cambridge was John Hatcher, brother of Harvey's friend Thomas Hatcher.

[25] See p. 69, below, for further discussion of the play.

[26] In *Strange Newes*, 1592 (*Works*, i. 303).

[27] See Moore Smith, ed., *Pedantius*, Louvain, Leipzig, and London, 1905, pp. xi–xx.

[28] Entered S.R. 30 June 1580 and probably published very shortly thereafter.

*Two Other very commendable Letters.*

[4] 'To the Worshipful his very singular good friend, Maister G. H. Fellow of Trinitie Hall in Cambridge.' This letter signed 'Immerito' has the Latin poem dated 5 October 1579 from Leicester House; it bids Harvey farewell before Spenser's expected departure to 'Gaul'.

[5] 'To my verie Friende, M. Immerito', dated from Trinity Hall '23 October 1579. In haste.' This letter of Harvey's provides further discussion and examples of versifying.

The volume opens with the preface by an unidentified 'Welwiller of the Two Authours' and is followed by the short introductory letter [1] of Spenser's to Harvey urging him, if he can spare the time from his legal studies, to impart some of his 'olde, or newe, Latine, or Englishe, Eloquent and Gallant Poesies'. Spenser, who was then in the Earl of Leicester's service, possibly as secretary, writes that there is little news stirring at Court 'but that olde great matter still depending [presumably the proposed Alençon marriage]. His honoure [Leicester] never better'. Spenser now introduces the subjects with which Harvey will deal in the two subsequent letters: the recent earthquake of 6 April 1580 (which is used in Harvey's letter as the subject of a satire on Cambridge 'scholarship') and the new 'reformed versifying' (based on quantitative measurement of syllables as in classical Latin poetry).

Introducing the topic of the earthquake, Spenser writes to Harvey at Cambridge: 'I thinke the *Earthquake* was also there wyth you (which I would gladly learne) as it was here with us: overthrowing divers old buildings, and peeces of Churches. Sure verye straunge to be hearde of in these Countries'. He next refers to Harvey's 'English Hexameters' which he likes 'so exceedingly well, that I also enure my Penne sometime in that kinde: whyche I fynd . . . neither so hard, nor so harsh, that it will easily and fairely yeelde it selfe to our Moother tongue'. And he alludes to its chief difficulty, that of accent. He requests that Harvey send him the Rules and Precepts of Arte, which you observe in Quantities, or else follow mine, that *M. Philip Sidney* gave me, being the very same which *M. Drant* devised, but enlarged with *M. Sidneys* own judgement, and augmented with my Observations, that we might both accorde and agree in one: least we overthrow one an other, and be overthrown by the rest.[29]

He encouragingly writes that Edward Dyer has praised Harvey's satirical verses:[30] 'Truste me, you will hardly beleeve what greate good liking and estimation Maister *Dyer* had of your *Satyricall Verses*, and I, since the

---

[29] A recent study of the quantitative meter problem is Derek Attridge's *Well-weighed Syllables*, Cambridge Univ. Press. 1975.

[30] Some of the verses extant in the 'Letter-book' (Sloane MS. 93) may have been a first draft for these. See especially fols. 58r–68r.

viewe therof, having before of my selfe had speciall liking.' Spenser tells
of his own writing plans, announcing that he is about to proceed with his
*Faerie Queene*, the sample of which he asks Harvey to return to him
promptly together with his judgement of it. In his letter of reply [3]
Harvey is far from enthusiastic about the verses which Spenser had sent
him. He writes his friend that if it is Ariosto he hopes to emulate and
overgo, Harvey finds Spenser's 'Nine Comoedies' nearer Ariosto's
comedies 'eyther for the finenesse of plausible Elocution, or the rarenesse of
Poetical Invention, than that *Elvish Queene*' is to the *Orlando Furioso*.
He further declares that if Spenser does not share his view, '*Hobgoblin*'
has 'runne away with the Garland from *Apollo*', i.e. that fanciful burlesque
has displaced the deeply moral epic poetry[31] which Harvey evidently
feels Spenser should be writing. Josephine Waters Bennett has convincingly argued that the portion of the *Faerie Queene* which Harvey
received was a part of Book III but probably in an early stage of its
present form.[32] Perhaps Harvey's criticism was not then wholly un-
justified; he certainly accords the finished epic poem the highest of
accolades.[33]

But let us return to the second of the *Three Letters* [2], a lengthy one
dealing with the recent earthquake and addressed by Harvey 'To my
looving frende, *M. Immerito*'. Harvey relates that at the time he was in the
house of a gentleman in Essex in the company of 'certaine curteous Gentlemen' and two 'shrewde wittie new marryed Gentlewomen'. As he sits

[31] For a comparison of Harvey's two terms, see William Nelson's *Fact or Fiction: The
Dilemma of the Renaissance Storyteller*, Cambridge, Mass., 1973, pp. 74–5 ff.

[32] *The Evolution of 'The Faerie Queene'*, New York, 1960, chapter I.

[33] For example, in *New Letter of Notable Contents* (1593), sig. A4$^v$ he writes: 'is not the
Verse of M. *Spencer* in his brave Faery Queene, the Virginall of the divinest Muses, and
gentlest Graces?' When the *Faerie Queene* finally went to press in 1590 it contained at the
end of Book III a commendatory sonnet by Harvey entitled 'To the learned Shepheard'
signed 'Hobynoll'. Here he expresses approval:

> Collyn I see by thy new taken taske,
>   some sacred fury hath enricht thy braynes,
> That leades thy muse in haughty verse to maske,
>   and loath the layes that longs to lowly swaynes.
> That lifts thy notes from Shepheardes unto kings,
> So like the lively Larke that mounting sings.
>
> .    .    .    .    .    .    .
>
> And fare befall that *Faery Queene* of thine,
>   in whose faire eyes love linckt with vertue sittes:
> Enfusing by those bewties fiers devyne,
>   such high conceites into thy humble wittes,
> As raised hath poore pastors oaten reede,
> From rusticke tunes, to chaunt heroique deedes.
>
> .    .    .    .    .    .    .

playing at cards, something he asserts he rarely does, tremors are felt. Harvey describes their onset in a light vein as follows:

The earth under us quaked, and the house shaked above: besides the mooving, and ratling of the Table, and fourmes, where wee sat. Whereupon, the two Gentlewomen having continually beene wrangling with all the rest, and especially with my selfe, and even at that very moment, making a great loude noyse, and much a doo: Good Lorde, quoth I, is it not woonderful straunge that the delicate voyces of two so propper fine Gentlewoomen, shoulde make a suddayne terrible Earthquake? Imagining in good fayth, nothing in the worlde lesse, than that it shoulde be any Earthquake in deede, and imputing that shaking to the suddayne sturring, and removing of some cumberous thing or other, in the upper Chamber over our Heades.

At this point the gentleman of the house comes stumbling into the parlour 'straungely affrighted' at the 'woonderous violent motion'. A servant, sent into the town to ascertain whether the same tremors have occurred elsewhere, soon returns with the report that the shaking of the earth is widespread.

The panic-stricken gentlewomen fall to praying and the gentleman of the house attempts to calm them by turning to Harvey and asking, 'What say you Philosophers . . . to this suddayne Earthquake? May there not be some sensible Naturall cause thereof, in the concavities of the Earth it self, as some forcible and violent Eruption of wynde, or the like?' Harvey replies that there well might be. He then mimicks the sort of explanation that a typical natural philosopher at Cambridge might have given: that the abundant rains which fell recently stopped the pores and crannies of the earth to such an extent that its natural vapours were unable to escape. But the 'verye Natures of things themselves so utterly unknowen, as they are to most heere, it were a peece of woorke to laye open the Reason to every ones Capacitie'. Then follows a pseudo-erudite discourse on 'poysonfull and venemous Hearbes, and Beastes, besides a thousand infective, and contagious thinges else', that sometimes 'the Evill (in the divels name) will . . . have his naturall Predominant Course', and that feverous spirits lurking in the earth cause an 'Ague' or 'fitte', which scholars called *Terrae motus*, a moving or stirring of the earth, while some learned gentlewomen termed it *Terrae metus*, a fear and agony of the earth. Harvey indicates that there has been considerable controversy as to which is the better term. Then, concluding this absurd discourse, he turns and asks, 'What think ye, Gentlewomen, by this Reason?' and one of them replies that there is neither 'Rime, nor Reason, out of any thinge I have hearde yet'. Although, she adds, 'me thinkes all should be Gospell, that commeth from you Doctors of Cambridge'.

The other gentlewoman suggests that this is a fanciful 'Tale of Robin-hood' and asks whether scholars really believe this. Harvey replies that it is indeed a ridiculous explanation, as though the 'Earth having taken in too much drinke, and as it were over lavish Cups . . . now staggereth, and reeleth, and tottereth, this way and that way, up and downe, like a drun-ken man, or wooman'. Some Cambridge scholars, he adds, even liken it to a fit of sneezing, or coughing, or sobbing on the part of the earth.

The ladies respond that, if this is doctorly learning, they have already had enough of it. Harvey suggests that he shares their reaction. Then, when asked what he himself believes is the cause of earthquakes, he pro-ceeds with a colloquy entitled 'Master H.s short, but sharpe, and learned Judgement of Earthquakes'. Although obviously limited in geophysical knowledge, it is a well-organized, lucid discourse on the earthquake and its possible causes. Using a Ramist type of analysis, he considers internal and external causes, the former subdivided into the traditional Aristotelian material and formal causes and the latter subdivided into efficient and final causes. God is conceded to play a part as creator, continuer, and corrector of nature, but Harvey's considered judgement is that earth-quakes are probably natural, not supernatural or preternatural phenomena.

This 'letter' with its two contrasting explanations of the earthquake is undoubtedly designed to differentiate between what Harvey considered as 'befuddled' and 'proper' methods of scientific inquiry. That this was recognized to be an attack on Cambridge pseudo-scholarship is evidenced by the fact that Harvey later feels the need of making 'a large Apology of my duetiful, and entier affection to that flourishing Universitie, my deere Mother'.[34] In 1589 John Lyly refers to 'One . . . that writing a familiar Epistle about the naturall causes of an Earthquake, fell into the bowells of libelling, which made his eares quake for feare of clipping'.[35] In analyzing this discourse of Harvey's, Gerald Snare points out that in the first part of the Earthquake Letter

The most absurd notions and windy rhetoric are constantly mentioned as examples of the 'great Doctorly learning' of 'you Doctors of Cambridge'. We know Harvey is writing tongue-in-cheek as this first part of his treatise consistently violates the rules of Ramist logic he is so careful to observe in the second part of the discourse, 'Master H's learned Judgment.'[36]

[34] In Foure Letters, 1592, sigs. C2r–C3r, he adds that although he wrote an apology he has until now suppressed it 'as unworthie the view of the busie world', but it is possible that 'extraordinarie provocations' may eventually stir him to 'publish many Traictes, and Dis-courses, that in certaine considerations I meant ever to conceale'.

[35] Pappe with an Hatchet, sig. B3r.

[36] Gerald Snare, 'Satire, Logic, and Rhetoric in Harvey's Earthquake Letter to Spenser', Tulane Studies in English, 18 (1970), 26.

When twelve years later in *Foure Letters* (1592) Harvey reminisces about the events of this period, he admits that he was then 'yong in yeares, fresh in courage, greene in experience, and as the manner is, somewhat overweeninge in conceit' and had been reading invectives and satires 'artificially amplifyed in the most exaggerate and hyberbolicall kinde'. Besides, he had been exasperated by some 'sharpe undeserved discourtesies', the particular ulcer being the Public Oratorship for which Harvey states he was well qualified and which he eagerly sought only to be defeated by Andrew Perne's wily manœuvring. Full of bitter resentment he withdrew to the background, but

some familiar friendes pricked me forward: and I neither fearing daunger, nor suspecting ill measure, (poore credulitie sone beguiled) was not unwilling to content them, to delight a few other, and to avenge, or satisfie my selfe, after the manner of shrewes, that cannot otherwise ease their curst hearts, but by their owne tongues, & their neighbours eares.[37]

He asserts that he had never intended publishing these 'infortunate Letters' but that they had happened to fall into the 'left handes of malicious enemies, or undiscreete friends: who adventured to imprint in earnest, that was scribbled in jest, (for the moody fit was soone over)'. Although at the end of the earthquake letter, he had requested Spenser to show it to the 'two odde Gentlemen you wot of [Dyer and Sidney[38]]', he had cautioned him that it was not to be shown to anyone else.

Josephine Waters Bennett has pointed out that contact with Sidney and the younger courtiers interested in the possibilities of the English language may have fired Harvey with the desire to distinguish himself in an English rather than a Latin publication.[39] Moore Smith finds it hard to believe that Harvey really did not want the 1580 volume of letters published.[40] It is possible, of course, that as a pose of modesty to make printing of such young men's letters more acceptable to the public, Harvey did feign reluctance to publish, but if one is to give credence to what he writes in the draft of a letter to Spenser in the summer of 1579 (Sloane MS. 93, fol. 35ᵛ), a project to publish Harvey's English poems was instigated by Spenser (perhaps with the help of E. K. or one of Harvey's other friends) and took him somewhat by surprise. Evidence that Harvey did not initially take an active part in the planning of the 1580 volume is to be found in its structure and in some later pages of his 'Letter-book'.

[37] *Foure Letters*, 1592, sig. C2.
[38] In *Three proper wittie familiar Letters*, 1580, Harvey's second letter identifies 'these two excellent Gentlemen' as Dyer and Sidney (sig. D4ʳ).
[39] 'Spenser and Gabriel Harvey's Letter-book' in *MP*, 29, 167.
[40] *Marginalia*, p. 31.

Let us first take a look at the 1579 letter. It is addressed to 'Magnifico Signor Benevolo', i.e. Edmund Spenser (for the addressee is identified on fol. 37ᵛ as 'mi belovid Imerito')⁴¹ and in it Harvey chides his friend for

publishing abroade in prynte to the use or rather abuse of others, and nowe bestowing uppon myselfe a misshapin ill favored copy of my precious poems. . . . Alasse they were hudlid, and as you know bunglid upp in more haste then good speede, partially at the urgent and importune request of a honest good-naturid and worshipfull yonge gentleman [John Wood?] who I knewe, beinge privy to all circumstaunces, and very affectionate towards me or any thinge of my doinge.⁴²

Harvey complains that his poems have been so poorly presented that they are suitable only for hawking at Bartholomew or Sturbridge Fair and even though Spenser has bestowed 'a delicate liverye upon them', he has foolishly christened them 'by names and epithetes, nothing agreable or apliante to the thinges themselves'. Harvey continues:

What greater and more odious infamye from one of my standinge in the Universitye and profession abroade then to be reckonid in the Beaderoule of Inglish Rimers, especially being occupied in so base an objecte and handelinge a theame of so slender and small importance?⁴³ Canst thou tell me or doist thou nowe begin to imagin with thyselfe what a wunderfull and exceedinge displeasure thou and thy Prynter [Singleton?] have wroughte me?

Nevertheless, he realizes that, however misguided the result, Spenser's intentions were of the best. Therefore, since the poems seem then on the verge of publication, he turns to practical matters and reminds Spenser that this work will need a commendatory letter which he presumes his friend will furnish to obviate the necessity of procuring one from a nobleman. Harvey's mood now changes to one of humorous teasing in which he commits Spenser to the obligation of covering Harvey's blushes of embarrassment by the loan of his own luxurious 'mustachyoes and subboscoes'.⁴⁴

One recalls E. K.'s urging (in his epistle of 10 April 1579) to Harvey 'to pluck out of the hateful darknesse, those so many excellent English poemes of yours, which lye hid, and bring them forth to eternall light'.⁴⁵ Perhaps E. K. and Spenser together approached the printer of *The Shep-*

⁴¹ See pp. 61–2, below.

⁴² In Harvey's marginalia in Domenichi's *Facetie* (1571), fol. 34ᵛ (p. 388) he refers to John Wood: 'Sir John Skinner when he speakes of mee, calles mee the great scholler. Sir Jon Woodes opinion of mee, greater & higher, then Sir John Skinners.'

⁴³ Probably these were rhymed poems in English about love, perhaps including polished versions of those found in Sloane MS. 93.

⁴⁴ 'Subboscoes' were sideburns. The draft of this letter is in Sloane MS. 93, fols. 35ᵛ–38ᵛ.

⁴⁵ From the postscript of the prefatory epistle to *The Shepheardes Calender*, 1579.

*heardes Calender*, Hugh Singleton, and began arrangements for an anonymous[46] edition of Harvey's English poems. Perhaps at some point Singleton had misgivings and backed out of the project, and Spenser, looking around for other possibilities, in the spring of 1580 approached Harvey's printer, Henry Bynneman, with the timely earthquake letter and Harvey's letter and poems related to the new quantitative unrhymed verse (putting aside for the moment Harvey's examples of rhymed verse). At any rate, on 30 June 1580 *Three Proper . . . Letters* was entered to Bynneman at Stationers' Hall, and was printed shortly thereafter.

Meanwhile, Harvey, although dilatory, had been making somewhat different plans for a publication with Edward Dyer as dedicatee. Evidence for this is found on folio 48ᵛ of Sloane MS. 93, where Harvey outlines a projected volume of his verse and prose, a far more ambitious work than that which actually appeared in the summer of 1580. At the top of the folio leaf Harvey inscribes the following dedication:

> To the right worshipfull Gentleman,
>      and famous Courtier,
>      Master Edward Diar,
>    in a manner owre onlye Inglishe Poett.
>    In honour of his rare Qualityes,
>      and noble Vertues./
>    *Quodvultdeus* Benevolo
>      J.W.[47] commendith the
>    Edition of his frendes
>      Verlayes: togither with certayne other of his
>                                    Poeticall Devises:
>    And instead of A Dedicatory Epistle,/
>      praesentith himself, and the uttermost
>      of his habilitye, and value
>    To his good worshippes curtuous, and favorable likinge./
>      This first of August. 1580.[48]

Below this Harvey enumerates the proposed contents:

1. The Verlayes.
2. The Millers Letter.
3. The Dialogue.
4. My Epistle to Imerito.

[46] On fol. 36ʳ of Sloane MS. 93 Harvey writes: 'if peradventure it chaunce to cum once owte whoe I am, (as I can hardly conceive howe it can nowe possibely be wholely kept in) . . .'

[47] Probably Harvey's courtier friend John Wood, nephew of Sir Thomas Smyth.

[48] Harvey had originally written 'March' but this was scored through and 'August' written below. Apparently the proposed publication date had been postponed.

He must have subsequently considered adding another composition and altering the sequence, for at the bottom of this page is the following revised list in his hand:

> The Verlayes/,
> My Letter to Benevolo./
> The Schollers Loove.
> The Millers Letter.
> The Dialogue.

The chief feature of the planned volume would thus have been his fine Verlayes',[49] short lyric poems which Harvey apparently esteemed but of which no extant samples have been identified, although it is possible that some of the pages of verse in Sloane MS. 93 are early drafts of these, for elsewhere in this 'Letter-book' are what seem to be preliminary drafts of portions of the other works which he lists: on folio 49$^v$ is 'An answer to a Millers vayne letter', on folio 51$^v$ a 'Dialogue in Cambridge between Master GH and his cumpanye at a Midsumer Comencement, togither with certayne delicate sonnetts and epigrammes in Inglish verse of his making'. 'My Epistle to Imerito' may be that on English reform versification [3] which is printed in the 1580 *Three proper wittie familiar Letters*, some verse of which (part of 'Speculum Tuscanismi') are found in Sloane MS. 93 on folio 52$^r$. 'My Letter to Benevolo' may have been an alternate title for this letter since, as previously noted, Spenser is frequently addressed by Harvey as 'Imerito' and at least once ironically as 'Benevolo'. 'The Schollers Love, or Reconcilement of Contraryes' is the long satiric poem beginning on folio 58$^r$ (referred to on pages 34–5, above). This and the 'Verlayes' were rhymed verse, as were parts of the 'Miller's Letter'. The brief sample of the 'Dialogue' contains a short prose passage with an unrhymed stanza, and the 'Epistle' (if I have properly identified it) was a treatise on versification. One therefore concludes that Harvey was planning a volume of English verse supplemented by some prose passages, but, undoubtedly, he had procrastinated, for, as has been pointed out, even as early as April 1579, E. K. had been urging him to publish his 'many excellent English poemes'. He and Spenser probably felt that if Harvey's English writing could gain recognition, this would increase his chances for a desirable

---

[49] See Sloane MS. 93, fol. 35$^v$ where he discusses these poems in a letter to Spenser. This type of poem originated in fourteenth-century France and was current from the Chaucerian period to the sixteenth century. It usually consisted of short lines arranged in stanzas with only two rhymes, the end-rhyme of one stanza being the chief one of the next. However, by Harvey's time the term 'Verlayes' may have been used in a much looser, more general sense.

position at Court. As we have seen, Edward Dyer had already made favourable comments on the 'satyricall verses' of Harvey's which Spenser had shown him; perhaps Philip Sidney, too, had shown interest. Although Harvey's letters continually admonish Spenser not to let anyone but Dyer and Sidney view his poems, he certainly seems desirous that they should see them.

While searching for some means of promoting his friend's career, Spenser may have received word of Harvey's very *avant-garde* discourse on the recent alarming earthquake. What more opportune time to approach Henry Bynneman who had published most of Harvey's Latin works! One can speculate that Spenser (probably with the help of another close friend of Harvey's like John Wood)[50] gave Bynneman the earthquake letter [2] and Harvey's epistle on versification [3] and that to accompany them Spenser wrote a brief epistle praising Harvey and introducing the subjects of the two main letters. The volume as published includes a preface by an unidentified 'Welwiller of the two Authours' (perhaps Bynneman himself) who tells of being 'made acquainted wyth the *three Letters following*, by meanes of a faithfull friende, who with muche entreaty had procured the copying of them oute, at *Immeritos* handes'. The 'foresaide faithfull and honest friende' (Wood?) had praised the substance and quality of these letters and added that he himself has written 'of the same stampe both to Courtiers and others'. The 'Welwiller' (Bynneman?) himself finds the discourses 'Twoo of the rarest and finest Treaties, as wel for ingenious devising as also for significant uttering, & cleanly conveying of his [the author's] matter, that ever I read in this Tongue'. The 'Welwiller' apologizes for not making the authors 'privy to the Publication' but feels they will not be too displeased with the result. The preface is dated 'XIX. of June. 1580'.

Although one wonders whether Harvey may not have had some inkling of what was going on, he claims that the printed letters were private ones that he had not intended to have divulged publicly.[51] When he found that they were about to be published, he may have tried to remedy what he felt was a lack of substance in the volume by adding two earlier letters, one of Spenser's [4] and one of his own [5]. If this is a correct surmise, it would explain why the *Two Commendable Letters*, although written earlier than the *Three Proper wittie familiar Letters*, were 'More lately delivered unto the Printer' and were added at the end of the volume, apparently as an afterthought, for they are not mentioned on the first

---

50 See n. 96 below, for evidence suggesting he had been contemplating publication of some 'unsatyricall Satyres'.

51 *Foure Letters* (1592), sig. C2.

title-page nor in the preface. As to Harvey's contemplated volume of verlayes, etc., dedicated to Edward Dyer, there is no evidence that Harvey proceeded farther than the draft of a title-page and contemplated publication date of 1 August 1580.[52] The disturbance caused by the early summer publication of the *Three Proper . . . Letters* (1580) would probably have put a damper on any inclination Harvey might have had to risk the issuing of a further volume at this time.

The last [3] of the *Three Proper . . . Letters* is Harvey's to Spenser 'with sundry proper examples, and some Precepts of our English reformed Versifying'. Harvey stresses that before setting up precepts of 'our Englishe Artificiall Prosodye' it is important to agree on a standard orthography conformable to natural pronunciation. In fact, he is noteworthy for his almost invariably consistent simplified phonetic spelling in an era marked by wide variation even within the work of any one writer. A few samples of Harvey's quantitative verse are given, most of which make apparent (perhaps unwittingly) the unsuitability of Latin metric rules when applied to English verse. One of these poems of Harvey's, 'Encomium Lauri', seems to me to be written with tongue in cheek, for its second line and its third from the last line cannot be anything but humorous in intent. The poem reads as follows:

> What might I call this Tree? *A Laurell*? O bonny Laurell:
> Needes to thy bowes will I bow this knee, and vayle my bonetto:
> Who, but thou, the renowne of Prince, and Princely *Poeta*:
> Th'one for Crowne, for Garland th'other thanketh *Apollo*.
> Thrice happy *Daphne*: that turned was to the *Bay Tree*,
> Whom such servauntes serve, as challenge service of all men.
> Who chiefe Lorde, and King of Kings, but th'*Emperour* only?
> And *Poet* of right stampe, overawith th'*Emperour* himselfe.
> Who, but knows *Aretyne*, was he not halfe Prince to the Princes.
> And many a one there lives, as nobly minded at all poyntes.
> Now Farewell *Bay Tree*, very Queene, and Goddesse of all trees,
> Ritchest perle to the Crowne, and fayrest Floure to the Garland.
> Faine wod I crave, might I so presume, some farther acquaintaunce,
> O that I might? but I may not: woe to my destinie therefore.
> Trust me, not one more loyall servaunt longes to thy Personage,
> But what says *Daphne*? *Non omni dormio*, worse lucke:
> Yet Farewell, Farewell, the Reward of those, that I honour:
> Glory to *Garden*: Glory to *Muses*: Glory to *Vertue*.
>                         Partim Jovi, et Palladi,
>                         Partim Apollini et Musis.

Harvey also includes in this letter a poem entitled 'Speculum Tuscanismi'

[52] Sloane MS. 93, fol. 48ᵛ, as transcribed on p. 61, above.

which he describes as a 'bolde Satyriall Libell lately devised at the instaunce of a certayne worshipfull Hartefordshyre Gentleman [perhaps Arthur Capel of Hadham Hall], of myne old acquayntaunce'. It depicts with ridicule the attire and mannerisms of an Italianate Englishman and was probably conceived as a veiled caricature of the Earl of Oxford whose recent insult to Philip Sidney[53] would have seemed insufferable to Harvey. The satirical verse was apparently brought to Lord Oxford's attention by John Lyly[54] (whose patron he was) and must have caused Harvey embarrassment although he eventually seems to have convinced the Earl that the description was not of him. In *Foure Letters*, 1592, Harvey, reminiscing about this episode, discreetly denies that he intended to satirize Oxford and he blames Lyly for having stirred up trouble for him:

[Lyly] would needs forsooth verye courtly perswade the Earle of Oxforde, that some thing in those Letters, and namely the Mirrour of Tuscanismo, was palpably intended against him: whose noble Lordeship I protest, I never meante to dishonour with the least prejudicial word of my Tongue, or pen: but ever kept a mindefull reckoning of many bounden duties toward The-same: since in the prime of his gallantest youth, hee bestowed Angels [gold coins worth about ten shillings each] upon mee in Christes Colledge in Cambridge, and otherwise voutsafed me many gratious favours at the affectionate commendation

---

[53] Sidney as spokesman for the opposition to the Queen's contemplated French marriage aroused the irascible de Vere's anger when the two met in 1579 at the tennis court. Oxford insultingly termed Sidney an insolent 'puppy' and demanded that he and his men withdraw from the tennis court. That Harvey's 'Speculum Tuscanismi' was related to this incident is suggested by three lines near the end which seem to contrast [Sidney's] gentlemanliness with Oxford's unattractive behaviour:

> None doe I name, but some doe I know, that a peece of a twelvemonth:
> Hath so perfited outly, and inly, both body, both soule,
> That none for sense, and senses, halfe matchable with them.

[54] In *Pierces Supererogation*, 1593, sig. I3, Harvey writes: 'Papp-hatchet [Lyly], desirous for his benefit, to currie favour with a noble Earle, and in defect of other means of Commendation, labouring to insinuate himselfe by smooth glossing, & counterfait suggestion, (it is a Courtly feate, to snatch the least occasionet of advantage, with a nimble dexteritie); some yeares since provoked me, to make the best of it, inconsiderately; . . . without private cause, or any reason in the world: (for in truth I loved him [i.e. Oxford], in hope praysed him [in *Gratulationes Valdinenses*, Bk. IV]; many ways favored him, and never any way offended him): and notwithstanding that spiteful provocation, and even that odious threatening of ten yeares provision, he [Lyly] had ever passed untouched with any sillable of revenge in Print, had not Greene, and this dog-fish [Nashe] abhominably misused the verbe passive, as should appeare, by his [Lyly's] procurement, or encouragement, assuredly most undeserved, and most injurious. For what other quarrel, could Greene, or this dogge-fish ever picke with me: whom I never so much as twitched by the sleeve, before I found miselfe, and my dearest frendes, unsufferably quipped in most contumelious, and opprobrious termes.' This intimates that Lyly had stirred up Greene to attack the Harveys in *Quip for an upstart courtier*, 1592. The passage of Harvey's quoted above immediately precedes his tract *An Advertisement for Pap-hatchet, and Martin Mar-prelate*. See pp. 85 ff., below.

of my Cosen, M. Thomas Smith, the sonne of Sir Thomas, shortly after Colonel of the Ardes in Ireland. But the noble Earle, not disposed to trouble his Joviall mind with such Saturnine paltery, stil continued, like his magnificent selfe. (sig. C4ʳ)

Oxford was apparently satisfied that Harvey meant no offence, but some notes within his 'Letter-book' suggest that Harvey's intentions may not have been altogether innocent. On folios 51ᵛ–52ᵛ of Sloane MS. 93 there is the draft of a discourse entitled a 'dialogue in Cambridge between Master GH and his cumpanye at a midsumer Comencement, togither with certayne delicate sonnets and epigrammes in Inglish verse of his makinge'. One of the gentlemen in the company quotes the first twenty-three lines of the satirical poem which in 1580 was published as 'Speculum Tuscanismi'. The discourse continues: 'Nowe tell me . . . if this be not a noble verse and politique lesson . . . in effecte conteyning the argumente of his [Master GH's] curragious and warly[k]e apostrophe to my lorde of Oxenforde in his fourth booke of Gratulationum Valdinensium'.[55] It would seem that Harvey was alluding to Edward de Vere, Earl of Oxford, in both 'Speculum' and in the verses of the 'dialogue in Cambridge betwene Master GH and his cumpanye' quoted above, in the printed version of the first more covertly, for there is no prose comment upon it and it is given the generalized title of 'Speculum Tuscanismi'.

Appended at the end of *Three Proper and wittie familiar Letters* are the *Two Commendable Letters* the first of which [4] is written by Spenser and contains a poem by him which is dated 5 October 1579 from Leicester House. It proves of especial interest for a biographical study of Harvey. Written just before Spenser's anticipated departure overseas, the long Latin poem bids a temporary goodbye to 'G. H., a man most eminent and long by many titles known to fame'. In it Spenser highly praises his friend yet offers him wise counsel. Among the more interesting lines are the following:

> A magnanimous spirit, I know, spurs you up to the summits
> Of honor and inspires your poems with emotions more solemn
> Than light-hearted love (yet not all love is light-hearted).
> Therefore you praise nothing so much as perennial fame.

.     .     .     .     .     .     .     .     .

[55] The apostrophe has a decided satirical undertone, praising Oxford's ability in warlike pursuits even though times are peaceful. It points out that he excels in letters and is more courtly and polished than Castiglione himself, but it advises him to put away his feeble pen, bloodless books, and impractical writings, for 'Steel must be sharpened. Who wouldn't swear you Achilles reborn? . . . It befits a man to keep the horrid arms of Mars busy even in peace. . . . Though there be no war, still warlike praise is a thing of great nobility.' Translation of Harvey's Latin is from Jameson's dissertation, op. cit., pp. 127, 129.

Other things which the giddy mob rabble adores as its gods—
Fat farmlands, gold, city freeholds, alliance of friends;
What gladdens the eye, pleasing forms, pageantry, paramours comely—
These you trample like muck and call a mocking of reason.
   Your scorn is worthy indeed of the Harvey I honor,
The orator copious, the heart that is noble; nor would
The wise Stoics of old fear to set their sanction, with seals
Everlasting, upon it. But men's tastes are not always the same.

.   .   .   .   .   .   .   .   .

Other things which the giddy mob rabble adores as its gods—
Fat farmlands, gold . . . etc.
Whatever flatters the ear or the palate—all these you despise.
Sapience, great indeed though your sapience be, is not sense;
And one who can in good season make light of trivial things
Often will easily triumph over supercilious wisemen.

.   .   .   .   .   .   .   .   .

Whoever strives to tickle the fancy of mighty patricians
Strives to be foolish, for follies multiply favor.

.   .   .   .   .   .   .   .

Nor, by your good leave, do you, O our age's great Cato,
Really deserve the name of time-honored poet,
However high-minded and noble the melodious verse you compose,
Unless you are willing to dote: the world is so laden with dullards.
   But the safe road still divides the abyss through the middle,
For you would describe as a wise man only one who wished to appear
Neither too much of a fool nor a mentor wise beyond measure.

.   .   .   .   .   .   .   .

Nor ought you to spurn voluptuous pleasures too much;
Nor a wife brought at long last to the altar;[56] nor, if you are wise,
An offer of gold . . . nor yet seek
These pleasures too often. Either course is fraught with reproach.
. . . he wins
Every vote of approval who mingles use with delight.

Spenser closes the letter: 'Farewell, fare very well, my sweetest Harvey,
to me of all my friends by far the dearest. On sea, on land, alive, or dead
. . . thy Immerito.'[57]

   The second of the *Two Commendable Letters* [5] contains further discussion of quantitative verse, especially of those samples of Spenser's

---

[56] Is Spenser here alluding to his own marriage to Machabyas Childe which was to take
place a few months later (27 October 1579)? If so, Harvey must have originally disapproved
although by 7 Apr. he had mellowed enough to address her in a friendly fashion: 'O mea
Domina Immerito, mea bellissima Collina Cloute.' (*Three Proper . . . Letters*, sig. Fiii<sup>r</sup>.)

[57] English translation is from the *Works of Edmund Spenser* (Variorum Edition), Baltimore, 1949, vol. ix, ed. Rudolf Gottfried, pp. 256–8.

that Harvey has seen. Spenser had evidently told him of the interest of
Sidney and Dyer in the movement and perhaps in the discussion of other
literary or philosophical problems,[58] for Harvey writes (sig. H4ʳ): 'Your
new-founded ἄρειονπάγον [areopagus] I honoure more, than you will
or can suppose: and make greater accompte of the twoo worthy Gentle-
menne, than of two hundreth *Dionisii Areopagitae*, or the verye notablest
Senatours, that ever *Athens* dydde affourde of that number.' He thanks
Spenser for his 'Latine Farewell' and calls it 'a goodly brave yonkerly
peece of work', telling Spenser that he is 'Marvellously beholding' unto
him for it and that he hopes to be able to reciprocate 'by that time I have
been resident a yeare or twoo in *Italy*' and become 'better qualifyed in
this kind'. He offers Spenser a wager that his own contemplated trip will
not materialize, 'Neither the next, nor the nexte weeke' and that Harvey
will be able to see him in person before long.

As it happened, Spenser's anticipated departure for France did not come
to pass, but he did leave for Ireland about ten months after the letter was
written (August 1580) when he received an appointment as secretary to
the Lord Deputy of Ireland, Arthur, Baron Grey of Wilton. In the autumn
of 1580 after Spenser left Leicester's service, Harvey for a brief time was
secretary to Leicester. His appointment to the Earl's service is confirmed
(as Moore Smith has noted[59]) by a few lines in Spenser's *Colin Cloutse
come home again*, 1595:

> For well I wot, sith I my selfe was there,
> To wait on Lobbin [i.e., Leicester] (Lobbin well thou knewest)
> As ever else in Princes Court thou vewest.[60]

Further confirmation is found in the Cambridge comedy *Pedantius*
which refers to the protagonist's going to Court, where a favourite pupil
had preceded him, and to his ignominious return. According to Nashe's
version in *Have with you to Saffron-walden* (*Works*, iii. 79), Harvey's
sojourn at Court was not a success and Leicester soon dispensed with his
services.

After returning to Cambridge and Trinity Hall in 1580/1, Harvey
resumed his briefly interrupted study of law and applied himself with
vigour. His marginalia in Oldendorf's *Loci Communes Juris Civilis*, 1551,
indicate that while relaxing from his arduous legal studies he amused
himself by listing and classifying astrological data.[61]

---

[58] See p. 74, below.                                        [59] *Marginalia*, pp. 39–40.
[60] *Spenser's Minor Poems*, ed. E. de Selincourt, Oxford, 1966, p. 328, lines 735–8.
[61] Blank pages belonging to what little remains of this volume are covered with closely
written astrological notes. At the top of one of the rectos is Harvey's inscription: 'Scripsi
in aula Trinitatis Cantabrigiae; cum forte e studio iuris civilis vacarem.'

Marginalia in another much annotated legal text disclose that at about this time Harvey enjoyed participating in a legal disputation at Trinity Hall. On this occasion he and John Gardiner, a fellow law student, presented the opposing points of view to Peter Wythipoll's[62] argument. In his marginalia he expresses his admiration for Wythipoll's impressive display of legal skill. The disputation must have taken place during 1579 or 1580, for Harvey did not enter Trinity Hall until 18 December 1578 and Withypoll resigned his fellowship in 1580. Harvey describes the forensic event: 'Problema Trinitense Petri Vithipoli, legum bacalaurii, illiusque aulae socii. Vithipoli respondens. Ego, et Gardinerus opponentes. Magna expectatio: satisfactio competens. Vithipolus se ipso paulo jurisperitior.'[63]

As mentioned before, Harvey's intensity, vanity, and ambitions became the butt of the comic satire *Pedantius* which was presented at Trinity College on 6 February 1580/1. Although the play's mood was more jovial than bitter,[64] it unmistakably parodied Harvey in the character of Pedantius, a Ciceronian rhetorician who aspires to be a lover, courtier, and man of affairs but succeeds only in making himself ridiculous. After purchasing an elaborate outfit of clothes, he has a brief trial at Court, but his hopes of preferment and his suit in love come to naught. Finding himself in financial straits, he faces the dire necessity of selling his cherished library with its books bejewelled with handsome marginalia.[65]

An Oxford lampoon written early in 1581, found in a commonplace book belonging to William Withie (a student of Christ's College) is entitled 'Uppon Harvyes vile arrogant English versifyinge'. Warren Austin believes that Withie had been irritated by *Three Proper . . . Letters* and by Harvey's pretensions to authority as rhetorician, critic, and poet.[66] The lampoon begins:

> Omnipotent Orator, famous Rhetorician Archpott
> whose front more clearly doeth shyne methincks, then a brasspott;[67]

---

[62] Gardiner became a fellow in June 1574, received his LL.B. in 1576, and his LL.D. in 1583; Wythipoll, a member of the Essex family referred to on p. 33 n. 2, above, graduated B.C.L. in 1572/3 and was a Fellow until some time in 1580. Both men were thus Harvey's seniors at Trinity Hall.

[63] It translates roughly: 'The Trinity Hall disputation topic of Peter Wythipoll, Bachelor of Laws and Fellow of that Hall. Wythipoll responding. I and Gardiner opposing. With great expectation, competing well. Wythipoll a little more skilled in law even than his usual self.' The marginalia appear on p. 373 of Harvey's copy of J. Hopperus, *In veram Jurisprudentiam Isagoges ad filium*, 1580.

[64] Sir John Harington in his discussion of comedies in *A Brief Apology for Poetrie*, 1591, writes: 'How full of harmeles myrth is our Cambridge *Pedantius*?'

[65] *Pedantius, a Latin comedy formerly acted in Trinity College, Cambridge*, ed. G. C. Moore Smith, London, 1905.

[66] 'William Withie's Notebook: Lampoons on John Lyly and Gabriel Harvey', *RES*, 23 (1947), 297–309.                                    [67] Sloane MS. 300, fol. 54ʳ.

At about this time Harvey's brothers also seem to have served as targets of ridicule. Nashe (whatever his report may be worth) writes that a 'shewe' was 'made of the little Minnow his [i.e. Gabriel's] Brother, *Dodrans Dicke*, at Peter-house, called *Duns furens*,[68] Dicke Harvey in a frensie'. Richard Harvey being of small stature was aptly termed 'Dodrans', a reference to three-fourths of anything. The title *Duns furens* is probably an allusion to his hot temper and to the fact that he was Praelector of Philosophy, a latter-day Duns Scotus. Nashe writes that Richard was always 'in hate, either with Aristotle, or with the great Beare in the firmament, which he continually bayted: or with Religion, against which in the publique Schooles he set up Atheistical Questions',[69] and according to *Pedantius*, Richard Harvey (like Gabriel an ardent Ramist) was persecuted for his belittling of Aristotle.[70] However, unlike his brother, Richard, while a Fellow at Pembroke, followed traditional procedures and trained himself in Divinity, eventually taking orders and gaining minor ecclesiastical posts.

In the early 1580s Richard became interested in the study of astrology and seems to have expended so much time on it that his brother (and tutor) Gabriel cautioned him either to desist or produce something of value in this field. The resultant pamphlet, *An Astrological Discourse upon the Conjunction of Saturne & Jupiter*, was published by Henry Bynneman in January 1583.[71] The author makes dire predictions of violence, or barrenness and sterility of the earth, and of the second coming of Christ because of the change from the watery to the fiery trigon, a change which had also occurred at the time of Christ's birth. The conjunction of the two major planets was to occur on 28 April 1583. *An Astrological Discourse* was almost immediately attacked and ridiculed: other books appeared controverting its predictions and a ballad was written (which Hyder Rollins believes was probably composed by William Elderton) beginning, 'Trust not the conjunctions or Judgementes of men when all that is made shalbe unmade againe'.[72] According to Nashe, Richard Tarleton made jests on the stage about Richard Harvey.[73] Another probable

[68] According to Nashe (iii. 80–1) *Duns furens* was acted at Peterhouse at a time when Dr. Perne was either Vice-Chancellor or Deputy Vice-Chancellor, i.e. in 1580/1 or 1586/7. This play (apparently a Latin satirical comedy) is now lost.

[69] *Works*, iii. 80; Nashe also refers to him as 'Pigmey Dicke' in iii. 81, 82.

[70] See also Nashe's *Works*, i. 196.

[71] On sig. Aii[r] of *An Astrological Discourse* Richard Harvey writes of Gabriel's advice to him.

[72] Entered S.R. on 3 May 1583 to Richard Jones. See Hyder Rollins, *Analytical Index to the Ballad Entries (1557–1709) in the Registers of the Company of Stationers of London*, 1967 (reprint of 1924 edn.), p. 237.

[73] *Works*, i. 196–7.

attack in lighter vein was a Cambridge show at Clare Hall alluded to by Nashe as mocking the three Harvey brothers. Nashe speaks of it as 'Tarrarantantara turba tumultuosa, Trigonum, Tri-Harveyorum, Tri-Harmonia'.[74]

Shortly after the printing of Richard Harvey's book, his younger brother John published *An Astrological Addition, or supplement to be annexed to the late Discourse upon the great Conjunction of Saturne, and Jupiter. Wherein are particularly declared certaine points before omitted, as well touching the elevation of one Plannet above another, with theyr severall significations: as touching Œconomical and household provision: with some Judicialls, no lesse profitable. Made and written this last March, by John Harvey, Student in Phisicke* (imprinted by Richard Watkins, 1583). Added at the end of this work is John Harvey's translation of Hermes Trismegistus' *Iatromathematica* which is described by the translator as 'A booke of especiall great use for all Studentes of Astrologie, and Phisicke'. The volume was entered at Stationers' Hall on 12 April and seems to have been put forth to support his brother's assertions by citing further authorities and answering his enemies' arguments. The work was dedicated to Master Justice Meade, but the text itself is addressed to Gabriel Harvey.

Justice Thomas Meade was a wealthy and learned judge who lived at Wendon Lofts, Essex, not far from Walden.[75] John Harvey had become friendly with his son, young Thomas, while the two were students at Queens' College, Cambridge. Subsequently, young Harvey became tutor to Meade's sister Martha and in the course of time married her. The couple moved to the coastal town of King's Lynn where he established a reputation as an able physician until his career was cut short by early death.

Richard Harvey had opened the text of his prognostication with an address to his brother Gabriel. John likewise invoked his prestigious scholarly name. Although eager to accumulate information about all matters, Gabriel, however, was very skeptical of the worth of astrological studies.[76] But when his brothers' work was attacked, fraternal considerations outweighed scholarly ones.

Among the assaults on Richard Harvey's prognostications was a small

[74] *Works*, iii. 80.

[75] On sig. A6[r] of the dedication of John Harvey's *An Astrologicall Addition* is an account of a visit to Justice Meade's made in 1580 by Gabriel and an Oxford friend who was a preacher. Although both young men were then strangers to Justice Meade, he entertained them courteously, 'partly by familiar discourse upon some chiefe pointes of learning, and partly by manifest declaration of good liking, & welwishing . . . as for the greatnes of the cheare, choise of companie, manner of welcome, and such like'.

[76] *An Astrologicall Addition*, sig. A5[r].

book by the learned Lord Henry Howard, Earl of Northampton, entitled *A Defensative against the poyson of supposed prophesies*. It was entered in the Stationers' Register on 13 June 1583 and was published a few weeks later, well after the conjunction of Saturn and Jupiter had taken place and proved innocuous. Although the ostensible purpose of Howard's tract was to denigrate judicial astrology, there may also have been a personal motive for attempting to injure Richard Harvey—at least so Gabriel seems to have thought, according to annotations he made in his own copy of the *Defensative*. At the bottom of sig. Hhi[r] one finds the following emotionally charged comments in his hand:

Iwis it is not the Astrological Discourse, but a more secret mark, whereat he shootith. *Latet anguis in herba: et per me latebit, etiam adhuc.* Patience, the best remedy in such booteles conflicts; God give me, and my Frends, Caesars memory, to forgett only injuries, offerid by other; and to remember especially such requisites, as especially concerne, and apperteine owrselves. An Ostridges stomack can digest harder iron, then this. *Qui seipsos confirmant, alios abunde confutant. Dabit Deus his quoque finem.*[77]

xx. Julii. 1583.   GH.

It is not known when Gabriel Harvey first met Henry Howard, but it is recorded in Cambridge archives that he was one of the two University examiners in 1569–70 when Harvey was placed ninth in the Ordo Seniori-tatis for his A.B. from Christ's College. Five years later Lord Howard would doubtless have heard of the Harvey family in the environs of Walden, for the Earl of Northampton was the uncle of Philip Howard, the young nobleman who in 1574 attempted a seduction of Gabriel's sister, Mercy Harvey. Henry Howard was then living at Audley End, for, after his brother's execution in 1572, he had retired there to oversee the education of Thomas Howard's sons.[78] As has been discussed above, when Harvey accidentally intercepted a *billet doux* to his sister from her wooer, he quickly put an end to the affair. Harvey had unquestionably tried to use the utmost tact, but the Howard family must have never-theless felt keen embarrassment. Harvey's marginalia in the *Defensative* suggest that he or his family had endured from the Howards affronts other than the thwarted seduction, but I have found no specific information as to what they may have been. It is known that Henry Howard and Harvey had widely divergent political views, and this may at various times have led to friction. In 1580 Henry Howard, a staunch Roman

---

[77] Freely translated the Latin sentences read: 'A serpent lies hidden in the grass: and it will remain concealed even now by me. . . . Those who assert themselves repress others exceedingly. God will put an end even to these men.'

[78] *D.N.B.*

Catholic, had circulated a manuscript tract advocating Elizabeth's marriage to the Duke of Anjou; in 1582 he had been arrested and held briefly on charges of correspondence with Mary Stuart. Although a cultured and erudite scholar 'reputed to be the most learned nobleman of his time', he seems throughout his life to have shown a lack of moral and political scruples. It was even said that the *Defensative* itself contained some well-disguised treasonable notions.[79] Throughout Harvey's copy of the book are underlinings and comments such as that on sig. Gg4[v]: '*Vincit qui patitur*'. Harvey seems continually to have needed to remind himself to practise patience and to 'turn the other cheek'.

In May of 1583 Harvey was appointed by Trinity Hall to fill a vacancy in the office of Proctor of the University for a period of five months.[80] His colleague as Senior Proctor was Anthony Wingfield, who in 1581 had become Public Orator. Two proctors were chosen every October by each of two colleges in turn nominating one person (who had to have been a member of the senate for three years and have resided in the University for the greater part of the three preceding terms). If the office for any reason became vacant before the expiration of the year's term (as occurred in 1583), Trinity Hall was empowered to nominate a Proctor for the remainder of the year.

The Proctorship was an important office of ancient tradition. Proctors played a leading part in the administration of Cambridge affairs and were maintainers of student discipline. Their duties included regulating the hours of lectures and disputations, of burials, of fairs, and other functions. They acted for the University in many kinds of business, patrolled streets to curb disturbances, exercised jurisdiction over improper persons, fined undergraduates for infringement of certain regulations, and, following John Whitgift's Vice-Chancellorship in 1571, even had the savage obligation of publicly whipping any undergraduate scholar who went into a local river or stream to wash or to swim. If Harvey had to exercise any such disciplinary functions, it is not likely to have increased his popularity with the students.

In June of 1583 Harvey may have paid a visit to Oxford University, for notes among his marginalia describe a philosophical disputation there between Giordano Bruno and the ecclesiastic Dr. John Underhill (who was to become Vice-Chancellor of Oxford the following year). According to Bruno's not necessarily impartial report in his *La Cena de le Ceneri*, 1584, Underhill made a poor showing in the disputation and had his fifteen syllogisms refuted fifteen times. In this passage Bruno gives

---

[79] John Strype, *The history of . . . Edmund Grindal . . .*, London, 1710, p. 157.
[80] Harvey filled out the expiration of Leonard Chambers's term.

an unflattering description of his antagonist and of the audience's rude behaviour.[81] They had become so incensed against Bruno that he subsequently broke off his course of lectures and left Oxford. Harvey's comments on the disputation read: 'Jordanus Neopolitanus, (Oxonii disputans cum Doctore Underhill) tam in theologia, quam in philosophia, omnia revocabat ad Locos Topicos, et axiomata Aristotelis: atque inde de quavis materia promotissime arguebat. Hopperi[82] principia multo efficiaciora in quovis Argumento forensi.'[83] As Moore Smith suggests, Bruno probably 'brought all questions back to Aristotelian axioms and commonplaces' because Oxford was then the stronghold of Aristotelianism.[84]

After leaving Oxford, Bruno seems to have joined the circle of Philip Sidney, Fulke Greville, and Edward Dyer. I. Frith believes that Bruno's account of their meetings suggests the existence of something like a philosophical club or 'Areopagus'.[85] In *La Cena de le ceneri* (London, 1584) he tells of their meeting in the eminent house of Sir Fulke Greville 'to draw appropriate topographies of a geographical, ratiocinative and moral order, and then make speculations of a metaphysical, mathematical and natural order.'[86]

There is no concrete evidence as to whether or not Harvey ever attended meetings of this sort, but it is clear that he had avid intellectual interests and the urge to comprehend and be informed in all fields of knowledge. He had already intensively studied languages and the verbal arts; he had read the salient works in the heritage of classical and medieval learning in the fields of history, metaphysics, astronomy, and cosmography; he had spent seven years in the difficult study of jurisprudence; he was now to educate himself in modern scientific subjects: medicine, chemistry, pharmacology, and navigation.

Having concluded his Cambridge legal studies in 1584 with an LL.B., he decided to accept a Medical Fellowship at Pembroke, for his annota-

---

[81] London: J. Charlewood, p. 93.

[82] Harvey here refers to the thought of Joachim Hopperus, author of *In veram Jurisprudentiam Isagoges ad filium* (1580) and *Seduardus, sive de vera Iurisprudentia* (1590), two compendious legal texts that Harvey owned.

[83] 'The Neapolitan Giordano (disputing with Doctor Underhill of Oxford) as much in theology as in philosophy, brought all questions back to Aristotelian axioms and commonplaces: and then very readily discussed any topic whatsoever. The principles of Hopperus are more powerful anywhere in a forensic argument.' (Ramus, *Oikonomia*, 1570, p. 162.)

[84] *Marginalia*, pp. 273–4.

[85] *Life of Giordano Bruno, the Nolan*, London, 1887, p. 128. See also p. 68, above, for Harvey's reference to a 'new founded Areopagus'.

[86] *The Ash Wednesday Supper: La Cena de le ceneri by Giordano Bruno*, ed. and trans. by Edward A. Gosselin and Lawrence S. Lerner, Archon Books, 1977, p. 68.

tions in a later acquired medical text state that he succeeded Lancelot
Browne (evidently when Browne left Pembroke for London to become a
Fellow of the Royal College of Physicians).[87] Harvey's marginalia (on sig.
h3ᵛ) of Hieronymus von Braunschweig's *Perfecte homish Apothecarye*,
1561, translate: 'And not by chance had this practical book been recommended to me by the very wise Doctor of Medicine, Launcelot Browne,
whom I was just then succeeding at Pembroke Hall in the particular
medical fellowship of his profession.'[88] As indicated by later lines in this
passage, Browne recommended this and a number of other medical
books to Harvey.

In 1584 he had acquired, probably from his physician brother John
(see page 222 below), Hugkel's *Semeiotice*, 1560, a medical text which
bases its diagnoses on Galenic theory and the four humours. At Cambridge, medical teaching was built on the classical principles of Galen, and
newer theories or developments were largely ignored. Most serious young
students of 'physick' endeavoured to go abroad for the more modern
and thorough medical training to be obtained at the Italian universities,
notably that of Padua. The inscription on the title-page of the *Semeiotice*
suggests Harvey's dissatisfaction with his progress in medicine. He writes
in Latin that the volume has not yet been (adequately) read, for its contents are not yet at his fingertips.[89] Marginalia in another medical volume
(Hieronymus von Braunschweig's *Homish Apothecarye*, 1561) are dated
'1590' by Harvey on sig. Hiiiʳ and show increased grasp of practical
medical knowledge. It was in the 1590s that Harvey made an inten-
sive study of medicine on his own, acquiring a number of reference
works in this field and annotating them copiously with aids to diagnosis
and memoranda about efficacious treatments. His interest in this later
period centred especially on the new Paracelsan methods of using chemical agents and pharmaceuticals.

In 1584 Harvey, having completed his legal training at Trinity Hall,[90]
was granted his LL.B. on 10 July with Dr. Henry Harvey, the Master,
acting as sponsor. However, according to *Grace Book Delta* (p. 389),
Gabriel Harvey was not subsequently inaugurated and therefore forfeited

[87] In 1584 Gabriel's brother John Harvey was also probably studying medicine, for by
1587 he was to receive from Cambridge a licence to practise as a physician (Venn, ed.,
*Grace Book Delta*, p. 494).

[88] 'Nec temere hic mihi practicus liber commendatus a sagacissimo Medicinae doctore,
Lanceloto Brouno fuit: cui iamtum aulae Pembrochianae medico succedebam in proprio
illius professionis sodalitio.'

[89] Harvey signed his name on the title-page and wrote: 'Nondum lectum, quod nondum
tuum ad unguem.'

[90] See Charles Crawley, *Trinity Hall: The History of a Cambridge College 1350–1975*,
Cambridge, 1976. Chapter IV, 'A Nursery for Civilians'.

twenty shillings. Eventually, he incepted as Doctor of Civil Law on 12 July 1585—but not at Cambridge—at Oxford.[91] His brother John and two Cambridge friends, Richard Wafeeld and John Barker, accompanied him there to depone for him, asserting that he had completed the Cambridge LL.B. requirements.[92] John Harvey was granted an Oxford M.A. during his few days' sojourn. (Although he had received an M.A. from Cambridge in 1583/4, an additional degree from another university would further enhance his prestige.)

Among Harvey's marginalia is a brief description of his own inauguration at Oxford: 'I did my Doctors Actes which a thousand heard in Oxford, & some knew to be done with as little premeditation, as ever such Acts were done, for I answered upon the questions, that were given me by D$^r$ Cathedrae but two days before, & read my cursory lecture, with a days warning'.[93] By way of exercises the candidate for the D.C.L. had to answer certain questions, read a lengthy text and be prepared to translate and make short comments on the subject-matter (this latter was known as the 'cursory lecture'). It was stipulated that the candidate was to be given at least three days to prepare for this examination; in Harvey's case this was cut down to two days for the questions and one for the cursory lecture. Harvey evidently was dissatisfied with his performance, for he noted in his Commonplace Book: 'at Cambridg, in my proctorship my default; at Oxford, in my Acts for the Doctorship'.[94] The ceremony of presentation included a curious rite known as 'nemo scit' in which the candidate had to produce a purse which he took an oath contained at least £6.13s. 4d. in gold and silver coins, but the precise amount of which he declared he did not know. He promised to make no complaint of whatever sum was removed. The money was then spread out, and the presenter and deponents helped themselves at their pleasure, returning the balance to the candidate.[95] As it was also necessary to purchase an expensive Doctor's robe for the ceremony, the obtaining of a Civil Law Degree must have involved Harvey in considerable expense.

After presentation the candidate's name was entered in the list for the year of those 'licentiati ad incipiendum in facultate juris'. The following

---

91 Thomas Baker wrote of his taking the civil law degree at Oxford rather than Cambridge that it 'being irregular, might be one thing (among others) that gave offense' (Baker MS. 36, 114).

92 Univ. Oxon. Arch., Register L 10, 1582–94 (an unfoliated MS.).

93 These annotations were transcribed by Thomas Baker from Harvey's own copy of his works and are entered on the verso of the flyleaf of Baker's copy (Bliss A. 110 at the Bodleian). Harvey repeats these lines in print on sig. E1$^v$ of *Pierces Supererogation*, 1593.

94 Add. MS. 32, 494, fol. 45$^v$.

95 *Register of the University of Oxford*, ii: 1571–1622, part i, 116 (5).

year on 10 July 1586 Harvey was incorporated as Doctor of Civil and Canon Law.[96]

There is a laudatory sonnet of Edmund Spenser's dated from Dublin, 'this xviii. of July, 1586' addressed 'To the right worshipfull my singular good frend, M. Gabriell Harvey, Doctor of the Lawes'. The salutation (although not the content of the sonnet which had probably been written some years earlier[97]) was very likely in response to the happy news that Harvey had at last been installed as Doctor of Civil and Canon Law.

In the previous year Harvey had undergone a distressing experience. In early February 1584/5 his relative and friend Dr. Henry Harvey, Master of Trinity Hall, had died. The candidate favoured to replace him was Gabriel himself since half the Fellows desired him as Master and the preferences of the other five were divided among three candidates none of whom was fully qualified for the office. It seemed obvious to Gabriel that he would be confirmed as the new Master. But, at this point, his old enemy Andrew Perne and some of his cohorts intervened to delay the vote by persuading the Fellows to defer the election out of respect for Dr. Henry Harvey's recent death. Then, according to the Cambridge historian Thomas Baker, with considerable manœuvring and 'by the cunning and conduct of some of the Heads' the Queen's mandate was obtained to install as Master of Trinity Hall the highly respected forty-eight-year-old Thomas Preston of King's College.[98] Gabriel was powerless to oppose the selection of this man who was over thirteen years his senior and one whom he much admired, but he must certainly have questioned the unusual procedure of bringing in as Master a man from another college. While the final choice was being made, a pathetic letter was delivered to William Cecil, Lord Burghley:

I was yesterday at Trinity Hall, when we universally agreed on the answere to your letters sent from your L: and M$^r$ Secretary, first to obey her Maj: commandment for the stay, & then to make humble supplication, that it might please her Maj: to vouchsafe a fuller cognisance of the cause—which humble supplication maketh exceedingly for me, considering how the statutes of the House make especially for me, how the suffrages of the company make especially for me & in truth how very favorable & charitable respect maketh especially

---

[96] Univ. Oxon. Arch. Register L 10, 1582–94 (unfoliated).

[97] The content of the sonnet seems designed as an accompaniment to a collection of satirical writings (whose publication had been contemplated but decided against by Harvey, probably because of the disturbing reception of the 1580 volume of letters). On sig. F2$^v$ of *Foure Letters* (1592) Harvey alludes to Spenser's 'overlooving Sonnet: A token of his Affection, not a Testimony of hys Judgement' and of its having 'long since embraced' Harvey's 'unsatyricall Satyres'. The sonnet was ultimately printed at the end of *Foure Letters*.

[98] Baker MS. 36, p. 114. Also see letter from the Heads to Lord Burghley (transcribed in Harleian MS. 7031, fol. 79$^r$).

for me. By our Statutes, none is eligible, but either a Fellow, sufficiently qualify'd or for want of such, a student in the town at this present. Whereunto these principall considerations are to be added, *ut non beneficiati* beneficiatis, pauperiores ditioribus *preferantur* [in order not to have benefitted those already benefitted, the poorer are placed ahead of the richer]. All which circumstances were supposed more agreable to me, then any my competitor. Then for voyces, I had five of ten, the other five being divided into three partialities for Betts, Whitcroft, & Berry. So that no man now is prejudiced & overthrown by her majesties mandat, but my poore miserable self—I never yet had anything bestowed upon me, having referred great part of my studies, to advance the honour of the greatest in authority—As for the judgment of any of our Heddes, the very truth is, not any one of them knoweth me to any purpose, but only D: Still, & not he so much as my L: of Rochester [John Young]—I stayed not the subscription to our answere, but provided my self for this journey, taking horse at three of the clock. Truly my Brothers & my self with my man, have nigh hard killed four good geldings about this cause—My self even for very shame to shew my face in the Town, am now constrained to go post, as I came Post.—Here in London raptissime this very Munday morning.

Your Lordship's ever most Dutifull at Commandment

Unhappy Harvey.[99]

While Harvey was journeying to London, Thomas Preston was elected Master of Trinity Hall and continued in this post until his death on 1 June 1598.

There are not many records of Harvey at Cambridge after 1585. One of the exceptions is to be found in a commemorative volume to Sir Philip Sidney published on 16 February 1587. It is entitled *Academiae Cantabrigiensis Lachrymae Tumulo Nobilissimi Equitis, D. Philippi Sidneii Sacratae* and contains a collection made by Alexander Neville[100] of Latin, Hebrew, and Greek verses written by various Cambridge scholars in tribute to Sidney whose premature death had occurred on October seventeenth of the previous year as the result of a wound received at the battle of Zutphen in the Netherlands. The title eulogy (on sig. A$^r$–A2$^r$) is a long Latin poem expressing Cambridge's grief for the loss of an outstanding man. At the end are appended in certain copies the initials 'G. H.' and in other copies the words 'Cantabrigiae xxx Novemb. 1586'.[101]

[99] Harleian MS. 7031, fols. 78$^v$, 79$^r$. These are Trinity Hall papers transcribed by Thomas Baker.

[100] Alexander Neville, unlike his brother Thomas (see pp. 17 ff.), was willing to honour Harvey's scholarly abilities, although one wonders whether it was his or Harvey's idea that the latter's poems should not be signed with his full name.

[101] The initials 'G. H.' are to be found in the British Library copy (C.34.h.1) and in the Huntington Library copy (31859); in a copy now at the Folger Library (STC 4473

In a letter to the *Times Literary Supplement* of 22 March 1947 (p. 127), Warren B. Austin definitively identified this and two other poems in the volume as being of Harvey's authorship and concluded that 'Gabriel Harvey had . . . been chosen, as the University's most noted Latin poet, to lead the procession of mourners with the titular elegy written on behalf of the whole body'. It is known that Harvey was an admirer and a friend[102] of Sidney's and had previously written other tributes to him.[103]

After this first poem there are two brief Latin elegies, one by Thomas Byng and one by Humphrey Tyndall, and then a quite personal tribute to Sidney in Latin verse (presumably Harvey's) titled 'De subito & praematuro interitu Nobilis viri, Philippi Sydneii, utriusque militiae, tam Armatae, quam Togatae, clarissimi Equitis: officiosi amici Elegia' (sigs. A3$^r$–A3$^v$) and a poem addressed to Sidney's uncle, 'Ad illustrissi-mum Dominum Comitem Leicestrensem protheoreticon' (sigs. A3$^v$–A4$^r$). At the conclusion of the latter are the initials 'G. H.' (and in other copies the words '29 Novemb. 1586'). Austin points out that in all three poems the 'rhetorical devices and characteristic phrases are redolent of Harvey throughout'. The poems apparently were composed about a month after Sidney's death. Only the first section of this volume contains tributes from Cambridge luminaries; following is a smaller section of eulogies from other contributors.

copy 1) the initials are replaced by the notation of place and date although the title-page does not vary. There was another issue of this work with a slightly different title-page. Part of it is bound together with the Folger copy, but I have not seen this issue in its entirety. The *STC* lists copies at the B.L., Bodleian, Huntington, and Cambridge. The Folger copies are apparently from Cambridge. On pp. 258 ff. of Appendix B, below, are transcripts of the three poems.

[102] In a 1598 letter to Robert Cecil (Salisbury Papers, 61, 5) Harvey alludes to his former 'inestimable dear friend Sir Philip Sidney' and in Harvey's copy of Sacrobosco (on sig. aii$^r$), his marginalia seem to refer to a conversation with Sidney: 'Sacrobosco, & Valerius, Sir Philip Sidneis two bookes for the Spheare. Bie him specially commended to the Earle of Essex, Sir Edward Dennie, & divers gentlemen of the Court. To be read with diligent studie, but sportingly, as he termed it.' A 1580 letter of Sidney's to Sir Edward Dennie (a copy of which is reprinted in James Osborn's *Young Philip Sidney*, New Haven and London, 1972, pp. 537–40) recommends 'Sacroboscus, and Valerius' but Harvey's comment on how they should be read probably is derived from Sidney's spoken words, perhaps to Harvey himself.

[103] Harvey's tributes include the 'Castilio, sive Aulicus', discussed on p. 43, above, and eighty-three Latin hexameters inscribed in James VI, *Essayes of a Prentise* (1585), and en-titled 'De tribus vividis scriptoribus Epigramma. Ad Astrophilum, vividiora delineantem.' After praising the pictorial quality of Homer, Livy, and Chaucer, the poem declares that Sidney's art surpasses even these.

# 6. To London: departure from academic life

HARVEY retained his Trinity Hall Fellowship until January 1591/2,[1] but it is probable that within a couple of years after 1586 (when he received his Oxford degree as Doctor of Civil and Canon Law) he moved to London to practise in the Court of Arches, the highest of the ecclesiastical courts. It was the court of appeal of the Archbishop of Canterbury from all diocesan courts and also a court of first instance in all causes under his jurisdiction, handling matters concerned with legacies, matrimonial contracts, defamation, perjury, usury, adultery, whoredom, debts, drunkenness, simony, blasphemy, and benefices. The Court of Arches received its name because of the archlike shape of the great stone pillars within the Church of St. Mary-le-Bow where the court met (on Cheapside near St. Paul's).

In Harvey's marginalia he alludes to The Arches when describing the great ecclesiastical court of the Rota at Rome: 'The Rota in Roome, much lyke the court of th'Arches.'[2] The Judge of this court was known as Dean of Arches. From 1583 until 1597 Richard Cosin was Dean; there are extant with Harvey's signature and annotations copies of two of Cosin's works.[3]

Becoming at last a 'Civilian' would have seemed to Harvey the fruition of a long-time dream. His marginalia reiterate his reverence for Jurisprudence as the queen of studies and the foundation of statesmanship. On one occasion he writes: 'Two imperial arts: the law and arms. Indeed, they are truly the art of arts, the science of sciences, and the souls themselves of flourishing states.'[4]

---

[1] On 22 Jan. 1591/2 Christopher Wivell was admitted to the Fellowship which 'Gabriel Havry [sic] nuper habuit'. (Smith, Marginalia, p. 61 n. 2.)

[2] Hopperus, In veram Jurisprudentiam Isagoges, 1580, sig. Ii 4ᵛ.

[3] An Answer to the two fyrst and principall Treatises of a certaine factious Libell, put foorth latelie, without name of Author or Printer, and without approbation by authoritie, under the title of An Abstract of certeine Acts of Parlement. London: Thomas Chard, 1584; An apologie for sundrie proceedings by Jurisdiction Ecclesiasticall. London: Christopher Barker, 1593. These titles of Cosin's are bound (presumably by Harvey) together with An abstract of certain Acts of Parliament for the peacable government of the Church, 1583 (authorship uncertain). See catalogue of Harvey's books, pp. 207, 208, and 241, below.

[4] Harvey's marginalia in Hopperus, sig. Aiʳ, are in Latin: 'Duae Imperatores Artes, Leges, et Arma; verae illae quidem Artes Artium, scientiae scientarum, ipsaeque Animae florentissimarum Rerumpublicarum.'

Harvey had confidence that with his new status as a civil lawyer he could gradually make his way toward an important position in government. But, as in other instances in his life, after hard study and earnest endeavour his glorious hopes were to be frustrated. We do not know the details of his legal career, but there are indications that he never attained great status as a lawyer nor was he able to broaden his base.

Perhaps, as had happened so often before, he was thwarted by petty personal jealousies or animosities, perhaps his own personality stood in the way; probably both factors played a part. Harvey's law career must be reconstructed largely by inference, for there is a dearth of documentation and his marginalia offer no suggestions of specific achievements in this field. Nashe, as might be expected, points to Harvey's legal practice as just one more cause for ridicule. He does, however, provide several references to the Court of Arches as the locale. For example, he writes:

If we were wearye with walking, and loth to go too farre to seeke sport, into the Arches we might step, and heare him plead; which would bee a merrier Comedie than ever was old Mother *Bomby*. As, for an instance: suppose hee were to sollicite some cause against Martinists, were it not a jest . . . to see him stroke his beard thrice, & begin thus? . . . O, we should have the Proctors and Registers as busie with their Tablebooks as might bee, to gather phrases, and all the boyes in the Towne would be his clients to follow him. Marry, it were necessarie the Queenes Decypherer should bee one of the High Commissioners; for else other-while he would blurt out such *Brachmanicall fuldde-fubs* as no bodie should be able to understand him.[5]   (*Works*, iii. 46. 11–18, 29–36.)

At the Court of Arches in the late sixteenth century there were about thirty-six advocates and thirty proctors[6] with functions corresponding roughly to those of present-day barristers and solicitors. The advocate, professionally independent (at least theoretically), was the pleader who spoke for the litigant in court and by way of remuneration received a gift fee from him but had no legal claim to a fee. The proctor, technically

[5] Another reference by Nashe to the Arches is quoted on p. 37, above.

[6] Cotton MS. Cleo F11, fol. 166ʳ states the following under the heading 'Court of Arches':

Judge: Deane of Arches. Dᵣ Cozen [Richard Cosin was Dean of Arches from 1583 to 1597]. Register or Examiner. Mʳ Edward Orwell.

Actuarie. John Lilley, or his Deputy now Jo: Clerk.

The Bedle or Crier of the Court, one which the generall apparitor to the Arches: doth appoynt.

There should be by usage and custome tyme out of mynd but: 12: Doctors of Law admitted (which be called Advocates) to plead, and tenne Proctors to be as Attourneys; But now they be trebled and after they be once admitted by warrant and commission directed from the Archbishop and by the Deane of the Arches they may then exercise as Advocates and Proctors there. . . .

an officer of the court, was paid a contracted salary. His function was to represent a person in the details of litigation. A qualified advocate might perform the functions of both advocate and proctor but the reverse was not possible as the training and status of the advocate were considerably higher. According to W. S. Holdsworth,[7] 'to become an advocate a candidate must have obtained the degree of civil law at Oxford or Cambridge, after performing the exercises prescribed by those universities; he must have obtained a rescript or fiat from the Archbishop of Canterbury; and he must then have been admitted during term time by the Dean of Arches, and have attended court for a year'. This 'silent year' was designed to acquaint the candidate with the art of pleading and the actual procedures of the court since the training received at the universities was largely theoretical rather than practical.

Harvey very likely became an advocate, for, according to Roscoe Pound,[8] 'it came to be the practice to admit as advocate anyone who had taken the degree of Doctor of Laws'. The quotation from Nashe on page 81, above, refers to hearing Harvey 'plead' in the Court of Arches.

But Nashe intimates that Harvey makes very little money in his law practice and wonders whether he has some other source of income:

Some maintenance of necessity he must have, or else how can he maintaine his peak in true christendome of rose-water everie morning? By the civil law peradventure you will alleage he fetches it in: nay, therin ye are deceivd, for he hath no law for that. I will not deny but his mother may have su'd *in forma pauperis*, but he never sollicited in form of papers in the Arches in his life. How then? doth he fetch it aloft with his poetrie? (iii. 71. 6–14)

Nashe also wrote:

Thou art a Civilian, and maist well fetch metaphors from the Arches, but thou shalt never fish anie monie from thence whilest thou liv'st. (iii. 46. 6–8)

Another mention of Harvey's meagre law practice is found in a passage that Nashe addressed to Gabriel's brother Richard:

I . . . hast backe to the right worshipfull of the Lawes, Master D. Garropius,[9]

[7] *A History of English Law*, London, 1922–52, iv. 236.

[8] *The Lawyer from Antiquity to Modern Times*, American Bar Association, St. Paul, Minnesota, 1953, p. 68.

[9] The reference to D[ominus[ Garropius suggests one who prates piously (Latin verb garrire' meaning to prate or chatter; Latin adjective 'pius' meaning 'conscientious', 'devout', or 'pious', the latter translation being doubly apt since Harvey was 'prating' in an ecclesiastical court, therefore 'piously'). The syllable 'rop' serves to remind Nashe's readers that the man of whom he is writing is the son of a rope-maker. 'Gabriel Garropius' also suggests a reference to the Paduan anatomist and chemical analyst Gabriel Fallopius who wrote a text on the distillation of medical waters entitled *De Medicatis Aquis atque Fossilibus*, 1569. It was translated into English in 1576. See pp. 84–5, below, for Harvey's interest in pharmaceuticals and the art of distillation.

thy brother, (as in everie Letter that thou writ'st to him thou tearmst him,) who, for all he is a civil Lawier, will never be *Lex loquens*, a lawier that shall lowd throate it with Good my Lord, consider this poore mans case. But thogh he be in none of your Courts Licentiate, and a Courtier otherwise hee is never like to be, one of the Emperor *Justinians* Courtiers[10] ... *malgre* he will name himselfe; and a quarter of a yeare since, I was advertised that aswell his workes as the whole body of that Law compleat, (having no other employment in his Facultie,) hee was in hand to tourne into English Hexameters.   (iii. 85. 29 ff.)

Nearly all the late sixteenth- and early seventeenth-century records of the Court of Arches were destroyed in the great London fire of 1660 but copies of a few documents are still to be found at the Bodleian Library in MS. Tanner 280, a group of papers on ecclesiastical courts. On folios 319ʳ to 320ʳ is a report clarifying the functions and historical origins of the offices of Dean of Arches and of Vicar-General (or Chancellor). This document is in the hand of the same scribe who has copied out a number of other papers in this collection. At the end is noted:

> Endorsed in another hand
> Doctor Harvy
> de
> Decano Arcub.
> Vicario Generali &c.

The tract consists of a little over two pages of closely written manuscript in English with a few phrases in 'Law Latin' and it is undated. The original of this copy has been attributed to Gabriel Harvey according to the listing in the Bodleian catalogue.

Harvey probably would have sought to join Doctors' Commons, a professional association composed of a small group of Trinity Hall graduates and other civil lawyers. Their headquarters were at Mountjoy House on Knightrider Street near St. Paul's. There most of them lived and took their meals and were afforded the opportunity of association with experienced lawyers as well as the facilities of a large reference library. Dr. Henry Harvey, former Master of Trinity Hall and Master in Chancery, had been a generous benefactor to Doctors' Commons and his arms were set up in the dining hall. In addition to its other advantages, membership seems to have been considered an almost essential prerequisite for advancement in legal practice.[11]

It would be conceivable that for reasons of social snobbery or personal animosities Harvey might not readily have been accepted into the group,

---

[10] Justinian was considered the chief founder of the Civil Law.

[11] Brian P. Levack, *The Civil Lawyers in England 1603–1641: a political study*, Oxford, 1973, p. 20.

and a listing from the Subscription Book containing the names of members from 1560–1696[12] does not include Harvey's name. However, while this purports to be a complete listing, G. D. Squibb finds this not to be the case and he notes that 'since there are some advocates who do not appear in the part of the Subscription Book devoted to the subscriptions and whose membership of Doctors' Commons is known only from casual entries elsewhere in the Subscription Book or in other records of the Society, it is possible that there were other members of whose membership no record has survived'.[13] There is presumptive evidence elsewhere that at least late in life Harvey was a member, for a leaf of an 18 June 1630 'Doctors Commons Treasurers Book' shows a 'Dr Harvy' to have made a 'contribution' (apparently annual dues) of ten shillings to the advocates' commons on that date.[14] This 'Dr Harvy' probably was Gabriel Harvey, for, although he was then nearly eighty years of age, he would have been in London at this time in connection with the settling of his brother Richard's estate.[15] And no other Harvey is known to have been a member of Doctors' Commons during this period.

After Harvey received his Oxford law degree in July of 1586, he had evidently returned to Cambridge, for his November 1586 *Lachrymae* poems are dated from there. One finds sparse documentation of his activities in the late 1580s but there is a provocative annotation on the title-page of his copy of Sextus Julius Frontinus's *Stratagemes, Sleightes, and Policies of Warre, 1539*.[16] Written vertically in Harvey's ornamental Italic hand the inscription reads: '1588. Revolutio meae Reformationis, seu Annus Assuetudinis'. In September 1588 the Earl of Leicester, Harvey's 'patron' (although a long neglecting one), died. Perhaps it was then that Harvey (equipped with his degree as Doctor of Civil and Canon Law and very possibly with some experience in teaching Jurisprudence at Cambridge) decided to leave the University and scholarship for a life of action in London and inscribed the above marginal comment: '1588. The revolution of my metamorphosis and the year of my habituation'.

One of Harvey's leisure time interests in London at this time is suggested by an interesting broadsheet[17] with his signature dated '1588', some

[12] MS. 139, fols. 129 ff. at Codrington Library, All Souls, Oxford.

[13] *Doctors' Commons: A History of the College of Advocates and Doctors of Law*, Oxford, 1977, p. 204.                                          [14] P.R.O. 30/26/8. fol. 141.

[15] Richard Harvey died early in June of 1630 leaving most of his estate to Gabriel and naming his brother-in-law and cousin as executors. On 10 June they relinquished their executorships to Gabriel and on 11 June he was confirmed as executor. Eight months before his own death he was apparently still able-bodied and legally competent. See p. 132, below.

[16] Lf. 18.54.8* at the Houghton Library.

[17] 'These Oiles, Waters, Extractions, or Essences, Saltes, and other Compositions; are at Paules Wharfe ready to be solde, by John Hester' (British Library: C.60.0.6).

manuscript underlinings of various items, and brief comments. The broadsheet lists the pharmaceuticals and chemicals which can be obtained at the shop of John Hester,[18] 'practitioner in the arte of Distillation, who will also be ready for a reasonable stipend, to instruct any that are desirous to *learne the secrets of the same in few dayes*, &c'. (Italics are Harvey's underlinings.) In the subsequent years Harvey became friendly with him and highly esteemed his special knowledge. Apparently he died in the early 1590s, for in *Pierces Supererogation* (sig. E4$^r$) Harvey wrote: 'Poules wharfe honour the memorye of oulde John Hester, that . . . would often tell me of *A Magistral unguent* for all sores'. In the marginalia of Harvey's copy of Sylvester's translation of Du Bartas[19] he noted: 'Hester's Chymical Epistle. An other Vade mecum'. He alludes to this epistle again in some Latin annotations in his Domenichi as one of a group of *'rara Opuscula'*.[20] Harvey's legal activities place him in the environs of St. Paul's so it is understandable that he may often have strolled into John Hester's apothecary shop on Paul's wharf. Hester was an ardent disciple of Paracelsus and perhaps was the one who converted Harvey to some of his medical theories[21] and stimulated Harvey to further reading and study of medical texts.[22] It is very likely that he was able to put this learning to practical use in later years.

However, in other respects Harvey may have found that life in London was disappointing. Probably his law career was not advancing as well as he had hoped. At any rate, he kept his Trinity Hall Fellowship and in 1589 was back in Cambridge, at least temporarily, for his 'An Advertisement for Pap-Hatchet and Martin Mar-prelate' is dated 'At Trinitie hall: this fift of November: 1589'.[23]

Martin Marprelate was the pen name of an anonymous writer or group of writers who were hostile to the existing episcopal system and were attempting to reform ecclesiastical practices by publishing racy,

[18] For a discussion of John Hester and other Paracelsians, see Allen C. Debus, *The English Paracelsians*, New York, 1966.

[19] Saluste du Bartas, *A Canticle of the victorie obteined by the French king, Henrie the Fourth, at Yvry*, 1590, p. 16.

[20] *Facetie, motti, et burle, di diversi signori et persone private*, Venice, 1571 (Folger Library: MS. H.a.2[1], fol. 33$^v$).

[21] In *A New Letter of Notable Contents*, 1593, sigs. A2$^v$ and D3$^v$ Harvey's references to various pharmaceuticals and their actions on the human body indicate that by then he was well acquainted with Paracelsan doctrines of chemical therapy.

[22] On the title-page of G. Bruele's *Praxis medicinae theorica*, 1585, is a note in Harvey's hand stating that he had bought it from his brother John on 15 Apr. 1589. There are numerous notes within it in Gabriel's hand. Hieronymus von Braunschweig's *Homish Apothecarye*, 1561, was acquired by Harvey in 1590, and is copiously annotated. J. Hugkel's *De Semeiotice Medicinae*, 1560, was acquired in 1584, see p. 75, above.

[23] It was published with *Pierces Supererogation* in 1593, see p. 105, below.

witty pamphlets in popular jargon designed to stir to action Parliament and the people at large.[24] The authorities had for some time been on the trail of Martin's itinerant secret press and on 14 March 1589 press and printers were captured near Manchester. Today it is believed likely that John Penry was author of the Marprelate tracts,[25] but in 1589 John Lyly published a vitriolic, coarse-languaged anti-Martin tract which insinuated that Harvey or one of his aides might perhaps be Martin Marprelate. Lyly titled his splenetic pamphlet *Pap with a Hatchet*, a proverbial phrase for giving nourishment in a harsh way. It is addressed 'To the Father and the two Sonnes, Huffe, Ruffe, and Snuffe' and specifically attacks Harvey in the following passage:

One we will conjure up, that writing a familiar Epistle about the naturall causes of an Earthquake, fell into the bowells of libelling. . . . If he join with us *periisti Martin*, thy wit wil be massacred . . . for this tenne yeres have I lookt to lambacke him. Nay he is a mad lad, and such a one as cares as little for writing without Wit, as Martin doth for writing without honestie; a notable coach companion for *Martin*, to drawe Divinitie from the Colledges of Oxford and Cambridge, to Shoo-makers hall in Sainct *Martins*. But we neither feare *Martin*, nor the foot-cloth,[26] nor the beast that wears it, be he horse or asse; nor whose sonne he is, be he *Martins*, sonne, *Johns*, sonne, or *Richards*, sonne; nor of what occupation he be, be a ship-wright, cart-wright, or tiburn-wright.[27]   (sig. B3)

Harvey's 'Advertisement for Pap-hatchet and Martin Mar-prelate' is an answer to John Lyly and a temperate and considered judgement on the Marprelate controversy. He identifies Lyly as the author of *Pap with a Hatchet* which had been published anonymously, signed only 'Double V'. Harvey refers to Lyly's Euphuistic style of writing and couples it with another awkward literary experiment, his own use of quantitative meters, and indicates that he has outgrown any enthusiasm for either. He writes ironically that Pap-hatchet's eloquence is 'as bleake and wan as the Picture of a forlorne Loover. Nothing, but pure Mammaday, and a few morsels of fly-blowne Euphuisme . . . . I long sithence founde by

[24] For a brief summary of the Marprelate controversy, see R. B. McKerrow, *Works of Thomas Nashe*, v. 34–50. Nashe seems to have been one of the writers on the anti-Martin side.

[25] See William Pierce, *An Historical Introduction to the Marprelate Tracts*, London, 1908, pp. 289–308 and Donald J. McGinn, *John Penry and the Marprelate Controversy*. New Brunswick, N.J., 1966, pp. 89–100, 194–201.

[26] A foot-cloth according to the *O.E.D.* was 'a large richly ornamented cloth laid over the back of a horse and hanging down to the ground on each side. It was considered as a mark of dignity and state'.

[27] A 'tiburn-wright' is presumably one who makes ropes for the gallows at Tyburn.

experience, how Dranting[28] of verses and Euphuing of sentences, did edifie.' (sig. K2$^v$).

He refers to the absurdity that at one time he had been suspected of being Martin:

He were a very simple Oratour, a more simple politician, and a most-simple Devine, that should favour Martinizing: but had I bene Martin, (as for a time I was vainely suspected by such madd Copesmates, that can surmize any thing for their purpose, howsoever unlikely, or monstrous:) I would have beene so farre from being mooved by such a fantasticall Confuter, that it should have beene one of my May-games . . . to entertaine such an odd light-headded fellow. (sig. K2$^v$)

Trying to make peace between the pro- and anti-Martin factions, Harvey appeals to both sides for tolerance, charity, and the maintenance of order:

Modesty is a Civil Vertue, and Humility a Christian quality: surely Martin is too too-malapert, to be discreet; and Barrow[29] too too-hoat, to bee wise: if they bee godly, God help Charity: but in my opinion they little wot, what a Chaos of disorders, confusions, & absurdities they breed, that sweat to build a reformation in a monarchy, upon a popular foundation, or a mechanicall plott. . . . (sig. M1$^v$)

He appeals to scholars and reasonable gentlemen:

You that be schollars, moderate your invention with judgement: and you that be reasonable gentlemen, pacify your selves with reason. . . . If Monarchies must suffer popular states to enjoy their free liberties, and amplest fraunchises, without the least infringement, or abridgment: is there no congruence of reason, that popular states should give Monarchies leave, to use their Positive lawes, established orders, and Royall Prerogatives, without disturbance, or confutation? Bicause meaner Ministers, then Lordes, may become a popular Cittie, or territorie; must it therefore be an absurditie in the majestie of a kingdome, to have some Lordes spirituall amongst so many temporall. . . . It is Tyrannie, or vainglorie, not reverend Lordship, that the Scripture condemneth. (sigs. M2$^v$–M3$^r$)

Stressing the need for order, he writes:

Let Order be the golden rule of proportion; & I am as forward an Admonitioner, as any Precisian [i.e. Puritan] in England. If disorder must be the Discipline, and confusion the Reformation, (as without difference of degrees, it

---

[28] Thomas Drant was an advocate of the quantitative verse movement.

[29] Henry Barrow, church reformer, who together with John Greenwood, adopted 'Brownist' tenets. In 1586 he had been arrested, examined, and imprisoned by legal and ecclesiastical authorities for denying their authority. He was ultimately hanged at Tyburn in 1593.

must needes) I crave pardon. *Anarchie*, was never yet a good States-man: and *Ataxie* [i.e. disorder, lack of discipline],[30] will ever be a badd Church-man. (sig. M4ᵛ)

Harvey, like Spenser and other Renaissance contemporaries, believed in a hierarchy of competence and that equality should apply only to things similar:[31]

Equality, in things equall, is a just Law: but a respective valuation of persons, is the rule of Equity: & they little know, into what incongruities, & absurdities they runne headlong, that are weary of *Geometricall proportion*, or distributive Justice, in the collation of publique functions, offices, or promotions, civile or spirituall. God bestoweth his blessings with difference; and teacheth his Lieutenant the Prince to estimate, and preferre his subjectes accordingly.   (sig. M4ᵛ–N1ʳ)

Harvey believed in being realistic:

Heat is not the meetest Judge on the bench, or the soundest Divine in disputation: & in matters of government, but especially in motions of alteration, that runne their heads against a strong wall. . . . Men are men, and ever had, and ever will have their imperfections: Paradise tasted of imperfections: the golden age, whensoever it was most golden, had some drosse of imperfections. . . . What reformation could ever say, I have no imperfections?  (sig. N3ʳ)

One wonders why Harvey did not publish this important tract until 1593, which in 1913 his biographer Moore Smith praised as follows:

Harvey's statesmanship, his independence of ecclesiastical prejudices, and his powers as a writer are seen to the highest advantage. He shows that a perfect system of Church Government is not to be had in a day, that the Primitive Church adapted itself to temporal circumstances, and that the creation of a theocracy represented by ministerial rule in every parish would be intolerable. . . . We must have mutual charity or Church and State will be overthrown. Perhaps nothing wiser or more far-sighted was ever written in the whole of the 16th century.[32]

Harvey apparently refrained from any involvement in the Marprelate controversy,[33] but his brother Richard took up the challenge by

---

[30] Harvey seems to have been the first to have used this word in English. He is probably responsible for many neologisms which have been absorbed into the English language (e.g., 'acumen').

[31] In *The Faerie Queene* (v. ii. 30–50) Spenser points out the dangers that would ensue if the egalitarian giant were allowed to equalize dissimilar things; in Shakespeare's *Troilus and Cressida* (i. iii. 101–26) Ulysses warns of the peril in allowing natural degree and hierarchy to be disturbed.                                    [32] *Marginalia*, pp. 58–9.

[33] Although it is possible that Harvey circulated the manuscript among his friends, he did not publish the tract until the Marprelate controversy was a thing of the past. On sig. I3ᵛ he asserted that he would never have published it had it not been for the attacks on him and his friends by Greene and Nashe.

writing and publishing two tracts: *Plaine Percivall* and *Lamb of God*. *Plaine Percivall*, printed without date or name of author probably made its appearance early in 1590. In it Richard Harvey stressed his role as a peacemaker between the two factions, but his native belligerence seems to have shown itself in an attack on anti-Martinist writers in general, a group which included Robert Greene and Thomas Nashe as well as John Lyly. In Richard Harvey's 1590 *A Theologicall Discourse of the Lambe of God and his Enemies* which he sub-titled *a briefe Commentarie of Christian faith and felicitie*,[34] *together with a detection of the old and new Barbarisme, now commonly called Martinisme* he again purports to assume the role of a mediator and peace-maker although in an Epistle (added after the original publication of the book[35]) he attacks the twenty-two year-old Nashe for having the effrontery, despite being a neophyte, to criticize in the preface to Robert Greene's *Menaphon* (1589) older highly regarded writers. Richard Harvey remarks:

It becummeth me not to play that part in Divinitie, that one *Thomas Nash* hath lately done in humanitie, who taketh uppon him in civill learning, as *Martin* doth in religion, peremptorily censuring his betters at pleasure. . . . Iwis this *Thomas Nash*, one whom I never heard of before (for I cannot imagin him to be *Thomas Nash* our Butler of *Pembrooke Hall*, albeit peradventure not much better learned) sheweth himselfe none of the meetest men, to censure Sir *Thomas Moore*, Sir *John Cheeke*, Doctor *Watson*, Doctor *Haddon*, Maister *Ascham*, Doctor *Car*, my brother Doctor *Harvey*, and such like. . . . Let not *Martin*, or *Nash*, or any such famous obscure man, or any other piperly make-play [peppery laughing stock] or makebate [breeder of strife], presume overmuch of my patience.

R. B. McKerrow (v, 76) suggests that Nashe's preface, incorporated in the work of the prominent anti-Martinist Robert Greene, and Richard Harvey's irritated reaction to it in his Epistle were the instigating sparks of the great Harvey–Nashe quarrel.

A genial satire on the Harvey brothers is found in Abraham Fraunce's *The Third Part of the Countesse of Pembrokes Yvychurch*[36] which was published late in 1592 (entered at Stationers' Hall on 2 October) but undoubtedly written before young John Harvey's death in July of that year. The satire, near the conclusion of Fraunce's pastoral colloquy, is a tale of three 'Academike Gardiners' who decide to mount to heaven to ascertain

---

[34] Richard Harvey was then Rector at Chislehurst, Kent. Until 1587 he had been at Peterborough as Deacon and Priest.

[35] This Epistle is found in only a few extant copies of the *Lamb of God*. It is reprinted by R. B. McKerrow in *Works of Thomas Nashe* (v, 176–83) from the Bodleian copy, Tanner 598. Quotation is found in the *Works*, v, 179.

[36] Gerald Snare has edited a critical edition of this work (California State University, Northridge, Calif., 1975).

whether or not the astrological prophecy of a disastrous flood is accurate and, if so, whether they can assuage the destruction by presenting to the gods gifts of fruit and flowers from their gardens.

The tale is preceded by a textual reference to Nemesis, the revenging goddess who punishes the insolence of those who in prosperity bear themselves too arrogantly. Her attributes are a bridle and a rule or measure to teach men to temper their tongues and deal justly.

The three university gardeners (there is passing allusion to Cambridge) are easily identifiable as the three Harvey brothers. 'Hemlock', who is called 'one Damoetas of the Deareless parck, *Factotum indeclinabile* to the Lady of the Lake', is John Harvey and is described as wearing a 'fryse bonnet'. It is he who is initiator of the outlandish scheme which ultimately gets the three in trouble; thus he is aptly termed 'hemlock', a poisonous weed. There is indication that John's wife Martha was an unpleasant person and that their relations were not of the fondest. Since they settled in the rather rural area of King's Lynn (where he practised medicine), their estate could punningly have been termed 'the dear-less park'. He was a competent young man and thus perhaps her 'factotum' but 'indeclinabile' because he was not subservient to her wishes.[37] The appellation 'Damoetas' may imply that he was somewhat of a ludicrous country bumpkin-like figure reminiscent of the Dametas of Sidney's *Arcadia*. One wonders whether his 'fryse bonnet' is a coarse cloth head-gear characteristic of the physician? 'Pasnip', 'a brave peece of a man, about foure and thirty yeares olde,[38] fayre, streight, and upright, so nimble and light, that he might well have walked on the edge of a sworde, or poynt of a speare', is Richard Harvey whose diminutive size is frequently alluded to by Thomas Nashe.[39] 'The Thistle' who 'was more auncient, as having passed full fortie yeares and was wholly addicted to contemplation' is, of course, the forty-two-year-old Gabriel Harvey. He is depicted as having great knowledge of astronomy and cosmography and is probably designated a 'thistle' because of the prickles of his ironical wit.

The three gardeners are guided to the upper reaches of heaven by *Intellectus* (Understanding) and his sister *Fantasie* or *Opinion* who speaks disparagingly of 'phantasticall and frantic astrologers' and the two express their view that 'stars rule fools, and wise men rule the stars'. Thistle agrees and adds that he makes 'no more account of these guessing astrologers, then of very asses'. The gardeners prematurely give away their fruits and flowers in order to please minor gods and goddesses. As

[37] See p. 99, below.
[38] Actually, Richard Harvey would then have been thirty-two years old.
[39] *Works*, i. 196, 262; iii. 82, 85.

a result, by the time that Hemlock is able to make his request to the major deities and informs them that he has brought baskets filled with fruit and flowers for them, he finds that the baskets have already been emptied of their contents. The gods now become angered at this and cast Hemlock and Pasnip down to earth to become humble weeds and roots. The Thistle is at first treated more kindly, but eventually the deities lose patience with him, too, because of his foolish petitions relating to his family.

This satire may have circulated in manuscript before publication. If so, Richard Harvey is probably commenting on it (along with other annoyances) when in an epistle dated '1592. the 14 of June' addressed 'To his most loving brother, Master Gabriell Harvey, Doctor of the Lawes' published in *Philadelphus* (1593), he writes:

I . . . heartily wish every man to take everie thing as it is, not as it is made of this and that scribler or pratler, which can tell better, howe to play the mocking Ape, then the just controller. Almightie God defend you dayly, and amend them one day: You know my minde in all matters, and that I would those petite Momes [mockers] had better manners: the schollers head without moderation is like the merchantes purse pennilesse without all credite: I desire that everie student may smell as the Lillies of Salomon, and that everie wilde Lilly may be set in his Gardens. I saye, out Hemlock, out Bramble, out Weedes, and let the bloud of furious *Ajax* himselfe, saith Ovid,[40] be turned into a pleasant herbe.    (sig. C1)

The sparks of angry controversy ignited by Nashe's 1589 Preface to Greene's *Menaphon* and Richard Harvey's subsequent censure of it in the *Lamb of God* did little more than smoulder and sputter until midsummer of 1592 when Greene caused them to flare up and spread with his slurs on John Harvey the elder, and his three Cambridge sons in *A Quip for an upstart Courtier: Or, A quaint dispute between Velvet breeches and Cloth-breeches*, entered in the Stationers' Register on 20 July and published promptly by John Wolfe. In a passage which, although it does not mention them by name, unmistakably alludes to the Harveys, the author meets a collier and a rope-maker and inquires where the latter is going:

Marry, sir [answers the rope-maker] . . . I dwel in Saffron Waldon, and am going to Cambridge to three sons that I keep there at schoole, such apt children sir as few women have groned for, and yet they have ill lucke.

The one, sir is a Devine to comfort my soule, & he indeed though he be a vaine glorious asse, as divers youths of his age bee, is well given to the shew of the world, and writte a late the lambe of God, and yet his parishioners say he

---

[40] A reference to Ovid's *Metamorphoses*, xiii. 389–98.

is the limb of the devill, and kisseth their wives with holy kisses, but they had rather he should keep his lips for madge his mare.

The second, sir, is a Physitian or a foole, but indeed a physitian, & had proved a proper man if he had not spoiled himselfe with his Astrological discourse of the terrible conjunction of Saturne and Jupiter.

For the eldest, he is a Civilian, a wondrous witted fellow, sir reverence sir, he is a Doctor, and as *Tubalcain* was the first inventer of Musick, so he, Gods benison light upon him, was the first that invented Englishe Hexameter: but see how in these daies learning is little esteemed, for that and other familiar letters and proper treatises he was orderly clapt in the Fleet,[41] but sir, a Hawk and a Kite may bring forth a coystrell,[42] and honest parents may have bad children.

Honest with the devil qd. the Colliar, How can he be honest, whose mother I gesse was a witch. For I have heard them say, that witches say their praiers backward, and so doth the ropemaker yerne his living by going backward,[43] and the knaves cheefe living is by making fatal instruments, as halters and ropes, which divers desperate men hang themselves with.[44]   (sig. E3ᵛ)

Next to this passage is a printed marginal note which cautions Richard Harvey to 'looke to it for all the Poets in England will have a blow at your breech for calling them poperlye [*sic*] makeplaies, and will if you reconcile not your selfe bring your worship on the stage' (sig. E4ʳ).

In all subsequent editions of *Quip* the section relating to the Harveys was deleted.[45] According to Gabriel Harvey, the deletions were made by Greene because he feared a lawsuit and therefore bribed his printer to remove the defamatory passage.[46] Nashe explains them as Greene's reluctance to include a derogatory remark about the physician John Harvey and thus risk offending the medical profession at a time when his own health was rapidly failing.[47] Both explanations may be true, for

[41] In *Foure Letters*, 1592 (sig. C3) Harvey categorically denies that he was ever imprisoned in the Fleet, and he also corrects Greene's allusion to Tubalcain, pointing out that it was 'Tuball' whom Genesis mentions as the father of music.

[42] A coystrell (or kestrel) was a small hawk; the term also had another meaning, that of knave or base fellow (*O.E.D.*).

[43] See description of rope-making on p. 5, above.

[44] For further discussion, see Edwin Haviland Miller's 'Deletions in Robert Greene's *A Quip for an Upstart Courtier* (1592)' in *HLQ*, xv, 1951–2, 277–82 and the same author's bibliographical analysis of the various editions in *Studies in Bibliography*, vi, 1954, 107–16. For ease in reading, I have paragraphed the quotation above and added occasional minor punctuation.

[45] The references to the Harveys appeared only in the first issue of the first edition. There is a unique copy of this issue at the Huntington Library and the passage has been reproduced in the Scolar Press facsimile of Greene's *Quip*.

[46] *Foure Letters*, 1592, sig. A4ʳ. John Wolfe, printer of *Quip* and of *Foure Letters* may have given Harvey this information.

[47] *Works*, I, 279–80.

during the summer of 1592 Greene suffered considerable physical pain and became frightened, contrite, and eager to make his peace with any whom his pen might have insulted. He died just before 5 September 1592.

Nashe, who was a close friend of Greene's, made his own attack on Richard Harvey in August of 1592 in *Pierce Penilesse*:

Some tired Jade belonging to the Presse, whom I never wronged in my life, hath named me expressly in Print (as I will not do him), and accused me of want of learning, upbraiding me for reviving, in an epistle of mine, the reverent memory of Sir *Thomas Moore*, Sir *John Cheeke*, Doctor *Watson*, Doctor *Haddon*, Doctor *Carre*, Maister *Ascham*, as if they were no meate but for his Maisterships mouth, or none but some such as the son of a ropemaker were worthy to mention them. (Works, I, 195)

This time Gabriel Harvey is omitted from the list. Although Nashe does not actually name Richard Harvey as the object of attack, he makes clear of whom he is writing by referring to Richard's works by title and later to his brother's (i.e. John's) interest in almanacs.[48] Nashe reminds Richard of the time at Cambridge that 'thou . . . hadst thy hood turned over thy eares when thou wert a Batchelor, for abusing *Aristotle*, & setting him upon the Schoole gates, painted with Asses eares on his head'. Then follows an embarrassing reference to his dogmatic prognostications in the *Astrological Discourse*, prognostications which proved erroneous and 'the whole Universitie hyst at him, *Tarlton* at the Theater made jests of him, and *Elderton* consumed his ale-crammed nose to nothing, in bearbayting him with whole bundles of ballets.'[49] Nashe then addresses the reader and asks, 'Would you, in likely reason, gesse it were possible for any shame-swolne toad to have the spet-proofe face to out live this disgrace?' Richard Harvey evidently made no reply to Nashe's attack. Perhaps by this time he had had second thoughts about astrological predictions and wished he had been more influenced by Gabriel's mistrust of the subject.

Gabriel, however, although he undoubtedly realized that there was justification for criticizing the *Astrological Discourse*, felt obliged to defend the honour of the Harvey family. In *Foure Letters*, 1592, he expresses his feelings that arrogant, irresponsible writers like Greene and Nashe should not be allowed to 'play fast and loose' with one's good name. Greene he treats with considerable harshness, but in the case of the youthful Nashe, Harvey merely counsels the misguided, wayward youth to show more character and make wiser use of his abilities.

[48] John Harvey had written *An Almanacke, or annual Calender, with a compendious Prognostication . . . for . . . 1589* (probably published in 1588).

[49] One of William Elderton's possible ballads about Richard Harvey is referred to on p. 70, above.

Harvey implies that Nashe is like the actor Tarleton who himself did not hesitate to indulge in the sin of lechery but hypocritically preached against it in 'his famous play of the seaven Deadly sinnes' moreover, that Nashe, like Tarleton, is 'of [Andrew] Perne's religion', for he changes his beliefs to please his audience (sig. D4$^r$).

In late July Harvey's favourite brother John, in whom he took pride almost as a father,[50] had become acutely ill when returning from Norwich to his home in King's Lynn and died suddenly at the age of twenty-eight. Gabriel had been with him at the end and movingly described his last moments.[51] Overcome with grief and conscious of strong familial devotion Gabriel was enraged to read Greene's ridicule of the rope-maker and his sons when it was published about the end of July.

Harvey's original intention seems to have been to take legal action against Greene because of the defamation of his family. But before the plan could be carried out, Greene died (at the beginning of September[52]). Harvey thereupon took recourse to print and chose as his publisher John Wolfe who had also put forth Greene's *Quip*. As Harvey explains in the dedicatory epistle to *Foure Letters* (sig. A2):

Venome is venome, and will infect. . . . Diffamation is intollerable: especially to mindes, that would rather deserve just commendation, then be any way blemished with unjust slaunder. . . . That is doone, cannot *de facto* be undone: but I appeale to Wisedome, how discreetely; and to Justice, how deservedly it is done: and request the one, to do us reason, in shame of Impudency: and beseech the other to do us right, in reproach of Calumny.

He explains that his aim is to 'mildly & calmely shew, how discredite reboundeth upon the autors'. Although he clearly demonstrates this in Greene's case, he must shortly after have ironically become aware that his own discredit of Greene (however justified in his opinion) was rebounding cruelly upon himself. He states in his epistle that he wants to be completely fair and show no distemper but (perhaps realizing his suppressed anger) he interjects, 'what Tounge or Pen may not slipp in heat of discourse?' He trusts that any such slips are minor and that the reader will be able to consider and judge the facts with fairness and impartiality. The dedicatory epistle, apparently written after completion of the work, is dated 'London: this 16. of September'.

The text of *Foure Letters* begins with 'The First Letter' from a friend,

---

[50] *Three Proper . . . Letters*, 1580, sigs. E3$^r$–E4$^r$.

[51] *Foure Letters*, 1592, sig. D1$^v$, written in August and September but not entered at Stationers' Hall until 4 Dec. 1592.

[52] 'The First Letter' (of 29 Aug.) assumes that Greene is alive but 'The Second Letter' (of 5 Sept.) describes Greene's death.

Christopher Bird of Walden, written 29 August 1592 introducing Harvey to Emmanuell Demetrius, a distinguished foreign merchant who lives 'by the Church in Lime-streete in London'.[53] In his letter Bird presents Harvey as 'a very excellent generall Scholler . . . desirous of your acquaintance and friendship, especially for the sight of some of your antiquities and monuments: and also for some conference touching the state of forraine countries'. He adds that he will be very grateful for any courtesy which Demetrius affords to Harvey.

The letter includes a postscript in which Bird comments upon Greene's 'famous new worke intituled A Quippe for an upstart Courtier':

as fantasticall and fond a Dialogue, as I have seene: and for some particulars, one of the most licentious, and intollerable Invectives, that ever I read. Wherein the leawd fellow, and impudent rayler, in an odious and desperate moode, without any other cause or reason; amongst sondry other persons notoriously deffamed, most spitefully and villainously abuseth an auncient neighbour of mine, one M. *Harvey*, a right honest man of good reckoninge; and one that above twenty yeres since bare the chiefest office in Walden with good credite: and hath mainetained foure sonnes in *Cambridge* and else where with great charges: all sufficiently able to aunsweare for themselves: and three, (in spite of some few *Greenes*) universally well reputed in both Universities, and through the whole Realme. Whereof one returning sicke from *Norwich* to *Linne*, in July last, was past sence of malicious injury, before the publication of that vile Pamphlet.

He concludes with a bitter sonnet about the brainsick, rakehell Greene who is 'a scurvy Master of Art' and for his *Quip* himself deserves the halter.

'The Second Letter' was written by Harvey to Bird telling him of his visit to the house of M. Demetrius, who was then away and of having met his wife who had been very courteous to Harvey. His next business, he recounts, was 'to enquire after the famous Author who was reported to lye dangerously sicke in a shoemakers house neere Dow-gate . . . of a surfett of pickle herringe and rennish wine, or as some suppose, of an exceeding feare'—a fear that he would be called to appear in the law courts because of his 'forged imputations' (sig. A4ʳ).

Harvey then expresses his views on invective and satire:

Invectives by favour have bene too bolde: and Satyres by usurpation too-presumptuous: I overpasse *Archilochus, Aristophanes, Lucian, Julian, Aretine,* and

---

[53] Christopher Bird (or Byrd) was a fellow townsman of Harvey's who came from a prominent Walden family of which Sir William Bird (see p. 188, below) was a younger member. The parish register lists a Christopher Byrde who died in October 1603. Bird's London friend Emmanuel Demetrius was a Dutch kinsman of Daniel Rogers and was an historian and merchant (see Van Dorsten, p. 22 and p. 40 n. 2).

that whole venemous and viperous brood, of old & new Raylers: even *Tully*,
and *Horace* otherwhiles over-reched: and I must needes say Mother Hubbard
in heat of choller, forgetting the pure sanguine of her sweet Feary [*sic*] Queene,
wilfully over-shott her malcontented selfe: as elsewhere I have specified at
larg, with the good leave of unspotted friendshipp . . .[54] if mother Hubbard in
the vaine of *Chawcer*, happen to tel one Canicular tale, father *Elderton*, and his
sonne *Greene*, in the vaine of *Skelton*, or *Scoggin*, will counterfeit an hundred
dogged Fables, Libles, Calumnies, Slaunders, Lies for the whetstone, what not,
& most currishly snarle, & bite where they should most kindly fawne, and
licke. Every private excesse is daungerous: but such publike enormities,
incredibly pernitious, and insuportable.   (sigs. A4v–B1r)

Harvey recalls his obligation to his dead brother and to his living brothers
and father as well as his determination to restore the good name and honest
reputation of his family,[55] and while musing on such matters, word reaches
him of Greene's death. He is distressed not only because Greene has departed 'without a charitable farewell' but also because now the 'remedy in
law' that Harvey had planned on behalf of his family's reputation no
longer is feasible.

As Harvey subsequently explains:

The persons abused, are not altogether unknowen, they have not so evell a
neighbour, that ever reade, or hearde those opprobrious villainies, (it is too-
mild a name, for my brother Richardes most abhominable Legend, who
frameth himself to live as chastely, as the leawde writer affected to live beastly)
. . . who ever heard such arrant forgeries, and ranke lies? A mad world, where
such shameful stuffe is bought, and soulde.   ('Third Letter', sig. D1r)

And he continues, '*Green*, vile *Greene*, would thou wearest halfe so
honest, as the worst of the foure, whom thou upbraidest: or halfe so
learned, as the unlearnedst of the three'.[56] He makes clear how much his
dead brother John had been esteemed by his neighbours: 'Hee was
reputed in Northfolke, where he practised phisicke, a proper toward man,
and as skilfull a Phisition for his age, as ever came there: how well
beloved of the chiefest Gentlemen, and Gentlewomen in that Shire,
themselves testifie' (sig. D1v).

In 'The Second Letter' Harvey states that he was not personally acquainted with Robert Greene but like many others had heard of his dissolute
and licentious style of living and his impious profanities. With sensational

[54] It is unclear whether Harvey is alluding to a previous spoken or written statement.

[55] In *Quip*, Greene upbraided John Harvey and only three of his four sons. The fourth
and youngest Thomas was perhaps a quiet yeoman who avoided the limelight.

[56] The 'foure' are obviously John Harvey and his three oldest sons; the 'three' are Gabriel,
Richard, and John the younger. Their father evidently could not be considered 'learned'.

details of the sordid misery of Greene's last days, Harvey depicts his death as an exemplum of deserved retribution evoked by sinful living.[57]

It may be hard to reconcile the Harvey we have viewed as a younger man (one more concerned with pragmatic than moral values) with the sanctimonious, puritanical writer of 'The Second Letter' who shows vindictiveness toward Greene and little tolerance for human weakness. But, if we review the events of the two previous decades, we note that time and again Harvey had suffered deep humiliation or frustration: in the struggle to obtain his M.A. degree, the near loss of his Greek lecturership, the failures to win the Public Oratorship and to be chosen Master of Trinity Hall, and more recently, in *Quip*'s gratuitous attack on Harvey's father and family at the poignant time when they were mourning the loss of young John Harvey. Such episodes left their mark on Gabriel Harvey's personality in his developing bitterness, the natural focus of which at this time would have been Greene. In the late summer of 1592 when he wrote *Foure Letters*, Harvey affirmed (sig. B1ᵛ): 'Patience hath trained mee to pocket-up . . . hainous indignities and even to digest an age of Iron'; but in point of fact the cumulative indignities by then were becoming so insufferable that they could really no longer be pocketed-up. At the end of 'The Second Letter' Harvey intimates that he is still yearning to be part of the world of valiant men and valorous deeds. That he was becoming more and more enmeshed in a web of pettiness and vulgarity must have been hard for him to take. Now at the age of forty-two he can be seen as a figure of pathos, for he could not have failed to realize that the glorious hopes of his youth for a position of political stature no longer had the slightest chance of attainment. His long, hard years of training for a pragmatic goal had not produced the expected reward.

Three prestigious patrons of his—Sir Thomas Smith, Sir Walter Mildmay,[58] and the Earl of Leicester—had died, as had his eminent friend,[59] Sir Philip Sidney. The demise of so many such 'flowers of knighthood' occasioned the lament of Sonnet XIII in *Foure Letters*. Entitled 'His intercession to Fame', it enumerates the many renowned knights who have recently died (Smith, Bacon, Walsingham, Hatton, Mildmay, Sir Humphrey Gilbert, Sir Philip Sidney, Sir William Sackville, Sir Richard

---

[57] For a full analysis of the text, see Janet Elizabeth Biller's 'Gabriel Harvey's *Foure Letters* (1592): A Critical Edition' (unpublished Columbia Univ. diss., 1969).

[58] Mildmay had died on 31 May 1589.

[59] In Harvey's 8 May 1598 letter to Sir Robert Cecil (see pp. 124–5, below) he refers to 'my inestimable dear friend Sir Philip Sidney'. There are also some marginalia references (e.g. that of p. 79 n. 102, above) which suggest their intimacy; notes on fol. 150ᵛ of Guicciardini tell of Harvey's esteem for Sidney.

Grenville, Walter Devereux [Earl of Essex]) and asks that their virtues
live forever enshrined by fame.[60] Further sense of loss must have been
felt by Harvey when his friend and confidant Edmund Spenser, who had
for some months in 1590 and 1591 been visiting England, returned
again to his home in Ireland.

Harvey's funds, always limited, had shrunken perilously due to his
inability to obtain preferment, a benefice, or an adequately remunerative
post. His connections with Cambridge had terminated in January 1591/2
with the relinquishing of his Trinity Hall Fellowship. As for his legal
practice, this apparently brought little return. Although his father still
owned considerable land, he had borrowed against it at the time of his
son John's marriage,[61] and like Gabriel now had financial worries.

At the beginning of 'The Third Letter' Harvey sadly observes, 'Ill-
lucke, I deny not: good lucke is not everie mans lotte: yet who ever heard
me complaine of ill-lucke, or once say, *Fortune my foe*?' But although he is
resigned to accepting the injustices of fortune, he will not tolerate attacks,
no matter by whom, which affect his reputation. He angrily warns:
'I can easely defie the proudest that dareth cal my credite in question: or
accuse me of any dishonest, or scandelous parte, either in deede, or in
word' (sig. C1$^v$–C2$^r$). During the summer of 1592 the 'credite' of the
Harvey family, and thus of Gabriel himself, was triply attacked: by
Robert Greene, by Thomas Nashe, and by John Harvey's widow Martha.

Greene's assault on the Harvey family in the first edition of *Quip*
and Gabriel's retaliation in *Foure Letters* has already been discussed, as has
Nashe's derogatory and abusive description of Richard Harvey in *Pierce
Penilesse* (sig. D4$^v$–E1$^v$). In *Foure Letters* (sig. E1$^r$) Harvey reprimands
Nashe: 'Better a man without money, then money without a man:
Penilesse is not his purse but his minde: not his revenue, but his resolu-
tion.'[62] Aware that the young man is an opportunist, Harvey cautions:
'Be a man, be a Gentleman, be a Philosopher, be a Divine, be thy resolute
selfe; not the Slave of Fortune, that for every fleabiting crieth out alas
& for a few hungry meales.'

The third attack on the 'credite' of the Harvey family came from
Martha Harvey. When her husband John died, she was determined to
collect what she considered her matrimonial right even if it involved
seizure of the Harvey homestead and properties in Walden. Within ten
months after John's death she had brought a lawsuit to recover what her
husband evidently had not wanted her to obtain, for he failed to make a

---

[60] It is worth noting that Leicester is not included in this list of the virtuous deceased.
[61] See p. 99, below.
[62] 'Pierce' and 'purse' were pronounced similarly at this time.

will although she had repeatedly urged him to do so. In trying to carry out what seem to have been his dead brother's wishes to protect his parents' source of livelihood, Gabriel is depicted by his sister-in-law as a malevolent conniver who has appropriated her property and turned it over to his father.

The story can be pieced together from the August and October 1592 entries in the Administrations Act Book[63] and a May 1593 petition by Martha Harvey of Lynn, Norfolk, to Sir John Puckering, Lord Keeper, against John Harvey of Walden, Essex, yeoman and Gabriel Harvey, his son. When John the younger married into the family of the eminent Justice Meade, certain monetary demands were made on him and his family. Martha brought £300 to the marriage and John contributed an unspecified sum which his father had been able to obtain by borrowing and offering as security for repayment his house and lands in Walden. In addition, he and Gabriel became bound, if John should die, to provide double the amount of marriage money that Martha had contributed. Her petition stated that 'she many times entreated her husband for a speedy accomplishment of a settlement in these terms in her favour. Her husband, his father, and Gabriel swore to do this but no such jointure was made'. She also stated that her husband 'did not do what he had previously resolved to do, that is, to make a will giving her all his goods and chattels and charging his father and Gabriel with his former promise'. One gathers that the relationship between husband and wife had for some time been a strained one and that John felt more obligation to his parents than to his wife and young daughters. Doubtless he bore in mind that his wife, by contrast, had an affluent father to whom she could always turn.

Gabriel, who was with his brother at the time of his death and subsequently stayed at his home to help handle his affairs, informed Martha Harvey that her husband had died intestate (perhaps lawyer Gabriel had advised John that it was shrewder not to draw a will in these circumstances) and that she might take out letters of Administration. This she did on 19 August represented by her proctor Thomas Ward but on 11 October, Administration was granted to Gabriel Harvey during the minority of Martha and John Harvey's two daughters (see note 63 below).

[63] In PCC Administrations Act Book 1592–8 (P.R.O. Prob. 6/5) are the following entries:

'Martha Harvey, relict of John Harvey, recently of Kings Lynn, Norfolk (diocese of Norwich) granted Administration. (She was represented by Thomas Ward, notary public, her proctor.) (fol. 27ᵛ 29 Aug. 1592)

Gabriel Harvey, doctor of laws, brother of John Harvey, late of Kings Lynn, Norfolk, granted Administration during the minority of Joan and Elizabeth Harvey, children of said John Harvey (Diocese of Norwich).' (fol. 31 11 Oct. 1592)

According to Martha Harvey's statement, after her brother-in-law had
been granted Administration he thereupon returned to the house and
'expulsed her from possession of all the goods and chattels and made sale
presently of a great part thereof and . . . took also all bonds and writings
and acknowledgments of debt, including a bond wherin he stood bound
to her husband for the sum of £80 for the payment of £40', which was
part of her unpaid marriage money.[64]

No further documents connected with this suit have been found, but
Nashe writes in *Have with you to Saffron-walden* (1596) of John's marriage
and its aftermath:

This *John* was hee that beeing entertained in Justice *Meades* House (as a Schoole-
Master) stole away his daughter, and, to pacifie him, dedicated to him an
Almanacke;[65] which daughter (or *Johns* wife), since his death, *Gabriel* (under
pretence of taking out an Administration, according as she in every Court
exclaimes) hath gone about to circumvent of al she hath: to the which effect
(about 3. yere agoe) there were three Declarations put up against him, & a
little while after I heard there were Attachments out for him: whether he hath
compounded since or no, I leave to the Jurie to enquire.   (*Works*, iii. 81)

Nothing more is heard of Martha Harvey or her two daughters. In the
1617 will of her brother, Sir Thomas Meade of Wendon Lofts,[66] he
leaves no bequest to his sister or her daughters although he leaves generous
sums to other members of his family and to his servants. Likewise,
with the 1625 will of Richard Harvey (Gabriel's brother) no mention is
made of his nieces Joan or Elizabeth although he includes bequests to his
other nephews and nieces.

In all probability it was problems connected with his brother's estate
which brought Harvey to London at the time of Greene's death. Such
problems added to all the other mounting pressures could easily have
made Harvey a more than usually bitter commentator. For over twelve
years he had scrupulously kept his resentments in check and adhered to a
policy of silence but, as he states in 1592, 'partlie the vehemente importu-
nity of some affectionate friends, and partly mine own tender regard of

[64] P.R.O. C.3/241/63, 9 May 1593. In this petition Martha Harvey makes no mention of
the first letters of Administration (which were in her name). She states only that she was
ill and had asked Gabriel to take care of the matter and that he hoodwinked her by taking
out papers in his own name. As a matter of fact, the deceased's brother or father (as blood
relatives) would take precedence over a spouse. Thus Gabriel's action in taking out Letters
of Administration in his name was legally proper though perhaps ethically questionable
according to modern standards. However, one must remember that Martha has not told
all the facts (e.g. that there was an earlier Admon. in her name).

[65] See p. 71, above.

[66] PCC 50 Meade (15–16 June 1617).

my fathers and my brothers good reputation, have so forcibly over-ruled me, that I have finally condescended to their passionate motion'. Yet he reminds himself and his readers that he really should not lower himself to involvement of this sort. He even minimizes his own injury, for he considers himself 'one born to suffer, and made to contemn injuries. He that in his youth flattered not himselfe with the exceedinge commendations of some greatest schollers in the worlde: cannot at these years, either be discouraged with misreporte, or daunted with mis-fortune'.[67]

One of the 'affectionate friends' whose 'passionate motion' so 'forcibly over-ruled' Harvey's natural inclinations to remain aloof may have been the printer John Wolfe. It is not known when the two first became acquain-ted but, if Nashe's repeated assertions[68] are true, Harvey lived at Wolfe's printing house (directly opposite the great South Door of St. Paul's)[69] and acted as publisher's reader[70] from early September of 1592 until the following April or later.

Although Harvey seems to have been convinced at this time that Wolfe was truly his friend, the printer–publisher was a man whose monetary interests usually took precedence in all his actions. He probably was a son of the fishmonger Thomas Wolfe, for he is known to have practised printing until 1583 as a member of the Fishmongers' Company, the likelihood being that he gained membership through patrimony.[71] In 1576 he was in Florence studying the trade of printing and apparently learned his craft well, for, as Harry R. Hoppe points out, 'Petruccio Ubaldini, a Florentine exile who was probably corrector of Italian books in his printing house [in London in later years], says in effect that "per studio, & diligenza di Giovanni Wolfio" Italian works can be printed as well in London as anywhere else'.[72] Returned to London, Wolfe began printing other men's privileged books. His piracy eventually led him to become leader of the rebellion against monopolies of printing rights. Iu 1583 he was bribed into submission, became a member of the Stationers'

[67] *Foure Letters*, sig. Ci[r].

[68] *Works*, iii. 87–90, 94 f., 102–3, 118.

[69] Harry R. Hoppe in 'John Wolfe, Printer and Publisher, 1579–1601' in *The Library*, fourth series, xiv, 3 (Dec. 1933), 264, points out that from 1590 Wolfe's imprints bear this address.

[70] At the Bodleian Library is a 1588 *Perimedes* of Robert Greene's (Malone 575[4]) with proof-reader's corrections in a hand that resembles Harvey's. If they are, as I suspect, in Harvey's autograph, the book was probably proof-read for John Wolfe in anticipation of bringing out a second edition and is corroboration of Harvey's activities as publisher's reader for Wolfe.

[71] Hoppe, pp. 242–3.

[72] Ibid. 243.

Company, and was awarded patents and printing privileges himself. He proceeded to assume respectability and by 1593 had been made Printer to the City of London.[73]

Early in his career he must have grown aware of the profit to be derived from publishing controversial works. Having in the 1580s printed what were probably the first English editions of Machiavelli and of Pietro Aretino,[74] he would quickly have realized the potential for a sensational exposé of the decline and death of the popular author Robert Greene written by the scholarly Doctor of Laws Gabriel Harvey. By urging Harvey to retaliate against Greene and Nashe, and it seems likely that Wolfe did just this, he may have had a share in fomenting the Greene–Nashe–Harvey altercation. At the end of *Foure Letters* (sig. H2$^v$) Harvey refers to himself as 'a Sheepe in Wolfes printe', and indeed the three of Harvey's books concerned with this controversy,[75] were printed by Wolfe.

When Harvey arrived in London, he needed to find living quarters and a source of livelihood. Wolfe apparently offered him both in return for Harvey's services at his printing house. As it was a time of bubonic plague, in addition to working on his own tracts and acting as printer's reader, according to Nashe, he wrote for Wolfe 'that eloquent *post-script* for the Plague Bills, where he talkes of the series, the classes, & the premisses, & presenting them with an exacter methode hereafter, if it please God the Plague continue'.[76] The Plague Bills were weekly lists of those who had died of the disease, and perhaps Harvey had applied his medical knowledge in an attempt to analyse occurrences of the plague. In *New Letter*, 1593 (sig. A2), Harvey refers to Wolfe's 'Indices of the Sicknesse'. By 14 July 1593 he was licensed to print 'the bills, briefs, notes and larges given out for the sickness weekly or otherwise'.[77]

Harvey seems also to have advised Wolfe about new works, for Nashe (*Works*, iii. 89–90) twits him for having recommended the printing of such worthless efforts as Barnabe Barnes' *Parthenophil and Parthenophe* (1593), Anthony Chute's *Shore's Wife*,[78] and Thomas Edwards's *Procris and Cephalus* (1595). Harvey while at Wolfe's was also busy writing against Nashe, 'ink squittring and printing against me', as the latter states:

He beganne to Epistle it against mee . . . in the ragingest furie of the last Plague,

[73] In *A New Letter of Notable Contents* which Harvey dates 'This 16. of September. 1593' he addresses John Wolfe as 'Printer to the Cittie'.

[74] They were printed, however, with spurious Italian imprints.

[75] *Foure Letters* (1592), *Pierces Supererogation* (1593), and *A New Letter of Notable Contents* (1593).

[76] *Works*, iii. 89.

[77] See F. P. Wilson, *The Plague in Shakespeare's London*, 1963 (reprint of 1927 ed.), p. 106.

[78] 'Shores Wife' is the sub-title for *Beawtie Dishonoured* (1593).

when there dyde above 1600, a week in *London*. . . . Three quarters of a yere thus cloystred and immured hee remained, not beeing able almost to step out of dores, he was so barricadoed up with graves, which besiedged and under-mined his verie threshold. (*Works*, iii. 87)

Nashe also gives another reason for Harvey's remaining cooped up at the printer's 'for seaven and thirtie weekes . . . never stirring out of dores or being churched all that while . . . after I had plaid the spirit in hanting him in my 4. Letters confuted, he could by no means endure the light, nor durst venter himself abroad in the open aire for many months after, for feare he should be fresh blasted by all mens scorne and derision.' (Ibid. 95.) Nashe writes of Harvey's mounting anger and efforts to take action against him:

his incensing my L. Mayor against me that then was, by directing unto him a perswasive pamphlet to persecute mee, and not let slip the advantage hee had against mee. . . . His inciting the Preacher at *Poules Crosse*, that lay at the same house in *Wood-streete* which hee did [had he stayed there on an earlier visit to London?], to preach manifestly against *Master Lilly* and mee, with *Woe to the Printer, woe to the Seller, woe to the Buyer, woe to the Author.* (Ibid. 95–6)

After having brought out his *Strange Newes* or 'The Foure Letters Con-futed' (entered S.R. January 1592/3), Nashe found it convenient to retire to the Isle of Wight. Harvey incensed returned to his writing-table and produced *Pierces Supererogation*.[79] If Nashe's report is to be credited, Harvey had promised to defray the cost of printing it, and this together with money owed for his board placed him in Wolfe's debt to the sum of £36. [80]

As the title-page informs us, *Pierces Supererogation* was conceived as 'A preparative to certaine larger Discourses, intituled NASHES S[t]. FAME', which Harvey intended shortly to publish against Nashe. Harvey points out that 'Pierce' (the protagonist of *Pierce Penniles*, viz. Nashe) is devoid of all morality and will use any means, no matter how knavish, to attain fame's limelight[81] and to line his purse. Harvey writes of the presumption and supererogation of Nashe who 'above all the writers that ever I knew, shall go for my money, where the currantest forgery, impudency, ar-rogancy, phantasticalitie, vanity; and great store of little discretion may

[79] The text is dated at the end '27. of Aprill: 1593'. Some of the introductory matter is dated July, but R. B. McKerrow believed that publication was delayed until early October (*Works of Thomas Nashe*, v. 102).

[80] *Works*, iii. 96.

[81] In *Strange Newes* at the end of the 'Epistle to the Reader' Nashe announces that this book begins the fray with Harvey and hurls his challenge: 'Heere lies my hatte, and there my cloake. . . . Saint Fame for mee, and thus I runne upon him.'

go for payment; and the filthiest corruption of abhominable villany passe unlaunced.'[82] Harvey intimates that were he himself not restrained by considerations of civil moderation, he would relish persecuting Nashe 'in his own vaine'.[83]

What Harvey chooses to do instead is to present as complete a case as possible and rely for judgement on the reader's sense of equity. He answers Nashe's semi-nonsensical jibes with a carefully reasoned and massive attack employing irony and subtlety and interlarding nearly every statement with evidence of his own learning and ability to vary and expand on an idea. What he has achieved is a document, of which many parts are cleverly and well written, but that in its inordinate bulk and copiousness tends sometimes to become tedious, especially when the erudite Harvey inappropriately sinks to the racy Nashe's level of coarse vulgarity. Despite the underlying intensity of the author's emotions and the justice of his cause, the reader is occasionally lulled into inattention and near insensitivity. Harvey has here chosen to write in Euphuistic style: filling each sentence with paratactic clauses and piling phrase upon phrase until the full possibilities of a subject have been included. One begins to long for a variety of sentence structure, for prose less continuously cadent and stylized, for thought more spontaneously expressed, but above all for a pause in the congested volume of words.[84] In Strange Newes (Works, i. 282) Nashe himself puts his finger on Harvey's weakness: 'his invention is over-weaponed; he hath some good words, but he cannot writhe them and tosse them to and fro nimbly, or so bring them about, that hee maye make one streight thrust at his enemies face'.

Nashe, by contrast, is a gadfly who is interested in stinging at random and disporting himself in the public eye. What he says, and whether it is true, is of far less importance to him than the wittiness with which he says it. Constantly aware of his reader's likely reactions, he concentrates on entertaining him and displaying his own virtuosity. Nashe's attitude is clearly expressed in the challenge he hurls at Harvey in Strange Newes (Ibid. 305): 'Gabriell, if there bee anie witte or industrie in thee, now I will dare it to the uttermost: write of what thou wilt, in what language thou wilt, and I will confute it and answere it. Take truths part, and I will prove truth to be no truth, marching out of thy dung-voiding mouth'. In addition to Nashe's deft handling of language, his text has another im-

---

[82] *Pierces Supererogation*, 1593, sig. D2ᵛ.

[83] Sig. E2ᵛ.

[84] Harvey advocates unaffected direct writing in *Ciceronianus* and he disparages Euphuism in his 'Advertisement for Pap-Hatchet' (see pp. 86–7, above) but chooses to use it against Nashe, perhaps because it was a style in sharp contrast to Nashe's casual mode of writing and perhaps because it enabled him to add epithets or illustrative examples.

portant asset conducive to readability: it is freely broken up into paragraphs, a practice unusual at this time but observed in other of Nashe's works. Harvey's long text of *Pierces Supererogation* has only a handful of paragraph indentations.

The volume is composed of three nearly independent sections which Harvey intended should be divided 'into three bookes'. Because of a printer's oversight, this was not done;[85] and it has resulted in an unfortunate loss in clarity. The first section is primarily an explanation of the need to refute the unsubstantiated railings of Nashe. Harvey regrets having to involve himself in this sort of contention because he considers it 'a base employment of precious time' which he would rather put to more worthwhile uses, for instance:

I could yet take pleasure, and proffite, in canvassing some Problems of naturall Philosophy, of the Mathematiques, of Geography, and Hydrography, of other commodious experimentes, fit to advaunce many valorous actions: and I would uppon mine owne charges, travaile into any parte of Europe, to heare some pregnant Paradoxes, and certaine singular questions in the highest professions of Learning, in Physick, in Law, in Divinity, effectually and thoroughly disputed *pro, & contra*: and would thinke my travaile as advauntageously bestowed to some purposes of importance, as they that have adventurously discovered new-found Landes, or bravely surprized Indies.[86]

The second section is the refutation of the arguments of another of Harvey's maligners, 'Pap-Hatchet' (John Lyly). This tract was written by Harvey in 1589 but not printed until 1593. In the third section, Harvey returns from discussion of 'Pappadocio' to an attack on the other arrogant braggart 'Nashe' whom he here terms 'Braggadoccio'.[87] Included is a paradoxical encomium of 'the Ass', the epithet repeatedly hurled by Nashe against Harvey. The encomium, although far briefer, is somewhat reminiscent of Erasmus's *Praise of Folly*. In all of Harvey's later writings he seems to take delight in the use of subtle ironies. It was his credo that this was the only permissible manner in which to express anger.[88]

*Pierces Supererogation* has been briefly but admirably summarized by

---

[85] This is noted on sig. Ff1ʳ among the many listed 'errours escaped in the printing'. The list continues for three pages; probably the volume was hastily printed.

[86] Sigs. A3ᵛ–A4ʳ.

[87] Harvey here employs the same name used by Spenser to personify vainglory in the *Faerie Queene*. The use of the Italian suffix augmentative may, however, have crept into Harvey's speech many years earlier as a result of his Italian studies in the late 1570s and early 1580s. Harvey acquired and annotated several Italian grammars in this period— Florio, Holyband, and Lentulo.

[88] In Harvey's marginalia in J. Foorth, *Synopsis Politica*, 1582, fol. I6ᵛ. See p. 161 n. 35, below.

R. B. McKerrow,[89] and an interesting paper has been written by David C. McPherson discussing the apparent paradox that in this text references to Pietro Aretino (and to Machiavelli) are derogatory ones whereas other of Harvey's writings, especially his marginalia, indicate that he was in many respects a greater admirer of both men.[90]

Preceding the text, there is an epistle of the author's addressed 'To my very gentle, and liberall frendes, M. Barnabe Barnes, M. John Thorius, M. Antony Chewt, and every favorable reader'. These three young writers,[91] associated with John Wolfe's publishing house, were supporters of Harvey's against Nashe. Also included in the prefatory matter are a group of sonnets and a letter from Barnes in which he refers to the author, of whom he first heard 'in Oxford, and elsewhere', as a commender and favourer of men of 'learning, witt, kinde behaviour, or any good quality' and as a loather of contention and invective. Barnes then mentions some sheets of Harvey's writings (presumably never published) which 'deserve immortall commendation': 'excellent Pages of the French King, the Scottish King, the brave Monsieur de la Nöe, the . . . Lord du Bartas, Sir Philip Sidney, and sundry other worthy personages' (sig. ★★★2ᵛ).[92]

An interesting feature of *Pierces Supererogation* is the introduction of a poetess whom Harvey describes as 'the excellent *Gentlewoman*, my patronesse, or rather Championesse in this quarrel [against Nashe]'. Harvey writes (on sig. ★★4ʳ and Dd3ᵛ) that she

is meeter by nature, and fitter by nurture, to be an enchaunting Angell, with her white quill, then a tormenting Fury with her blacke inke. . . . She was neither bewitched with entreaty, nor juggled with persuasion . . . but only moved with the reason which the Equity of my cause, after some little communication, in her Unspotted Conscience suggested.

Harvey describes her learned reading in 'notablest Historyes of the most-

---

[89] *Works of Thomas Nashe*, v. 88–95.

[90] Harvey had previously admired Aretino's originality, his use of hyperbole, and his political cleverness; the specific targets here are Aretino's opportunism and his hypocrisy (he wrote religious tracts quite as readily as bawdry) which Harvey realized were models for Nashe. Since Nashe saw himself as the English 'Scourge of Princes', it was appropriate for Harvey to attack his *persona*. It is also possible that as Harvey grew older he became more conservative in his views. ('Aretino and the Harvey–Nash Quarrel' in *PMLA*, 84 (1969), 1551–8.)

[91] Barnes was author of *Parthenophil and Parthenophe* (1593), mentioned above, *A divine century of spiritual sonnets* (1595), *Four bookes of offices* (1606), and *The divils charter* (1607); Thorius was translator of Antonio de Corro's *Spanish Grammar* (1590) and compiler of a dictionary published with it; Chewt (or Chute) was author of *Beawtie dishonoured, written under the title of Shores wife* (1593). All were published by Wolfe, and Harvey probably owned copies of most of them.

[92] Perhaps these are among the tracts to which Harvey alludes in his 1598 letter to Sir Robert Cecil. See pp. 124–5, below.

singular woomen of all ages, in the Bible, in Homer, in Virgill, (her three soverain Bookes . . .); in Plutarch, in Polyen, in Petrarch, in Agrippa, in Tyraquell'.[93] Harvey also writes of her: 'She is neither the noblest, nor the fairest, nor the finest, nor the richest Lady: but the gentlest, and wittiest, and bravest, and invinciblest Gentlewoman, that I know'.   (sig. Dd3ᵛ)

> A Dame, more sweetly brave, then nicely fine;
> Yet fine, as finest Gentlewomen be:
> Brighter, then Diamant in every line;
>                     (sig. Ff4ʳ)

'I dare not Particularise her Description . . . without her licence, or per-mission, that standeth upon masculine, not feminine termes; and is respectively to be dealt withall, in regarde of her courage, rather than her fortune. And what, if she can also publish more workes in a moneth; then Nash hath published in his whole life; or the pregnantest of our inspired Heliconists can equal? . . . Yet she is a wooman; and for some passions may challenge the generall Priviledge of her sexe . . . full of spirite, and bloud, but as full of sense, and judgment, that she may rather seeme the marrow of reason, than the froath of affection. . . . Her pen . . . runneth like a winged horse, governed with the hand of exquisite skill' (sig. Dd4). 'She hath in meere gratuity bestowed a largesse upon her affectionate servaunt; that imputeth the same, as an excessive favour, to her hyperbolicall curtesie, not to any merite in him-selfe' (sig. Ee1ᵛ).

In the three sonnets of hers prefixed to the text of *Pierces Supererogation* Harvey's 'Championess' asks:

> Ist possible for puling wench to tame,
> *The furibundall Champion of Fame?*
>
> .   .   .   .   .   .   .
>
> Shall Frend put-up such braggardous affrontes?
> Are milksop Muses such whiteliverd Trontes?[94]
> Shall Boy the gibbet be of Writers all,
> And none hang-up the gibbet on the wall?
> . . . for dread of Danters scarecrow Presse:[95]

She answers, 'Truth feares no ruth, and can the Great Div'll tugg.——

---

[93] Polyaenus was the second-century Greek author of *The Stratagemes*; André Tira-queaux ('Tyraquell') was a French sixteenth-century writer of legal and antiquarian works.

[94] 'Trontes' sounds like a dialect word but I have been unable to identify it.

[95] John Danter was the publisher of Nashe's *Strange Newes*, 1592, and of a 1593 reissue of *Pierce Penniles*; in 1596 he was to publish *Have with you to Saffron-walden*.

Ultrix accincta flagello [the female avenger equipped with a whip]' and concludes:

> If nothing can *the booted Souldiour*[96] tame,
> Nor Ryme, nor Prose, nor Honesty, nor Shame:
> But *Swash* will still his trompery advaunce,
> Il'e leade the *gagtooth'd fopp* a newfounde daunce.
>
> .    .    .    .    .    .    .    .
>
> See how He brayes, and fumes at me poor lasse,
> That must immortalise the killcowe Asse,
>                                   (sig. ***1ʳ)

One wonders whether Harvey's 'Championesse' may be the same gentlewoman who in 1578 inspired him to further his career by writing *Gratulationes Valdinenses?*[97] Not yet satisfactorily identified, she evidently was someone on the outskirts of the Court who was sympathetic to Harvey and his struggles and impressed him, perhaps unduly, with her poetic abilities. I have found no mention of her in his marginalia.

In July 1593 John Harvey the elder, Gabriel's father, died.[98] Gabriel inserted an apostrophe to his dead father and dead brother at the end of *Pierces Supererogation*. The work includes a group of commendatory letters and poems, a feature found also in *Foure Letters* but not typical of Harvey's other works. He was trying to counter Nashe's irresponsible attacks by standing on his own dignity and reputation and pointing out his esteem by others. For instance, he ironically but rather heavy-handedly retorts to Nashe's accusation that he is an ass: 'If I be an Asse, what asses were those curteous frendes, those excellent learned men, those worshipfull, & honorable personages, whose Letters of undeserved, but singular comendation may be shewen?' Harvey then lists the names of those who think highly of him and have at one time or another given him letters of extraordinary commendation: Christopher Bird, Edmund Spenser, Jean Bodin, Thomas Watson (the poet),[99] Thomas Hatcher

---

[96] On sig. Dd4ᵛ, Harvey refers to Nashe as 'the booted Shakerley', one who 'is always riding, and never rideth; always confuting, and never confuteth'. The allusion is to a bragging half-wit, Peter Shackerley, who was the joke of London for his vainglory. Harvey recalls an occasion when Shackerley ridiculously 'advaunced the triumphall garland upon his owne head, before the least skermish for the victorie' (sig. S4ʳ), apparently completely unaware of his absurdities; he seems to have subsisted upon the charity of those he amused. See McKerrow, iv. 155.

[97] See p. 45, above.

[98] The Saffron Walden Parish Register (a seventeenth-century copy of the original one) records the burial of 'Mʳ John Harvey' on 25 July.

[99] Nashe retorts to Harvey (*Works*, iii. 127) that once at the Nag's Head in Cheapside, Watson spoke of Harvey's vanity and quoted hexameters derogatory of him. Thomas Watson died in Sept. 1592.

('a rare antiquary'), Daniel Rogers ('of the Court'), Doctor Griffin Floyd (Queen's Professor of Law at Oxford), Doctor Peter Baro (Professor of Divinity at Cambridge), Doctor Bartholomew Clerke ('late Dean of the Arches'), Doctor William Lewin ('Judge of the Prerogative Court'), Doctor John Thomas Freigius ('a famous writer of Germany'), Sir Philip Sidney, Secretary Thomas Wilson, Sir Thomas Smith, Sir Walter Mildmay, John Young (Bishop of Rochester), William Cecil (the Lord Treasurer), and Robert Dudley (Earl of Leicester).[100]

At one point Harvey analyses Nashe's various attacks on him and makes clear that they boil down to but two accusations: Harvey's habits of dress and his pride. As he very humanly comments:

Who is not limed with some default; or who reddier to confesse his own imperfections, then miselfe? . . . yet can he [Nashe] not, so much as devise any particular action of trespas, or object any certaine vice against me, but only one grevous crime, called Pumps and Pantofles, (which indeed I have worne, ever since I knewe Cambridge,) & his own deerest hart-root, Pride:[101] which I protest before God, and man, my soule in judgment as much detesteth, as my body in nature lotheth poyson. . . . It is not excesse, but defecte of pride, that hath broken the head of some mens preferment. (sig. E4)

One understands why Nashe who chose to rail in a needling but jocular, only half-earnest manner with multifarious thrusts of skilful sword-play was easily the winner in this war of words with a methodical man who persisted in painstakingly rebutting the substance of every attack. At his best, Nashe concentrated verbal skill on clever and colourful caricatures of his adversary, amusingly exaggerating his observations and embellishing them with graphic details which sounded plausible even though some of them were far from accurate.[102] Nashe had a gift for seeing and depicting the ridiculous in human behaviour. His mixture of amusing fact and colourful fiction proved readily saleable to a public only too glad to deflate a Doctor of Law so much preoccupied with abstruse and erudite matters.

Harvey's effort to defend himself was not successful, for he found Nashe an elusive antagonist who was always changing weapons and finding new points of vulnerability. As for Nashe himself, he was thoroughly enjoying the fray since it enabled him to gain public notice,

---

[100] Sig. E4v–F1v.

[101] There is evidence of Nashe's own self-conceit in the *Menaphon* preface and in the 1594 epistle preceding *Christs Teares* (*Works*, ii. 181).

[102] Among Nashe's innuendoes that are falsehoods are Harvey's imprisonment in the Fleet in the 1580s; and identification of Christopher Bird, of the 'Gentlewoman', and of Gabriel Frende (author of Almanacs) as actually Harvey himself. See *Works*, i. 300 & 273; iii, 111 & 70.

the 'St. Fame' he was seeking. At times he congratulates himself on his own abilities with language, as for instance, after verbally thrashing Richard Harvey in *Pierce Penilesse* (*Works*, i. 199), he turns to his audience and asks:

Have I not an indifferent prittye vayne in Spurgalling an Asse? if you knew how extemporall it were at this instant, and with what hast it is writ, you would say so. But I would not have you thinke that all that is set downe heere is in good earnest . . . but only to shewe howe for a neede I could rayle, if I were thoroughly fyred.

What happened to Harvey as a result of his reluctant involvement with Nashe is aptly expressed in an epigram by John Harington, the Queen's godson:

> To Doctor Harvey of Cambridge
> The proverbe sayes, Who fights with durty foes,
> Must needs be soyld, admit they winne or lose.
> Then think it doth a Doctors credit dash.
> To make himselfe Antagonist to Nash?[103]

During the summer of 1593 Nashe suddenly purported to have a change of heart and prefaced his current book, *Christs Teares over Jerusalem*, with an expression of contrition to Harvey. This pamphlet consists of two sections: Part I deals with the crimes of the Jews and the fall of Jerusalem, Part II with the crimes of Londoners which, unless checked, might draw similar divine vengeance upon their city. Nashe infers that the plague now raging uncontrollably may be a manifestation of that vengeance. To guard against it, Londoners should repent of their sins and reform. Two sins which Nashe especially emphasizes are 'Pride' and 'Atheism', the latter being a broad term for any sort of religious deviance. Like his prototype Pietro Aretino, Nashe easily turns from the writing of satire and bawdry[104] to the production of a homiletic religious tract.

*Christs Teares* (1593) was not entered at Stationers' Hall until 8 September, but Harvey obviously knew of its subject-matter and of the contents of its preface when he composed *A New Letter of Notable Contents* which is dated by him 'This 16. of September, 1593'. Harvey had the

---

[103] Published in *The Most Elegant and Witty Epigrams of Sir John Harington, Knight*, 1618 (sig. E8ᵛ) but probably written about 1593.

[104] In *Pierces Supererogation* (sig. F4ʳ misprinted E4) Harvey refers to a group of unprinted, bawdy obscene verses which Nashe had apparently circulated (perhaps his 'Choise of Valentines'): 'I will not heere decipher thy unprinted packet of bawdye, and filthy Rymes, in the nastiest kind: there is a fitter place for that discovery of thy foulest shame, & the whole ruffianisme of thy brothell Muse, if she still prostitute her obscene ballatts.'

ability to work fast so it is barely possible that he had seen Nashe's published work before writing *New Letter*. It seems more likely, however, that Harvey had read Nashe's work in manuscript or had been apprised of its contents before publication.

This letter of Harvey's to his printer John Wolfe is a twenty-six page discourse on genuine and sham values and an attempt to appraise Nashe's intentions in these respects. Like *Pierces Supererogation* it is written in modified Euphuistic style and makes elaborate use of figurative language. It is more pithy and better organized than the earlier work although it still seems excessive to modern taste. The reader must proceed slowly, stopping frequently to interpret and interpolate, for each phrase is pregnant with meaning. In fact, reading this letter and the curious sonnet 'Gorgon' appended to it becomes at times a difficult intellectual exercise, but it is a rewarding one because of the gratification derived by the reader as he unravels clue after clue to grasp the essence of thought, the author's discernment, and his prodigious background of learning. One is fascinated by observing Harvey's mind at work and one senses the writer's obvious enjoyment of his own rhetoric.

In *New Letter* Harvey expresses his feeling that Nashe's contrition is but another evidence of his volatility and hypocrisy, that he is shedding 'crocodile tears' of insincerity. In the address 'To the Reader' of the 1593 edition of *Christs Teares* Nashe had written:

A hundred unfortunate farewels to fantasticall Satirisme. . . . Nothing is there nowe so much in my vowes, as to be at peace with all men, and make submissive amends where I have most displeased. . . . Even of Maister Doctor *Harvey*, I hartily desire the like, whose fame and reputation (though through some precedent injurious provocations, and fervent incitements of young heads)[105] I rashly assailed: yet now better advised, and of his perfections more confirmedly perswaded, unfainedly I entreate of the whole worlde, from my penne his worths may receive no impeachment. All acknowledgements of aboundant Schollership, courteous well governed behaviour, and ripe experienst judgement, doe I attribute unto him. Onely with his milde gentle moderation, heerunto hath he wonne me.

Take my invective against him in that abject nature that you would doe the rayling of a Sophister in the schooles, or a scolding Lawyer at the barre, which none but fooles wil wrest to defame. As the Tytle of this Booke is *Christs Teares*, so be this Epistle the Teares of my penne. Many things have I vainly sette forth whereof now it repenteth me.   (*Works*, ii. 12)

To Harvey this apology smacked of hypocrisy and hyperbole in the

---

[105] Was one of the inciters perhaps Nashe's friend Christopher Marlowe? Marlowe had died on 30 May, a few months before this apology was made.

typical Aretine vein.[106] In *New Letter* Harvey seeks justification for believing that his antagonist may truly intend peace; but his naturally sceptical
mind cannot fail to evaluate and question Nashe's sincerity in the light
of his customary opportunistic behaviour. As a precursor to such analysis,
Harvey examines pretences of all kinds: recent truces on the political
scene, camouflage and wiliness in animal lore, adulterate or inappropriate
medicines, allurements by perfume in cosmetics and by sugar and honey
in cooking, bravado as contrasted with true valour, unsound and pre-
tentious writing as compared to true eloquence. (Harvey's paragons of the
latter were Sidney, Spenser, Cheke, Ascham, Harington, and Mary
Sidney, Countess of Pembroke.) He suggests that Nashe's 'pangs of
September' (the apologies in *Christs Teares*) if compared to his 'fits of
April' (the publication of *Strange Newes*) are seen to be merely enticing
counterfeits. Harvey notes that Nashe's favourite reading is by flamboyant
writers like Aretino and Rabelais and that more recently he has immersed
himself in the lugubrious religiosity of Southwell's *Marie Magdalens
funeral teares* (1591). With their help Nashe has achieved his new work of
supererogation, *Christs Teares*.

One can speculatively reconstruct the background events of September
1593. Friends of Harvey's and of Nashe's had evidently tried to make
peace between the two, and some tentative expressions of goodwill had
been exchanged through intermediaries although Harvey had unsuccess-
fully sought to meet his adversary face to face.[107] For some time Harvey
had been eager to put an end to this destructive, time-consuming altercation, and Nashe on his part was probably glad to forestall the publication
of *Pierces Supererogation*, the text of which had been completed on 27 April
and the preliminary matter on 16 July. Harvey insisted that Nashe's
private protestations of amity be backed up by public confession in
print of the injurious wrong he had done.[108] Harvey may have agreed,
if Nashe guaranteed to do this, to withhold publication of his current
tract. Harvey's arrangement with Wolfe had been to pay for the book's
printing cost, and he therefore would have assumed that he could control
its publication. Wolfe, however, eager to recoup some of the funds he had
advanced to Harvey probably decided to publish the work on his own,
with or without Harvey's approval.[109]

---

[106] In *New Letter* (sig. A4$^r$) Harvey writes to John Wolfe: 'He that hath read, and heard
so many gallant Florentine Discourses [the reference here seems to be both to Aretino
and Machiavelli], as you have done, may the better discerne, what is what: and he that
publisheth so many books to the world, as you do, may frame unto himselfe a private, &
publique use of such conference'.

[107] *New Letter*, sig. C4$^r$.                                              [108] Ibid.

[109] This is essentially R. B. McKerrow's view (v, 103–5). It is supported by the many

When Nashe finds that notwithstanding his apology in *Christs Teares* (1593), *Pierces Supererogation* has come forth in print and shortly thereafter (probably during October)[110] *New Letter* as well, he completely retracts the apology and inserts a vicious new assault on Harvey in a revised preface to *Christs Teares* in the edition the following year. The lengthy 1594 epistle 'To the Reader' includes the passage:

Whereas I thought to make my foe a bridge of golde, or faire words, to flie by, he hath used it as a high way to invade me. *Hoc pia lingua dedit* [this is what devout language has produced]. This it is to deal plainly. An extreme gull he is in this age, and no better, that beleeves a man for all his swearing. Impious *Gabriell Harvey*, the vowed enemie to all vowes and protestations, plucking on with a slavish privat submission, a generall publike reconciliation, hath with a cunning ambuscado of confiscated idle othes, welneare betrayed me to infamie eternall, (his owne proper chaire of torment in hell.) I can say no more but the devill & he be no men of their words. Many courses there be (as *Machiavell* inspiredly sets downe) which in themselves seeme singular and vertuous. . . . This course of shaking hands with *Harvey* seemed at the first most plausible and commendable, and the rather because I desired to conforme my selfe to the holy subject of my booke; but afterwards (being by his malice perverted) it seemd most degenerate and abject.   (*Works*, ii. 179–80)

He further writes:

I have tried all wayes with mine adversary. Heretofore I was like a tyrant which knowes not whether it is better to be feared or loved of his subjects. First I put my feare in practise, and that housed him for a while, next into my love and my favour I received him, and that puft him up with such arrogance that he thought him selfe a better man then his maister, and was ready to justle me out of all the reputation I had. Let him trust to it Ile hamper him like a jade as he is for this geare, & ride him with a snaffle up & down the whole realm. . . . I have heard there are mad men whipt in Bedlam, and lazie vagabonds in Bridewell; wherfore me seemeth there should be no more difference betwixt the displing of this vaine *Braggadochio*, then the whipping of a mad man or a vagabond.   (*Works*, ii. 181)

Robert Greene and Christopher Marlowe were two close companions of Nashe's who likewise were scoffers and 'plagued' their tongues 'with

printer's errata in *Pierces Supererogation*. Harvey seems customarily to have done his own proof-reading and overseen the publication of his works. The many errors in the printing of this volume suggest that he was not in close touch with its publication. If there was a breach of faith, it seems more likely that it was Wolfe's doing rather than Harvey's, because neither the discourses titled 'Nashes S. Fame' which on sig. Dd3ʳ he states 'are already finished and attend the publication' nor 'The Gentlewomans Reply' were ever put into print.

110 Registered at Stationers' Hall on 1 Oct. 1593.

desperate blasphemies in jest'. Harvey counsels Nashe to break away from
perfidy of this sort. 'Though Greene were a Julian [a despiser of Christi-
anity] and Marlow a Lucian [a scoffer at religion], yet I would be loth he
[Nashe] should be an Aretine [one who taints and perverts religious
feeling for his own purposes]'. Harvey points out that Greene and Marlowe, like Nashe, admonished others about religious values[111] but they
would have been wiser to have admonished others 'to advise themselves'.[112]
As Harvey will show in the poem appended to this work, Marlowe
(like Greene) met his deserved end and Nashe should guard against a
similar fate.

The closes with the statement that although his '*Affection* is ready
to subscribe to any indifferent articles of *accorde* . . . *Reason* hath reason
to pawse awhile'. Yet, he says he is still open to any truce that is sincere.
He adds that his final request and affectionate prayer is 'that howsoever
poore men be used, the deare *Teares of Christ*, and the cheap *Tears of Repen-
tance*, be not abused'. He closes the letter with a medical metaphor which
illustrates the advantage of healing with pure medications (oil of roses and
mercury of bugloss), three drops of which will strengthen the brain and
heal the heart far more than will six ounces of their dilute common syrups.

The curious and difficult poem which concludes *New Letter of Notable
Contents* is the following:

Sonet.

*Gorgon,* or the wonderfull yeare.

St Fame *dispos'd to cunnycatch the world,*
    *Uprear'd a wonderment of* Eighty Eight:
    *The* Earth *addreading to be overwhurld,*
    *What now availes, quoth She, my ballanceweight?*
    *The Circle smyl'd to see the Center feare:*
    *The wonder was, no wonder fell that yeare.*

*Wonders enhaunse their powre in numbers odd:*
    *The fatall yeare of yeares is* Ninety Three:
    Parma *hath kist;* De-Maine *entreates the rodd:*
    *Warre wondreth,* Peace and Spaine in Fraunce *to see.*
    *Brave* Eckenberg, *the dowty* Bassa *shames:*
    *The Christian* Neptune, *Turkish* Vulcane *tames.*

Navarre *wooes* Roome: Charlmaine *gives* Guise *the* Phy:
    *Weepe* Powles, *thy* Tamberlaine *voutsafes to dye.*

---

111 Presumably, Greene in *Groatsworth of Wit* (1592) and Marlowe in *Doctor Faustus* or
in his oral discourses as described by Richard Baines in Harleian MS. 6648, fols. 185–6.
112 *New Letter,* sig. D1ʳ, D2ʳ.

L'envoy.

*The hugest miracle remaines behinde,*
The second Shakerley Rash-Swash to binde.

*A Stanza declarative*: to the Lovers
of admirable Workes.

*Pleased it hath a* Gentlewoman rare,
*With Phenix quill in diamont hand of Art,*
*To muzzle the redoubtable Bull-bare,*
*And play the galiard Championesses part.*
*Though miracles surcease, yet Wonder see*
The mightiest miracle of Ninety Three.

*Vis consilii expers, mole ruit sua.*

The Writers Postscript: or a frendly *Caveat*
to the *Second Shakerley* of Powles.

Sonet.

*Slumbring I lay in melancholy bed,*
*Before the dawning of the sanguin light:*
*When* Eccho *shrill, or some* Familiar Spright
*Buzzed an* Epitaph *into my hed.*

Magnifique Mindes, bred of Gargantuas race,
In grisly weedes His Obsequies waiment,
Whose Corps on Powles, whose mind triumph'd on Kent,
Scorning to bate Sir Rodomont an ace.

*I mus'd awhile: and having mus'd awhile,*
*Jesu, (quoth I) is that* Gargantua minde
*Conquerd, and left no* Scanderbeg *behinde?*
*Vowed he not to Powles A Second bile?*

What bile, or kibe? (*quoth that same early Spright*)
Have you forgot the Scanderbegging wight?

Glosse.

*Is it a Dreame? or is the* Highest minde,
*That ever haunted Powles, or hunted winde,*
*Bereaft of that same sky-surmounting breath,*
*That breath, that taught the Timpany to swell?*

*He, and the* Plague *contended for the game:*
*The hawty man extolles his hideous thoughtes,*
*And gloriously insultes upon poore soules,*
*That plague themselves:* for faint harts plague themselves.

*The tyrant Sicknesse of base-minded slaves*
*Oh how it dominer's in* Coward Lane?
*So Surquidry rang-out his larum bell,*
*When he had girn'd at many a dolefull knell.*

*The graund Dissease disdain'd his toade Conceit,*
*And smiling at his tamberlaine contempt,*
*Sternely struck-home the peremptory stroke.*
*He that nor feared God, nor dreaded Div'll,*
*Nor ought admired, but his wondrous selfe:*
*Like Junos gawdy Bird, that prowdly stares*
*On glittring fan of his triumphant taile:*
*Or like the ugly Bugg, that scorn'd to dy,*
*And mountes of Glory rear'd in towring witt:*
*Alas:* but Babell Pride must kisse the pitt.

L'envoy.

Powles steeple, *and a* hugyer *thing is downe:*
*Beware the* next Bull-beggar *of the towne.*
——— Fata immatura vagantur.

*FINIS.*

The poem *in toto* consists of two sonnets and amplifying stanzas. The first sonnet entitled 'Gorgon, or the wonderfull yeare' points out that because of dire astrological predictions 1588 had been anticipated as a year of awesome events.[113] But the earth need not have feared, for it had been tricked by 'St. Fame':[114] 'no wonder fell that yeare'. The truly 'wonderful' year, one of incredibly amazing events (for the most part welcome rather than dire ones) has been the current year. During 1593 the following occurred: Alexander Farnese, Duke of Parma died; the Duke of Mayenne sought but did not attain the rule; it went instead to Henry of Navarre who surprisingly converted to Catholicism; although Spain

[113] The threatening prognostications for 1588 were discussed by John Harvey in his *A discoursive probleme concerning prophesies,* 1588.

[114] 'St. Fame' was Harvey's nickname for Nashe (Nashe's *Works,* iii. 52), undoubtedly because of the latter's remark in *Strange Newes* (1592) at the end of the 'Epistle to the Reader': 'Saint Fame for mee, and thus I runne upon him.' Evidently Nashe's prime motive in attacking Harvey was to be able to publicize himself.

In the above sonnet the inference is that just as St. Fame was 'dispos'd to cunnycatch the world' with threatening prophecies, Nashe is tricking the public with his dire warnings in *Christs Teares,* which was just about at this time appearing in print.

That 'St. Fame' was Harvey's nickname for Nashe is further corroborated by the title-page of *Pierces Supererogation,* 1593 which informs the reader that this work is 'A Preparative for certaine larger Discourses, intituled NASHES S. FAME.' On sig. Dd3 Harvey alludes to this massive attack on his enemy as already finished and awaiting publication; apparently it was never printed.

retained a port in France, peace was achieved between the two countries; the formidable military commander of the Turks was defeated by the brave Christian warrior Eggenberg; Charlemagne was shamed by his descendants, the House of Guise, who failed to live up to his greatness.[115] But, most miraculous of all, in 1593 the frightening bogy 'Tamburlaine' (i.e. Christopher Marlowe) consented to die. Harvey cynically adds that St. Paul's churchyard, hub of stationers, writers, and controversialists, can now weep for its 'Tamburlaine'.

The title 'Gorgon' has at least two connotations. It can, of course, refer to the terrifying Medusa, the sight of whom petrified those who gazed upon her. Near the end of *Pierces Supererogation* (sig. Dd2) Harvey, discussing what traits in a man are deserving of fear, writes: 'He were very simple, that would feare a conjuring Hatchet, a rayling Greene, or a threatening Nash'.[116] 'Gorgon' surely also alludes to the underworld potentate Demogorgon. One recalls a relevant line in Part I of Marlowe's *Tamburlaine* (IV. i. 18): Zenocrate's father, the Soldan of Egypt, has asserted that he will not retreat even 'were that Tamburlaine as monstrous as *Gorgon*, prince of Hell'. Since the Soldan is overwhelmingly defeated by the Scythian shepherd, the implication is that Tamburlaine is at least as monstrous as Gorgon.

The sonnet is capped by a two-line 'L'envoy':

> The hugest miracle remaines behinde,
> The second Shakerley[117] Rash-Swash to binde.

Marlowe's bravados[118] have succumbed to death; the second braggadoccio, Nashe, is yet to be stifled.

Now follows a stanza addressed 'to the Lovers of admirable Workes'. It describes another miracle of 1593: the muzzling of the redoubtable

---

[115] I am indebted to Hale Moore's 'Gabriel Harvey's References to Marlowe' in *SP*, xxiii, July 1926, 337–57 for clues to some of Harvey's historical allusions, although I am not in agreement with his conclusion that the 'wonderment of eighty-eight' is an allusion to the Spanish Armada; I believe it is a reference to astrological forecasts of cataclysms which failed to occur.

[116] Harvey adds that there are other traits in men which are very much to be feared: 'the flattering Perne, or pleasing Titius [Roberto Tizio, teaching at Bologna in the late sixteenth century] . . . notable men in their kinde, but pitch-branded with notorious dissimulation; large promisers, compendious performers; shallow in charity, profounde in malice; superficiall in theory, deepe in practise; masters of Sophistry, Doctors of Hypocrisie; formall frends, deadly Enemies; thrise-excellent Impostours. These, these were the Onely men, that I ever dreaded . . . other braggardes, or threatners whatsoever I feare, as I feare Hobgoblin, & the Bugges of the night. When I have sought-up my day-charmes, and night-spelles, I hope their power to hurt, shal be as ridiculously small, as their desire to affright, is outragiously great'.

[117] See p. 108 n. 96, for description of Peter Shackerley.

[118] See *Pierces Supererogation*, sig. H4ᵛ.

'bullbear' Thomas Nashe by the 'Gentlewoman'. She has played the part
of a valiant championess for Harvey. The final Latin verse advises us
(and Nashe) that force bereft of judgement tumbles to its own destruction.

Harvey now turns directly to Nashe and writes the second sonnet, a
friendly warning to the 'Second Shakerley' of St. Paul's.

An epitaph for Marlowe has occurred to Harvey as he lies abed dream-
ing. It is a lament for a magniloquent mind 'bred of Gargantua's race',
for a man who strutted in Paul's churchyard and triumphed even over the
Archbishop of Canterbury who scorned to involve himself with a
braggart of this sort. While musing on this epitaph Harvey realizes that,
although Marlowe's 'Gargantua minde' is now conquered by death, he
has left the dissembling rascal Nashe behind to follow in his footsteps.
The eruptions of Marloweism and Nasheism Harvey likens to boils and
to chafing irritations.

Next follows the 'Glosse', a stanza of moralizing explanation which
Harvey begins in the typical rhythm of Marlovian blank verse:

> *Is it a Dreame? or is the* Highest minde,
> *That ever haunted Powles, or hunted winde,*
> *Bereaft of that same sky-surmounting breath,*
> *That breath, that taught the Timpany to swell?*

Harvey asks himself whether it can really be true that this 'Highest
minde' so well known around St. Paul's is now bereft of breath (the
breath of life and of soaring speech). In the fourth line Harvey ends on a
sardonic note with his allusion to 'the Timpany'. The term can be used
figuratively to apply to inflated, bombastic style or to self-conceit as
diseased tumorous swelling. The stanza continues with a statement that
Marlowe and the Plague have been contending for prey: Marlowe by
swaying weak-minded folk with his contaminating 'atheistic' ideas (as
expressed in his plays and verbal harangues), the bubonic plague by
claiming the weak with its own dangerous brand of infection. So arrogance sounded its bell of alarm. Previously the haughty Tamburlaine
had only bared his teeth at its many doleful tollings. Now, however, the
ultimate leveller Death, disdaining Marlowe's intolerable presumption,
smilingly struck home and annihilated the man who had feared neither
God nor Devil nor admired anything but his wondrous self. His towering
wit that had once enabled him to mount to glory, this overweening
verbalizing pride, has because of its morbid disease been doomed to
perish miserably, to be cast into a pit of anonymous filth and decay.

The poem closes with a pungent 'envoy':

St. Paul's tall steeple crumbled (in 1561) and now a huger thing has fallen. Let this be a warning to London's next terror-inspiring bugbear (Nashe)! A not-yet-accomplished fate is abroad.

Commentators have incorrectly, I am convinced, taken the Glosse to mean that Harvey believed Marlowe had died of the bubonic plague but it seems to me that Harvey is here only using the prevalent plague as a metaphor of disease much in the way that he employed medical imagery in *New Letter*.[119] One must remember that Harvey was in a position to know how Marlowe had actually died since Richard Harvey was then serving as Rector in Chislehurst at St. Nicholas's, the parish church attended by Marlowe's patron, Thomas Walsingham, at whose manor Marlowe had probably been sojourning shortly before his death.[120] The facts seem to be that on 30 May 1593 Marlowe was killed either accidentally or intentionally at Deptford in the company of three men (Poley, Frizer, and Skeres) who were associated with or employed by Walsingham.[121]

If Harvey within a few days' time managed to ferret out the details of Robert Greene's death in September of 1592, it is not likely that he would have been completely misinformed in the case of Christopher Marlowe, for when in September 1593 Harvey wrote of his death, he had had over three months to assess the facts with his brother stationed in a position to hear behind-the-scenes gossip. Harvey's motivation to learn whatever was the sordid truth would have been very similar in both cases. Like Greene, Marlowe had been a friend and somewhat of a patron of Nashe's[122] and possibly encouraged his baiting of the Harveys.[123] In fact, Nashe quotes Marlowe as having once termed Richard Harvey an ass fit only to preach of the Iron Age.[124]

Marlowe entered Cambridge in 1580/81 so he still would have been an undergraduate when Gabriel Harvey became Proctor in 1583. Since Marlowe was not inclined to be subservient to authority, it is quite possible that there may have been friction between them as early as this.

[119] See sigs. A2ᵛ, A3ʳ, D2ᵛ, D3ʳ.

[120] The Privy Council warrant of 18 May 1593 to one of the Messengers of her Majesty's Chamber directs him 'to repaire to the house of Mʳ Tho: Walsingham in Kent, or to anie other place where he shall understand Christofer Marlow to be remayning, and by vertue hereof to apprehend and bring him to the Court. . . .' Presumably Marlowe was wanted for questioning (PC 2/20, p. 374).

[121] The Coroner's Inquisition brought to public notice by Leslie Hotson (Chancery Miscellanea, Bundle 64, File 8, No. 241b) states that the killing was accidental and an attempt of Frizer to defend himself. The document leaves a number of unanswered questions, some of which are discussed by John Bakeless in his *Tragicall History of Christopher Marlowe*, Cambridge, Mass., 1942, chap. VI.

[122] See *Pierces Supererogation*, sig. Dd4ᵛ.

[123] See p. 111 n. 105, above.

[124] *Works*, iii. 85.

The two men's religious attitudes were certainly considerably at variance. As we learn from 'Advertisement for Pap-hatchet, and Martin Marprelate', Harvey felt that church organization and discipline were important, and he deplored any attack on them. A firm moderate in religious matters, he thought that the political good of the state was the paramount consideration, and he was convinced that this required a shying away from extreme views of all sorts whether fanaticism or religious nonconformity. Marlowe, by contrast, was a persistent questioner of traditional Biblical teachings.

*New Letter* was written from Walden in September 1593, whither, it seems, Harvey retired to lick his wounds. Nashe recounts another humiliation that Harvey underwent (in 1594?) when John Wolfe attempted to have him imprisoned for an outstanding debt which Harvey, he attested, had made no effort to repay. According to this version (*Works*, iii. 97 f.) although protesting violently, Harvey was jailed at Newgate but was almost immediately bailed out by the Revd. Robert Harvey, Rector of St. Alban's, Wood Street, who, if Nashe is reporting accurately, was no relative but paid the required amount 'for the name's sake'.[125] The minister procured a room for Harvey after his release 'at one *Rolfes*, a Serjeants in Wood-streete'.[126] According to this account, Harvey shortly thereafter fled without paying his board and 'coopt up himselfe invisible' at Saffron Walden.[127]

One of his most constant problems was a lack of funds. Nashe notes, 'His discontented povertie (more disquiet than the Irish seas) hath driv'n him from one profession to another'.[128] Harvey's marginalia contain comments in which he bitterly implies that life might have been very different for him had it not been for his financial insufficiency. An early example is a note of Harvey's dated 1577 (in his copy of Erasmus's *Parabolae*, 1566, sig. L8ᵛ) which translates: 'A learned man without money is the same as a little Brucus worm [a type of wingless locust] which flies without wings'.[129] A later example probably written after 1596

---

[125] There were two Robert Harveys at Trinity Hall while Gabriel was in residence there. One was the nephew and heir of Dr. Henry Harvey and was a Fellow who later became a civil lawyer; the other, described in Henry Harvey's will as born in Coventry and a scholar of Trinity Hall (no kinship is mentioned) was left a bequest of Dr. Harvey's 'winter gown' and forty shillings. Perhaps it was this younger man who later became a minister and befriended Gabriel.

[126] Another mention of Harvey's sojourn in 'Woodstreete' is found in Nashe's *Works* (iii. 96) where he alludes to Harvey's having incited 'the Preacher at *Poules Crosse*, that lay at the same house in Wood-streete which hee did, to preach manifestly against *Master Lilly* and mee, with *Woe to the Printer, woe to the Seller, woe to the Buyer, woe to the Author*'.

[127] iii. 101.        [128] iii. 61.

[129] 'Vir doctus, sine pecunia; idem quod Brucus vermiculus, qui volat sine alis.'

(the date of Jean Bodin's death) is found in Harvey's copy of Guicciardini's *Detti, et Fatti Piacevoli, et Gravi*, 1571 on sig. *2ᵛ (fol. 74ᵛ):

Professors of generalized learning without specialized fields or assured provision for their needs die as beggars: as did not only Agrippa and Turnebus (what fine philologians!) but also Machiavelli and Bodin (what great political analysts!). It is important to provide for one's personal needs so that one is self-sufficient and not dependent upon friends, for there are no friends to the indigent. The most deserving men are despised if they are poor and the most unworthy are esteemed if they are rich.[130]

I am inclined to believe that an epigram by Thomas Bastard in his *Chrestoleros* (1598) is an allusion to Harvey and his monetary problems especially those with John Wolfe:

### De Publio

*Publius* sweares he is not false nor wicked,
Free from great faults, and hath no other lett,
Save this great fault he is in debt.
This is the greatest sinne he hath committed.
This is a great and hainous sinne indeede,
Which will commit him if he take not heede.
(Liber Primus: Epigramma 12)

In 'Liber Secundus', 'Epigramma 1' seems to deal with the attack on Harvey by Thomas Nashe who in 'deluding raisest up a fame'. A second epigram about 'Publius' (Liber Quartus: Epigramma 11) may perhaps refer to Harvey and his two surviving brothers (Richard and Thomas).[131]

Dubious loyalty to Harvey by John Wolfe is suggested by the fact that on 17 September 1593 there is entered to him in the Stationers' Register a work of Nashe's, *The Unfortunate Traveller*. It was not, however, published by Wolfe but by T. Scarlet for C. Burby in 1594. This fictional tale contains a brief passage about the 'verie solemne scholasticall entertainment of the Duke of Saxonie' at the University of Wittenberg. The scene is somewhat reminiscent of Queen Elizabeth's visit to Audley End in 1578. A speech is delivered by the University Orator, who seems to me intended as a parody of Harvey. The section is as follows:

The chiefe ceremonies of their intertainment were these: first, the heads of

---

130 'Generales Professores, sine speciali praxi, aut certa necessariarum opum provisione; moriuntur Mendici. Ut non modo Agrippa, et Turnebus, quanti philologi? sed etiam Machiavellus et Bodinus, quanti politici? Tanti refert, solide fundare rem familiarem: et a teipso pendere; non ab amicis: qui nulli sunt egentibus. Contemnuntur pauperum dignissimi: aestimantur divitum indignissimi.'

131 See Appendix C, p. 263, below for text of these two stanzas.

their universitie (they were great heads of certaintie) met him in their hooded
hypocrisie and doctorly accoustrements, *secundam formam statuti*; where by the
orator of the universitie, whose pickerdevant was very plentifully besprinkled
with rose water, a very learned or rather ruthfull oration was delivered (for it
raind all the while) signifieing thus much, that it was all by patch & by peece-
meale stolne out of Tully . . . a thousand *quemadmodums* and *quapropters* he came
over him with; every sentence he concluded with *Esse posse videatur*: through
all the nine worthies he ran with praising and comparing him; *Nestors* yeeres
he assured him off under the broade seale of their supplications, and with that
crowe troden verse in Virgil, *Dum iuga montis aper*, hee packt up his pipes and
cride *dixi*.[132]

The besprinkling with rose water and use of 'esse posse videatur' ('it
seems to be possible') as a Ciceronian clausula are elsewhere pointed out
by Nashe as Harveian trademarks.[133]

    An allusion to the antipathy between Doctor Harvey and Master Nashe
is found in William Covell's *Polimanteia: England to her three Daughters,
Cambridge, Oxford, Innes of Court, and to all her Inhabitants*, 1595. The text
and printed marginal comments are the following:

| | |
|---|---|
| *Cambridge* make thy two children | |
| friends, thou hast been unkinde to | |
| the one to weane him before his | |
| time; & too fond upon the other to | |
| keepe him so long without preferment | |
| the one is ancient, & of much reading, | *D. Harvey* |
| the other is young but ful of wit: | *M. Nash.* |
| tell them both thou bred them, and | |
| brought them up: bid the ancient for- | *Doctores liberi* |
| beare to offer wrong; tel the yonger | *sunto.* [May |
| he shall suffer none: bid him that | teachers hence- |
| is free by law,[134] think it a shame to | forth be unen- |
| be entangled in small matters: but | tangled!] |
| tell the other, he must leave to medi- | *Others of that* |

---

[132] *Works*, ii. 246–7.

[133] Nashe writes of Harvey: '*Some there be (I am not ignorant) that, upon his often bringing
it in at the end of everie period, call him by no other name but* esse posse videatur' (*Works*, iii. 66);
Nashe elsewhere asks, 'How can he maintaine his peak in true christendome of rose-water
everie morning?' (iii. 71).

[134] 'Him that is free by law' is evidently Harvey, for in the 1597 *Trimming of Thomas
Nashe* it is stated (on sig. G3ᵛ) that Nashe 'sundrie and oftentimes hath been cast into manie
prisons (by full authoritie) for his mis-behaviore' and has been only 'lately set at libertie'.
If Harvey was actually imprisoned on account of debt (as Nashe states), he had been almost
immediately set free by the Revd. Robert Harvey's payment of bail. The subsequently
mentioned 'adversarie' who '(to learnings injurie) lives unregarded' is apparently Harvey.

| *Great* | tate revenge, for his adversarie (and | *name, as fit for* |
|---|---|---|
| *pittie.* | let that suffice for al revenge) (to | *a Scholler to* |
| | learnings injurie) lives unregarded. | *inveigh against.* |

(sig. Q4)

Harvey and Nashe seem to have remained subjects of public interest throughout the 1590s and a number of literary works of this period allude to one or the other of them.[135]

In 1596 Nashe published his *chef d'œuvre*, the satirical biography, *Have with you to Saffron-walden*, sub-titled 'Gabriel Harveys Hunt is up. Containing a full Answere to the eldest sonne of the Halter-maker', and in it Nashe recurrently needles Harvey for being the son of a rope-maker.

The epistle dedicatory is addressed to Dicke of Lichfield who is the barber at Trinity College, Cambridge[136] and is equipped to shave immediately 'a certain kinde of Doctor of late very pitifully growen bald . . . to trie if that will helpe him'. The woodcut caricature on sig. F4ʳ (*Works*, iii. 38) is entitled 'The picture of Gabriell Harvey, as hee is readie to let fly upon Ajax'. It depicts a rather short-statured man with tufts of white hair framing his thin cheeks which are accented by a moustache and small beard. He is nattily attired in a large hat (perhaps to cover a bald pate), oversized ruff, jerkin, and trunkhose with one hand hidden somewhere within their folds as if indeed he is about to make use of a 'jakes'. In the adjacent text Nashe explains:

Those that bee so disposed to take a view of him, ere hee bee come to the full Midsommer Moone and raging *Calentura* of his wretchednes, here let them behold his lively counterfet and portraiture, not in the pantofles of his prosperitie,[137] as he was when he libeld against my Lord of *Oxford*, but in the single-soald pumpes of his adversitie, with his gowne cast off, untrussing, and readie to beray himselfe, upon the newes of the going in hand of my booke.

If you ask why I have put him in round hose, that usually weares Venetians; it is because I would make him looke more dapper & plump and round upon it, wheras otherwise he looks like a case of tooth-pikes, or a Lute pin in a sute of apparell.

R. B. McKerrow states his belief that 'the "portraiture" here given was simply an old wood-cut which the printer happened to have by him,

---

[135] For instance, George Peele's *Old Wives Tale* (1595), Shakespeare's *Love's Labour's Lost* (1598), and the Cambridge University play, *Returne from Parnassus* (produced 1599/1600).

[136] In *Works*, iii. 33, there is the identification 'one *Dick Litchfield*, the Barber of *Trinity Colledge*, a rare ingenuous odde merry Greeke, who (as I have heard) hath translated my *Piers Pennilesse* into the *Macaronicall* tongue . . .'

[137] Compare with this the implied portrait of a handsome young man in 1578 (see p. 42, above).

probably a piece cut from a larger one', and he adds, 'So far as I am aware, however, it has not been identified'.[138]

About a year later in the autumn of 1597 a reply to Nashe's epistle in *Have with you* was published as *The Trimming of Thomas Nashe, Gentleman, by the high-tituled patron Don Richardo de Medico campo, Barber Chirurgion to Trinitie Colledge in Cambridge.*[139] The epistle 'To the gentle Reader' is signed 'Yours in all curtesie, *Richard Lichfield*'. *The Trimming* contains a woodcut of a smirking, tousle-haired Nashe with his ankles shackled by iron chains. On sig. E4ᵛ he is referred to as 'the moth of fame'[140] who scorns oblivion and so has recently augmented his deeds with 'that most infamous, most dunsicall and thrice opprobrious worke *The Ile of Dogs*'.

At the time that *The Trimming* was published it was automatically thought by many to be the work of Gabriel Harvey, but there is every indication that this was not so. Not only is the style completely unlike any extant examples of his prose but also the text itself alludes to 'this my first work and offspring'. Although the actual author has never been pinpointed, both McKerrow and Moore Smith concur that Harvey's authorship of this work is most unlikely.

Little is known about Harvey's activities after he left London late in 1593. However, in 1598 he seems to have made one more attempt for a post at Cambridge. Having heard that Thomas Preston, Master of Trinity Hall, was then at the point of death, Harvey wrote a long letter from Walden to Sir Robert Cecil (Principal Secretary to the Queen) earnestly requesting his assistance in the same way that Cecil's father, the Lord Treasurer Burghley, and the then Principal Secretary Sir Francis Walsingham, had assisted Preston to the Mastership in 1585: by first staying the election and then obtaining her Majesty's mandate. Harvey lists his various qualifications: his academic background and that he has always put his time to good use

in reading the best Autors extant, aswell in Lawe, as in other emploiable faculties; or in writing sum Discourses, either of privat use, or of publique importance . . . I had ever an earnest & curious care of sound knowledg, & esteemed no reading, or writing without matter of effectual use in esse: as I hope shoold soone appeare, if I were setled in a place of competent maintenance, or had but a foundation to build upon.

---

[138] McKerrow justifies his view by the condition of the block itself, especially the unsupported leaves on the left hand and because Nashe apologizes for not putting Harvey in the type of costume he customarily wore (*Works of Thomas Nashe*, iv. 321).

[139] Entered S.R. 11 Oct. 1597.

[140] It is worth noting that the name 'Moth' is used for the Nashe-like character in Shakespeare's *Love's Labour's Lost*.

He recalls his many writings in honour of the Lord Treasurer's 'weightie and rare vertues, and in memorie of his tru-honorable name', which he is ready to show his son whenever convenient. The Lord Treasurer was then critically ill;[141] Robert Cecil had just returned to England from his ambassadorship to France.

Harvey also mentions the 'sundrie royall Cantos . . . in celebration of her Majesty's most . . . glorious government'.[142] He plans to transcribe and revise some of these and would like to publish them together with manie other mie Traicts & Discourses, sum in Latin, sum in Inglish, sum in verse according to the circumstance of the occasion, but much more in prose; sum in Humanitie, Historie, Pollicy, Lawe, & the sowle of the whole Boddie of Law, Reason; sum in Mathematiques, in Cosmographie, in the Art of Navigation, in the Art of Warr, in the tru Chymique without imposture (which I learned of your most learned predecessour, Sir Thomas Smith, not to contemne) & other effectual practicable knowlage, in part hetherto un-revealed, in part unskilfully handeled for the matter, or obscurely for the forme; with more speculative conceit, then industrious practis, or Method, the two discovering eies of this age. I speak it not anyway to boste, (that loath the follie of any such vanity) but to certifie the truth. For I can in one yeare publish more, then anie Inglishman hath hetherto dun: I hope with the allowance of the sharpest & deepest Judgments in Ingland: whose censure I am not only willing, but desirous to underlie: as one that woold be lothe to divulge anything with-out hope of life, & continuance in it. But thereof more at fitter opportunitie.

In closing he again begs for Cecil's help and assures him that he will always remain grateful and will serve him and his family in whatever way he can. (Cecil Papers, vol· 61, 5 at Hatfield House.)

The long missive is endorsed: '1598. 8° Maii. D$^r$ Harvey to my M$^r$. The M$^r$ of Trinitie Hall dangerouslie sycke. Desires the eleccion may be stayd for him'. Whether Cecil himself ever read this is not known. There is no record that Harvey ever received an answer.

During 1598 Harvey's interest must have strongly turned to literature and drama, for he annotated with handsomely penned appraisals of contemporary writers a newly purchased folio volume containing Geof-frey Chaucer's life and works. (It also included John Lydgate's 'Story of Thebes', and a catalogue of Lydgate's works.) Harvey signed his name and the date of acquisition '1598' on the title-page and on the last page below the 'Finis' he inscribed: 'gabrielis harveii, et amicorum. 1598'. Ap-parently he circulated this volume among his friends as was his practice with some of his books of general interest.[143] His elegant edition of

<hr>

[141] William Cecil died on 4 Aug. 1598.    [142] See p. 51, above.

[143] Other volumes which Harvey inscribed in this way are: Demosthenes, *Gnomologiae* [1552]; [John Eliot], *The survey; or topographical Description of France*, 1592; Luca Gaurico,

Chaucer was edited by Thomas Speght and published in 1598. Harvey's copy (B.L. Add. MS. 42518) has a number of black ink underlinings and brief manuscript comments related to the text, and at the end of the Chaucer and Lydgate poems on two half-pages not covered by text (fols. 421ᵛ and 422ᵛ) are the closely written literary appraisals referred to above.

Chaucer was Harvey's especial delight, for he admired his fine craftsmanship and the astronomical and other learning which underlay his imagery. In his marginalia in another of his books Harvey praises Chaucer and observes: 'It is not sufficient for poets, to be superficial humanists: but they must be exquisite artists, and curious universal schollers.'[144]

On fol. 421ᵛ, Harvey laments that too few writers of 'those dayes' lived up to the high standards set by Chaucer, Lydgate, Gower, Occleve, Surrey, or Heywood and that there are today all too few Aschams, Phaers, Sidneys, Spensers, Warners, Daniels, Silvesters, or Chapmans. He asks, 'when shall wee tast the preserved dainties of Sir Edward Dier, Sir Walter Raleigh, M. Secretarie Cecill, the new patron of Chawcer; the Earle of Essex, the King of Scotland, the soveraine of the divine art; or a few such other refined witts, & surprising spirits?'

Harvey now comments on his own literary efforts. An interesting feature of these annotations is Harvey's use of personae to express certain aspects of his personality or interests. 'Axiophilus' is his name for himself as a writer or lover of worthy poetry,[145] 'Chrysotechnus' is one whose technical excellence is golden, and 'Anonymus' is obviously one who cannot gain recognition in his own name. Harvey's observations are the following:

No marvell, thowgh Axiophilus be so slowe in publishing his exercises, that is

Tractatus Astrologicus, 1552; [Richard Grafton], A brief treatise conteinyng many proper Tables, 1576; Hieronymus von Braunschweig, A most excellent . . . apothecarye, or homely physicke booke for all the grefes and diseases of the bodye, 1561; James VI, His majesties poeticall exercises at vacant houres, [1591[; Livy, Romanae Historiae Principis, 1555; Oldendorf, Loci Communis Juris Civilis, 1551; Sleidanus, De statu religionis et reipublicae, 1568. Although the practice of adding 'et amicorum' to ownership notes is observed elsewhere (e.g. in Dr. Henry Harvey's copy of Libri De Re Rustica, 1528, he inscribes 'Sum Henrici Harvey & amicorum'), the fact that only certain of Gabriel Harvey's books bear this inscription is likely to have some significance.

[144] In Dionise Alexandrine's Surveye of the World (trans. by Thomas Twine), 1572, on the verso of the second of seven flyleaves preceding the title-page Harvey writes of Chaucer and Lydgate: 'I specially note their Astronomie, philosophie, & other parts of profound or cunning art . . . wherein few of their time were more exactly learned. It is not sufficient for poets to be superficial humanists: but they must be exquisite artists, & curious universal schollers.' These notes were written about 1574.

[145] In James VI, Essayes of a prentise, 1585, Harvey has inscribed a number of original poems which he signs 'Axiophilus'.

so hastie in dispatching them: being one, that rigorously censures himself; unpartially examines other; & deemes nothing honorable, or commendable in a poet, that is not divine, or illuminate, singular, or rare; excellent, or sum way notable. I dowbt not, but it is the case of manie other, that have drunk the pure water of the virgin fountaine. And Chrysotechnus esteemes a singular poet worth his weight in gould: but accountes a mean versifier a Cipher in the algorisme of the first philosopher: who imitated none, but the harmonie of heaven; and published none but goulden verses. . . . More of Chaucer, & his Inglish traine in a familiar discourse of Anonymus.

On fol. 422ᵛ Harvey expresses his esteem for Petrarch, Ariosto, Tasso, and Bartas. Among 'owre best Inglish' he lists 'the Countesse of Pembrokes Arcadia, & the Faerie Queene . . . & Astrophil, & Amyntas'.¹⁴⁶ The lines which follow have often been cited as evidence of the date of Shakespeare's *Hamlet*, for Harvey writes:

The Earle of Essex much commendes Albions England. . . . The Lord Mountjoy makes the like account of Daniels peece of the Chronicle. . . . The younger sort takes much delight in Shakespeares Venus, & Adonis: but his Lucrece, & his tragedie of Hamlet, prince of Denmarke, have it in them, to please the wiser sort. Or such poets: or better: or none . . .

Next to the mention of Shakespeare's works three stars are drawn in the margin—apparently part of Harvey's comment.

Since the Earl of Essex is referred to as alive it has been assumed that the Shakespeare allusions were written before Essex's execution on 25 February 1600/1.¹⁴⁷ Although the Chaucer folio was acquired in 1598, I am

¹⁴⁶ Presumably Harvey is referring to Sidney's *Astrophel and Stella*, 1591, and Watson's *Amintae Gaudia*, 1592.

¹⁴⁷ For further discussion of the significance of Harvey's comment on *Hamlet*, see especially the following: Moore Smith, *Marginalia*, pp. vii–xiii; E. K. Chambers, *William Shakespeare* (1963), ii. 196–8; E. A. J. Honigmann, 'The Date of *Hamlet*' in *Shakespeare Survey* (1956), pp. 24–34; and George Iam Duthie, *The 'Bad' Quarto of Hamlet* (1941), pp. 78–84. Geoffrey Bullough in *Narrative and Dramatic Sources of Shakespeare* (1973), vii. 5, concludes that the play was mainly written between 1598 and 1601 and that alterations were probably made in 1601 or 1603. 'A booke called the Revenge of Hamlett Prince of Denmarke as yt was latelie acted by the Lord Chamberleyne his servantes' was entered at Stationers' Hall on 26 July 1602.

The dating of Harvey's comments presents two problems: 1. Harvey includes in a group of 'our flourishing metricians' Thomas Watson who had died in 1592. However, this can be explained by assuming that Harvey is referring to writers whose *works* are still 'flourishing' and in popular demand. 2. Harvey states, 'I have a phansie to Owens new Epigrams, as pithie as elegant, as pleasant as sharp, & sumtime as weightie as briefe.' Although John Owen's *Epigrammata* were not published until 1607, it is possible that Harvey could have seen at least some of them before 1601 in manuscript. One of them (that addressed to Lord Burghley) is dated as early as 1596. Another, and likely, explanation is that, since the mention of Owen is one of the last two sentences on the page, it may have been inscribed considerably later than the Shakespeare allusion. In fact, these last two sentences seem to

inclined to believe that the marginalia in question were probably written after 1 June 1599, for Harvey's notes on the previous page (fol. 421ᵛ) in the same hand and apparently part of the same literary comment sound as though they were entered after the Archbishop had banned all future publishing of Harvey's work.[148] He seems to be bemoaning the fact that he ('Axiophilus') has been so slow in publishing his many accumulated poems and to be implying that as an author he will now have to be anonymous. On fol. 422ᵛ Harvey concludes his literary comments with the following wistful observation:

amongst so manie gentle, noble, & royall spirits methinkes I see sum heroical thing in the clowdes: mie soveraine hope. Axiophilus shall forgett himself, or will remember to leave sum memorials behinde him: & to make use of so manie rhapsodies, cantos, hymnes, odes, epigrams, sonets, & discourses, as at idle howers, or at flowing fitts he hath compiled. God knowes what is good for the world, & fitting for this age.

In Harvey's copy of Guicciardini's *Detti, et Fatti*, Venice, 1571, on sig. K2ʳ written vertically in the inner margin is another reference to *Hamlet*. The Guicciardini volume was acquired by Harvey in 1580. He entered annotations at that time, in 1590, and again later. The marginalia in question are not dated but the mention of *Hamlet* is included with *Richard III* in a listing of fifteenth- and sixteenth-century works which are of especial interest to Harvey. The inscription, which begins on the inside margin of sig. K1ᵛ, reads as follows:

Now Domenichi, & the 4. of Guazzo [i.e. Book IV of his *Civile Conversation*], super omnes: & for miself, ante omnes, Argutissimae altercationes, et rotundus Logismus [above all, the keenest debates and polished reasoning] in Senecae Tragaed. Eliots dialogs: Gascoignes steel-glasse: Greenes quip for an upstart Courtier; & his art of Conniecatching: Diets drie dinner; a fresh supplie of Mensa philosophica; the Tragedie of Hamlet: Richard 3.

As for Harvey's own writings, nothing further seems to have been published after 1593—at least in his own name. His marginalia had always abounded in advice to himself that time should not be wasted in writing.[149] More and more he seems to have taken these precepts to heart. Yet, one

have been written with a different pen or with different ink, for they are somewhat blurred whereas the writing on the rest of the page is clear and well defined with clear cut edges. This seems to argue for the likelihood that the last two sentences were written at a different time from the top part of the page.

[148] See p. 129, below.

[149] For instance, in his Commonplace Book (Add. MS. 32, 494) on fol. 16ʳ he wrote: 'Avoyde all writing, but necessary: w[hi]ch consumith unreasonable much tyme, before you ar aware: you have alreddy plaguid yourselfe this way: Two Arts lernid, whilest two sheetes in writing'.

would expect that he must have intended eventually putting into print some of his many discourses and unpublished poems. If he did have any such notions, he was shortly disabused of them, for on 1 June 1599 an order was issued by Archbishop John Whitgift and Bishop Richard Bancroft who had had their fill of scurrilous and vituperative satires. The order included the stipulation that 'all Nasshes and Doctor Harvyes bookes be taken wheresoever they maye be found and that none of theire bookes bee ever printed hereafter'.[150] A comedy of Nashe's, *Summer Last Will*, was nevertheless published in 1600.

By 1601 Nashe had died[151] at the age of not more than thirty-four. His frequent imprisonments had probably taken their toll of his health.

[150] E. Arber, ed., *Transcripts of the Stationers' Registers* (1913), iii. 677.

[151] In Charles Fitzgeffrey's *Affaniae* (1601), sig. N3ʳ, there is a Latin epigram on Nashe's death and in *The Second Part of the Returne from Parnassus*, lines 310 ff. (first acted in the Christmas season of 1601/2) there are a few elegiac lines to Nashe.

# 7. In later years

HARVEY left London in 1593. Thereafter he seems to have retired to Walden, collecting rents on the extensive lands inherited from his father, continuing his studies, being consulted on legal matters, perhaps practising some medicine among the townsfolk,[1] and occasionally visiting Cambridge.

As for the University, it had not forgotten Harvey and his colourful eccentricities. Certain speeches of the character 'Luxurio' in the Cambridge comedy *The Returne from Parnassus* (First Part) which was produced during the Christmas season of 1599/1600 are undoubtedly take-offs of Harvey. The following in Act I, scene i, are examples:

*Luxurio* [to *Ingenioso*, a Nashe-like character]: To London Ile goe, for there is a great nosde ballet-maker deceaste [William Elderton died *c.* 1592], & I am promised to be the rimer of the citie. Ile fit them for a wittie *In Creete when Daedalus.*[2] I have alwaies more than naturallie affected that poeticall vocation.

*Ingenioso*: Wilt thou leave Parnassus [Cambridge] then?

*Luxurio*: Is it not time thinkest thou? I have served here an apprentishood of some seaven yeares, and have lived with the Pythagorean and Platonicall Δίαιτα [manner of diet][3] as they call it. Why a good horse would not have endured it. Adew single beare and three qua of breade, if I converse with you anie longer, some Sexton must toll the bell for the Death of my witt. Here is nothing but levelinge of colons, squaringe of periods, by the monthe.[4] My sanguin scorns all such base premeditation. Ile have my pen run like a spigot & my invention answerr it as quick as a drawer. Melancholick art, put down thy hose, here is a sudaine wit, that will lashe thee in the time to come.[5]

Very little is known about Harvey after the age of fifty except that he managed to outlive his friends as well as foes. William Lewin died in

---

[1] Marginalia in medical texts by Gualterus Bruele and Hieronymus von Braunschweig strongly suggest that this was so.

[2] Nashe writes (*Works*, iii. 67) in an imaginary letter from Harvey's tutor to his father that this song [a sixteenth-century ballad] is to Harvey 'food from heaven, and more transporting and ravishing than *Platoes* Discourse of the immortalitie of the soule was to Cato'.

[3] Harvey's marginalia evince his belief that a sparing diet was conducive to mental and physical health, e.g. see fol. 12ᵛ of Add. MS. 32, 494 (Harvey's 'Commonplace Book').

[4] In *Pierces Supererogation* (sig. Z4ʳ) Harvey writes: 'It is for Cheeke, or Ascham, to stand levelling of Colons, or squaring of Periods by measure, and number: his [Nashe's] penne is like a spigot; and the Wine-presse a dullard to his Inke-presse.'

[5] J. B. Leishman, ed., *Three Parnassus Plays* (1598). Reprinted in London, 1949, pp. 155–7.

1598; Edmund Spenser succumbed in January 1599. During his later years Harvey probably turned more and more toward his family, but they were to prove a mixed blessing.

On 31 March 1600, according to the Parish Register, Gabriel's sister Marie (Mary) was married in Walden to the yeoman Phillipp Collyn. She must have been about thirty-three at the time of her marriage, for as has been previously noted she was baptized in May of 1567. Three documents are extant, a deposition and two court decrees, which show that in 1608 husband and wife brought a lawsuit in Chancery against Gabriel Harvey claiming that he had not paid his sister the sum of sixty pounds which their father's will stipulated should be paid within four years after his death. The will provided that, if the sum was not paid, Gabriel would forfeit to her the greater part of the inheritance which was to come to him after his mother's death,[6] viz., the family mansion and the major part of the family lands. Since the value of the property being bequeathed[7] must have been far in excess of sixty pounds, Marie may have preferred that the sum not be paid. At any rate, she waited not four years but fifteen years after her father's death to complain that Gabriel was delinquent. It is stated in the decree (Chancery Decrees and Orders, 33/115, fol. 145) that she applied to the court in 1608 because the only witness to her father's will, the notary public James Crofte, was old and she feared that he might die leaving her without a witness for her claim. A deposition was thereupon taken from James Crofte who averred he was 'of the age of fiftie fower yeeres or ther aboutes', that he knew the complainants and the defendant and had known John Harvey of Walden for many years before his decease and was acquainted with his written will which he recited for the court (C.24/346/31).

As an outcome of the suit the court ordered a personal attachment against Gabriel Harvey, and the Sheriff of Essex was enjoined to seek him out. Harvey evidently went into hiding or fled Walden, for the Sheriff reported that he was unable to find him. For this reason the court issued a Commission of Rebellion directing Henry Collen, Kenelme Collen (perhaps relatives of Phillipp's), Henry Taylor, and Zerobabell Veale jointly and severally to attach the said defendant (C.33/115/fol.149).[8]

---

[6] Gabriel's father died in July 1593, his mother Ales Harvey was buried 14 Apr. 1613, according to the Walden Parish Register. It is hard to understand how Gabriel's father could have expected him to be able to pay his sister £60 before he had inherited anything. Nothing seems to have been willed to him until after his mother's death.

[7] The deposition by the notary public James Crofte recites the will of John Harvey the elder and includes a description of his properties (P.R.O. C.24/346/31). See pp. 6–7, above.

[8] These documents are fully described by Irving Ribner in 'Gabriel Harvey in Chancery—1608', RES, ii. 6 (Apr. 1951), 142–7.

No further documents have been found to indicate whether or not Harvey was apprehended or forced to pay, but the likelihood is that the suit eventually was settled amicably, for in the will of Richard Harvey,[9] who throughout his life remained close to his elder brother, Gabriel is made heir, and designated as executors are their cousin John Gyver and brother-in-law Phillipp Collin. In Collin's own will (dated 15 June 1625)[10] he specifies that he and his wife dwell in Goldstreet, Walden, and he describes his messuage as having buildings, barns, stables, yards, orchards, and gardens, but no mention is made of the mansion on Walden's Market Square[11] or of the other properties that had belonged to John Harvey (i.e. the properties as described in James Crofte's deposition).

What happened to Gabriel Harvey after the 1608 lawsuit? Did he flee to London or possibly manage to go abroad as he had always yearned to do? His later marginalia are found chiefly in Italian texts and have many annotations in Italian.[12] But our query must remain unanswered, for nothing concrete is known about Harvey during the next twenty years.

At the time of his brother Richard's death in 1630 he was in England, for the two executors named in the will renounced their execution of it and asked that this function be granted to 'Gabriel Harvey doctor of lawes'. The document which is written in English is dated '10th daie of June 1630' and is signed by John Gyver and Phillipp Collin and witnessed by John Ayer and Richard Lyon, probably in Walden, for all four were local residents. A second document in Latin dated 'xi Junii 1630' states that Mr Wyan as proctor for Gabriel Harvey appeared before the Bishop of Rochester's Chancellor, Dr Edmund Pope, and requested that execution be granted to Gabriel Harvey since the named executors had deferred

[9] A verified copy of the will (dated 25 Aug. 1625) is at the Kent Record Office (DRb Pw28). Richard Harvey died in June 1630, apparently at Chislehurst.

[10] D/ACW/11/158 at the Essex Record Office.

[11] According to the *Essex Review*, 7 (1898), 22, the Walden Churchwardens' books have an entry dated 1623: 'Paid to Thurgood for a bellrope iij.ij.', and the item occurs again later on. A. R. Goddard, writer of this article on 'The Harveys of Saffron Walden', takes these entries as evidence that the rope-maker's shop at the back of the Harvey mansion passed to a new name even during Gabriel's lifetime. There is a will of 18 March 1638 at the Essex Record Office for a Thomas Thurgood of Stansted Mountfitchet, rope-maker. Since Stansted Mountfitchet was the town adjoining Walden, it probably was Thomas Thurgood who operated the Walden rope-making establishment. The business was a highly specialized one and rope-makers generally serviced quite a wide area. The mansion itself was eventually owned by the Thurgood family; in 1855 when the old Harvey homestead was finally demolished, it was in the ownership of Robert Driver Thurgood, one of the eminent mayors of Saffron Walden.

[12] For example: Guicciardini, Domenichi, Porcacchi.

to him.[13] Harvey was granted the executorship of his brother's estate but never concluded the legal matters connected with it as he himself died shortly thereafter on 7 February 1630/1 at the age of about eighty years.[14] The 1630 burial entry in the Walden Parish Register is prominently written in exceptionally large script and reads:

'M<sup>r</sup> Doct<sup>r</sup> Gabriel Harvey        11 Feb.'

Thomas Baker, the seventeenth-century Cambridge historian, wrote of Harvey's death: 'He must have liv'd to a great age, for I have seen an elegy on D<sup>r</sup> Harvey of Safron Walden, compos'd by W<sup>m</sup> Pearson dated an: 1630: whereby it appears, he died that year'. (Baker MS. XXXVI, 114.) A further comment has been incorrectly attributed to Baker, in Baker MS. XXXVI, 107 although at no place in this manuscript is there any such amplifying statement. The traditionally repeated remark is a substitution for the final phrase in the above ('whereby it appears, he died that year'). Instead there is interpolated the sentence: 'By that it would seem he practised physic and was a pretender to astrology'.[15] There was a William Pearson who matriculated at Pembroke in 1580/1 and received his B.A. from Christ's in 1584/5 and his M.A. in 1588.[16] Very likely it was he who composed the elegy, but the poem itself is lost.

It is quite possible that Harvey may have 'practised physic' in Walden, as his marginalia give evidence of detailed interest in and practical knowledge of medical treatment. However, although profound interest in cosmology and astronomy is clearly manifested in his marginalia, it seems totally unlikely to me that with his scorn for astrological predictions he would ever have been a 'pretender' in what must even then have seemed to him a pseudo-science.

The impressive burial entry in the Parish Register and the long-standing

[13] These two documents are filed with the will and bear the same classifying number, see p. 132 n. 9, above.

[14] According to Moore Smith's *Marginalia*, p. 75 n. 1: 'On 20 April 1631 administration of Richard's goods "de bonis non admin. per Gabrielem Harvey etiam defunctum" was granted to Alice Lyon, natural sister of the deceased.'

[15] Even such a customarily careful scholar as Moore Smith quotes this on pp. 75–6 of the *Marginalia*. The partial canard seems to derive from Philip Bliss's 1815 revision of Anthony Wood's *Athenae Oxonienses*. It is found on p. 230 of the 'Fasti' in vol. 5 of the 1815 edition, although it does not appear either in the first (1691) or second (1715) editions. However, Wood himself makes some questionable statements, e.g. he credits Harvey with the writing of '*Almanacks* much in esteem in the reign of Qu. *Elizabeth*'. Here Wood is either confusing Gabriel Harvey with his brother John or with Gabriel Frende or else he is relying too much on Nashe. Wood also misreports the content of 'An Advertisement for Pap-hatchet, and Martin Mar-prelate' by stating (from hearsay) that it is an attack on Robert Greene.

[16] A family named Person (or Pearson) resided in Walden at this time, but I have been unable to ascertain whether it included this William Pearson.

respect that Waldeners have had for his name suggest that he was held in high esteem by his native community.[17] At the time of his death, his sister Alice Lyon, then in her mid seventies, was appointed Administrator for his estate,[18] for he left no will nor has any inventory been found. His huge library probably was sold and dissipated almost immediately, for no provision seems to have been made for keeping it intact.

Gabriel Harvey's life story cannot be concluded without the record of a final sardonic touch of irony. The man who had tried so hard and suffered so much was commemorated in 1631 by the publication in 12° of *Pedantius*. Just two days after Harvey's death (and two days before his burial) it was entered at Stationers' Hall[19] and was published shortly thereafter. Apparently, even at the end of his life Harvey was still a figure of public interest—but to many Englishmen not one to be treated seriously.

[17] Richard Lord Braybrooke in his *History of Audley End*, 1836, alludes (p. 195) to the notices of natives of Walden at the time of Harvey's death. To my knowledge these notices are no longer extant.

[18] See P.R.O. PROB. 6/114A, fol. 8ᵛ which is dated from London 1632, 18th day of 'Ascension' (one of the four yearly law terms). That the probate was in London probably indicates that Harvey owned property in several counties.

[19] Janet Biller, in *Notes and Queries*, 1968, p. 49, remarks on the closeness to Harvey's death of the S.R. entry: 9 Feb. 1630/1.

PART II

# MARGINALIA

# 1. Harvey's Habits in Annotating

THE foregoing biographical section has outlined Harvey's interaction with the world around him: his achievements, disappointments, and failures. But there is more to search out about the inner man: what Harvey thought, his changes of mood, devices he used to cope with the cruelties and injustices he encountered, his views of himself and of others, his self-admonitions and precepts, his methods of scholarship, and above all his unceasing efforts towards self-improvement. Glimpses of this sort are obtainable by investigating his personal notes to himself: the annotations which he systematically inscribed in the margins and blank spaces of his cherished books. As he matured and aged there were changes in his personality and attitudes, but his habits of recording marginalia remained amazingly constant over a span of at least forty years.

His extravagant aim seems to have been to acquire for his own use the sum total of human knowledge—learning, however, not for learning's sake but rather for pragmatic purposes. An omnivorous reader and student, he was intent upon being able to recall and use the fruits of his reading and study. How he achieved his prodigious scholarship can best be understood through analysing typical examples of his marginalia. His annotating methods seem to have gradually developed from the youthful inscriptions in his first serious books to the copious manuscript reflections which fill nearly every blank space in some of his favourite volumes. The earliest annotations which I have seen are those in Erasmus's *Parabolae* (1565),[1] acquired by Harvey in January of 1566 when he was about sixteen and just entering Christ's College, Cambridge. Probably used as a college text, it is an anthology of parables and significant quotations from classical authors: Plutarch, Seneca, Aristotle, Pliny, and Theophrastus. Such florilegia were widely used in Harvey's day. Although during the course of his life he intensively studied the original writings of most of the great thinkers and authors, he nevertheless seems to have treasured such collections as a stimulant to his own thought, and they evidently proved to be so, for most such volumes in his library are thickly annotated with his comments.

---

[1] A brief description of each of Harvey's extant books is to be found in Part III which has a listing of his library.

Even this early he must have sensed that his books and their manuscript notes would play an important role in his life, for on the title-page of the Erasmus he clearly inscribed his name and the date in black ink, as he made a practice of doing in the books he subsequently acquired: for instance, in 1567, in Quintilian's *Institutionum oratoriarum Libri XII* and in 1568 in Livy's *Romanae Historiae Principis, Decades Tres*.

On the *Parabolae* title-page (see Plate A) in the space to the right of the printer's device, Harvey wrote in Latin in a slanted, angular, somewhat immature hand: 'Gabriel Harvejus. mense Januario: 1566.'[2] And below the device he added 'Â quibus nihil boni spero, quià nolunt: ab iis nihil mali metuo, quià no[n] possunt',[3] and at the bottom of the page appended, 'Vel Arte, vel Marte'. Perhaps he had been reflecting on Gascoigne's 'Tam Marti, quam Mercurio' and was experimenting with a motto of his own.[4]

Although the title-page also bears another signature of Harvey's (in a large florid Humanist hand) and a comment written in his later rounded Italic script,[5] those I have quoted above were written early, probably at about the time that he acquired the little book. There are certain characteristics which help to distinguish Harvey's early from his later hand (i.e. after about 1580). In the 1560s and the early 1570s his letter formations, whether in English Secretary, Humanist, or Italian (Italic) script, are far more angular and pinched; pen strokes are usually narrower, and there is less evenness of script and less judicious spacing than is subsequently found; a gradual development toward greater control takes place in the late 1570s. The examples of Harvey's Italic hand written in the early 1570s in Sloane MS. 93, although extremely neat and fairly regular in formation, are far more pinched, angular, and slanted than those that are found, for instance, in the beautiful and decorative 1580 inscriptions of his Livy or the 1598 Chaucer inscriptions.[6] The later marginalia are typified by broad pen strokes, rounded and free-flowing forms, deep intensity of ink, and usually an Italic script. Being skilfully inscribed and

---

[2] In these lines the 'i's' and the 'j's' are dotted with curled, comma-like marks. This is a practice which Harvey occasionally follows, probably with decorative intent.

[3] 'From those whom I hope nothing good, because they are unwilling: from these I fear nothing evil because they are unable.' As was common practice at this time, Harvey uses various accent marks on his Latin script. The above is given as a sample of this use but, for the sake of simplicity, such diacritical marks have not been included in the other transcriptions in this study.

[4] This motto is also found on the title-page of Harvey's copy of Frontinus's *Stratagemes*, translated by Richard Morysine. Published in 1539, it was acquired by Harvey in 1578.

[5] The later comment reads: 'Quae ipse profitetur esse exquisitas Gemmas.' ['He declares that they are *exquisite jewels*.'] Italics represent Harvey's underlinings.

[6] For comparative samples of Harvey's early and late hands, see Plates A–H.

spaced, they present a truly ornamental appearance on the page.[7] There is a discernible change in the darkness and opacity of the script about 1579 when Harvey apparently adopted a more permanent type of black ink which tended to maintain its deep intensity and is easy to read even today despite the ravages of time (although occasionally the black has turned to brown).

Harvey probably used both a Secretary and an Italic hand throughout his life, but the former is found far more frequently in his early than in his later marginalia, being used primarily in his workaday notes within his university texts. Harvey's beautiful legible Italic hand is employed for summarizing remarks on the flyleaves at the end of such texts or elsewhere for observations of a generalized or philosophical nature, and in nearly all of his annotations after 1582. His handwriting is usually moderately large but can at times vary to quite minute. Its size is usually determined by page dimensions[8] or other kinds of space limitation but an extra large hand is sometimes used for the sake of emphasis.

Harvey's favourite books were read and annotated a number of times. The initial (1566) annotations of the title-page of the Erasmus are in a thin-lined immature hand which may be an early precursor of Harvey's Italic but is far too irregular and uncontrolled to resemble it clearly as yet. A number of other notes throughout the volume are in a similar hand and were probably written at about the same time. On sig. L4ᵛ at the conclusion of the text is a large inscription in an angular but firm and assured Secretary hand which reads: 'Relegi mense Septembri. 1577: Gabriel Harveius'. In this same 1577 hand are other marginalia within the volume and a few in Harvey's slanted, somewhat pinched, early form of Italic which probably also dates from this period. Not until about 1580 does his Italic hand attain its mature and final form.

As stated above, many of Harvey's volumes bear evidence of successive readings and annotations; he sometimes signs and dates re-readings. For example, his Livy folio has marginalia from perusals in 1568, 1580, and 1590. Here the earliest notes are in an angular, somewhat irregular hand of Secretary type, while those dated '1580' or '1590' are in a rounded,

---

[7] Harvey's marginalia are remarkable for being almost totally without corrections. The only one that I can recall seeing is in Domenichi on fol. 11ʳ where Harvey changes the word order from 'virtus adamantina' to 'adamantina virtus' by crossing out the first 'virtus'.

[8] A very tiny Italic is found in parts of Quintilian and on the final flyleaves of two of Harvey's octavo law texts (Humphrey and Hopperus) where the annotations are of a generalized nature closely and copiously inscribed; an especially minute Italic is found on the tiny pages of the 12° Il Pastor Fido by Guarini. Compare the handwriting here with the large Italic script in the folio Livy (Plates C, F, and G).

cursive, broad-stroked Italic. On some pages two different periods of handwriting are represented with the Secretary script above and the Italic following directly below or superimposed on the page after the Secretary inscription had been written. A similar sequence of Secretary or Humanist first and then Italic is also found on various pages of Frontinus's *Strategemes* (1539) and Bruele's *Praxis Medicinae Theorica* (1585).[9] In his copy of Frontinus, there is evidence of an initial reading in 1578 with marginalia from this period as well as from 1580 and 1588. Though it would seem possible to distinguish between Harvey's very early annotations and those made after about 1580, the foregoing dating criteria must be used with discretion unless one has the opportunity to study a wide enough range of marginalia.[10]

A number of studies of Harvey's marginalia have, of course, been made, one of the first and most comprehensive being that of G. C. Moore Smith in 1913. He listed fifty-three extant books of Harvey's and published selected annotations from twenty-six of them together with notes from two Harvey-owned manuscripts (Add. MS. 32,494 and 36,674) and thus provided an excellent sampling of the 'treasures' to be found. Smith did not, however, make mention of the many marginal symbols which Harvey customarily appends to his notes nor of the coloured slashes found through some of his pages.

In 1948 Harold S. Wilson added to our comprehension of the marginalia by describing Harvey's well-organized system of annotation: his use of planetary and other symbols or brief comments to classify subject matter, and his careful cross-referencing.[11] Wilson mentions Harvey's yet-to-be understood practice of making broad, coloured (usually red or green) chalk-like markings through the text of some of his books.[12] Wilson also observed that there were not fifty-three but at least a hundred of Harvey's books extant.[13] To date I have found one hundred and eighty, and more will undoubtedly turn up. Other writers, notably Caroline Ruutz-Rees, Gregory Smith, Caroline Bourland, John Lievsay, and

---

[9] For example, in the Livy folio the flyleaf just preceding the title-page and at the beginning of the Frontinus text there is such a sequence of hands. The final page of Cicero's *Topica* can be used for a comparison of Harvey's 1570 'Humanist' hand and 1579 'Italic'. These three pages are reproduced on Plates C–E.                 [10] See Plates A–H.

[11] When he is referring to a passage in the same book, Harvey uses the folio number following a symbol for *supra* (s$^{a}$) or *infra* (j$^{a}$); when referring to another volume, he uses its title together with the folio number.

[12] Suggested explanations are that these slashes represent sections studied by him or alternatively that they represent passages assigned to his students for study. However, neither explanation seems to fit all the many instances of the use of these coloured markings.

[13] 'Gabriel Harvey's Method of Annotating His Books' in *Harvard Library Bulletin*, 2 (1948), 344–61. However, Wilson seems to have made no listing of Harvey's books.

Eleanor Relle have printed informative studies of one or more individual annotated volumes.[14] The present study will survey briefly Harvey's methods in a number of characteristic volumes (especially richly annotated ones which have recently come to light) and will then examine in more detail one of his favourite and most copiously annotated books.

Harvey's annotations are roughly of three types. First, there are concise notations or marginal symbols[15] which classify the subject matter so as to make it quickly available to him for future use. For the most part he uses abbreviations, planetary, and diagrammatic symbols. For example, the symbol for Mars ♂ designates matter relating to military affairs or strategy, an asterisk-like star denotes astronomy, a large and small contiguous circle is used for controversy or opposition, a bisected circle for the earth or natural history, the sun for kingship.

For instance, in the preface to Thomas Wilson's *The arte of Rhetorike, for the use of all such as are studious of Eloquence* (1567), the text recounts God's granting the gift of eloquence so that men

might with ease, win folk at their will, and frame them by reason, to al good order. And therefore, where men lived brutishly in open feldes . . .[16] these appoynted of God, called them together by utteraunce of speache, perswaded with them what was good, what was bad, & was gainfull for mankinde.

Next to this passage Harvey inscribes his symbol for eloquence, the planetary sign of Mercury ☿. Other symbols are used in a similar manner in other texts and the notation 'J. C.' (*Jurisconsultus*) or 'LL.' (*leges*) are frequently employed to indicate a passage which has legal relevance. The term 'nota' or opening quotation marks are used to mark a notable or quotable section. Succinct captions of one or two words placed in the margin often summarize a fairly lengthy textual discussion, e.g., in Thomas Blundevill's *The foure chiefest Offices belonging to Horsemanship* (1580), on fol. 1r [17] at the beginning of a chapter entitled 'Of the colours of Horses, and which be best' Harvey writes 'Semeiotice', i.e. 'diagnosis'. On fol. 5r he inscribes at the top 'How the Rider ought to sitt in his Saddle'; on fol. 11v next to a discussion of Musroll and Martingale, Harvey captions 'Extraordinari instruments'; and on fol. 51v he summarizes with the inscription, '*The parts of the bit*: as they ar termed by their *proper names*'. Italics represent Harvey's underlinings.

[14] See list in Appendix E, p. 272, below.

[15] For further discussion of his marginal symbols, see Wilson, pp. 354–8.

[16] Ellipsis here and in some of Harvey's subsequent quoted marginalia represents my deletions of the original.

[17] In some of Harvey's heavily annotated volumes the signature markings have been completely covered over by inscriptions, in some the markings have been removed due to trimming. In these cases I have designated passages by folio number.

Second, there are his critical and supplementary comments on ideas in the text, often pointing out the relevance to historical or current problems. Examples include the following: In Wilson's *Art of Rhetorike*, on sig. C2$^r$ next to the printed marginal caption: 'He that will stirre affections to other must first bee moved hymselfe'. Harvey exemplifies: 'Checus. Templaeus. Gardinerus. Valsinghamus. Vilsonus. omnes regii Cantabrigienses.'[18] On sig. K3$^r$ which Harvey captions, 'Ironies enlarged', the text deals with dissembling and, illustrating a mock encomium, Wilson writes of a man to beware, he who

hath more *honesty with hym then he nedes*; and therefore both is *able* and *will lende, where it* pleaseth hym beste. Beware of hym above all menne that ever you knewe. He *hath no fellowe*, ther is none suche, I thinke he will not live long, he is so honest a manne; the more *pitye that suche good fellowes shoulde know what death meaneth*. But it maketh no matter, *when he is* gone, all the *worlde will speake of hym*, his name *shall never dye*, he is so well knowen universallie.* Thus we maie mockingly speake well of hym, when there is not a *naughtier fellowe within al England* again. Italics represent Harvey's underlinings.

Next to the asterisk (*) Harvey notes, 'Green, & Nash at this instant'. At the bottom of this page Harvey inscribes, 'Julian in his Misopogon[19] finely commendes his foes: ut s: 65'. This cross-reference to (fol.) 65 (sig. Ii$^r$) is to another annotation of Harvey's: 'The Roman prudence in extolling the valour of their mightiest enemies: as Pyrrhus, Annibal, Mithridates &c.' On sig. K4$^v$ the text, dealing with the power of jesting, points out that when an evil fellow wrongfully accuses a man, far more harm is caused when the accusation is made in a witty and humorous manner. Here Harvey notes, 'Nash the rayler'.

In Sextus Julius Frontinus's *The strategemes . . . of warre*, translated by Richard Morysine (1539) and purchased by Harvey in 1578, there are marginalia inscribed at the time of purchase and again, as previously mentioned, in 1580 and 1588. Many of these cite successful or unsuccessful manœuvres used in classical times and compare them with contemporary military problems in the Netherlands or liken them to the conduct of the Spanish and English forces at the time of the Armada. On sig. H8$^v$ the text treats of various stratagems that have been used to gain entry into a heavily fortified enemy camp, one being the donning of the apparel of a slain enemy leader. Harvey comments, 'How easely might S$^r$ Humfry Gilbert, or Captain Forbusher, or Captain Drake, have gained sum lyke opportunity? The Spaniards with bribes, have greatly advancid

---

[18] (John) Cheke. (William) Temple. Stephen Gardiner. (Francis) Walsingham. (Thomas) Wilson. All were Regius Professors of Cambridge.

[19] Harvey is referring to 'The Beard Hater' a bitter satire addressed by Roman Emperor Julian to the citizens of Antioch in which he commends them for hating him.

his [Philip II's] procedings in the low cuntrys, & other places. Corruption, the great stratagem of Philip of Macedonia: and now of this Philip of Spain'.

On sig. N2ʳ is the account of Cassius' setting fire to certain large but not too valuable ships of his own and allowing them to be driven by the wind into the midst of the enemy's ships thus igniting them. Harvey writes in the margin: 'Owr Inglish pollicy against the Spanish Armada, this other day'.

On sig. H6ᵛ the text tells of the capture of a formidable Spanish town by Marcus Cato's forces, a victory achieved because in two days they managed to travel a four days' journey and thus caught the enemy unaware. Here Harvey interposes in Latin, which may be freely translated: 'But nothing of the sort have I accomplished. [I have progressed] to some extent towards a man of learning but what [have I accomplished] towards the man Gabriel Harvey? Without industry of one's own there are no topics for discussion'.[20]

In A. P. Gasser's *Historiarum, et Chronicorum Totius Mundi Epitome* (1538) purchased by Harvey in 1576, are brief manuscript characterizations of some of the historical figures mentioned. For instance, in Harvey's early hand at the top of sig. Q8ᵛ he writes: '*Tamerlane* of a lusty stowt Heardman, a most valiant, & invincible Prynce', and at the top of sig. R4ᵛ: 'Georg Scanderbeg, Prynce of the Epirotes. An other Pyrrhus'.

In George North's *Description of Swedland, Gotland, and Finland* (1561) on sig. G2ʳ next to a textual discussion of the 'Swecian Language' and a copy of 'The Lordes Prayer' in Swedish, Harvey adds: 'The same radical of owre Inglish, & Scottish: notwithstanding sundrie dialects, or idioms, even amongst owrselves'. At the bottom of the prayer he notes: 'Ut prosodia differt, sic orthographia [As the prosody differs so the spelling]'.

Harvey evidently kept abreast of current scientific writing even as he got older, for in Thomas Hill's *Schoole of Skil* (1599), a text on the sphere, Harvey inscribes (probably in 1599, the date of purchase) on sig. S6ᵛ: 'Astronomical notes readily reducible to the spherical method of Scribonius, or to Freigius's physical questions about the sphere. [This is] also the sort of practice of our Blundevill in his clear tract on the sphere, recently published'.[21]

---

[20] 'At nihil tale feci. Ad polyhistorem, ista aliquid: sed quid ad Gabrielem Harveium? Nihil Loci Communes, sine propria Industria'.

[21] 'Scholia astronomica, ad methodum Scribonii Sphaericam facile redigenda: aut ad Freigii quaestiones physicas de Sphaera. Qualis etiam nostri Blundevili usus in perspicuo tractatu de Sphaera, nuper edito'. The works to which Harvey is referring are Adolphus Scribonius (Wilhelm Adolph Schreiber) of Marburg's *Physica et spherica doctrina*, published in a third edition in Frankfurt in 1593, Johann Thomas Freigus's *Quaestiones physicae*, 1579,

The third type of marginal annotation is not closely connected with the text and consists of Harvey's personal reflections, introspections, and precepts. From the first and second kind of annotation, classification of subject matter, and critical and supplementary comments on the text, one derives an understanding of Harvey's scholarly methods and erudition, but it is from the second and third kind that one learns particularly of his attitudes and feelings. Lodovico Domenichi's *Facetie, motti, et burle* (1571), a cherished volume of Harvey's which will be discussed later, has notes which belong almost wholly to the third category. An Italian collection of short, miscellaneous observations and anecdotes which are printed with wide margins, its shrewd wisdom stimulated Harvey to jot down a variety of musings and random philosophical reflections.

Before proceeding further, however, it would be wise to say a word about Harvey's unusual but quite logical phonetic spelling[22] and to comment on the evidences of scholarship that are displayed throughout his notes.

Harvey's spelling is far more consistent than that of most of his contemporaries, for Harvey is aware of the importance of orthographical standardization. In a letter to Spenser on reformed versifying[23] he writes:

I am of Opinion, there is no one more regular and justifiable direction . . . to bring our Language into Arte, and to frame a Grammer or Rhetorike thereof: than first of all universally to agree upon one and the same Orthographie, in all pointes conformable and proportionate to our Common Natural Prosodye: whether Sir Thomas Smithes in that respect be the most perfit, as surely it must needes be very good: or else some other of profounder Learning and longer Experience, than Sir Thomas was, shewing by necessarie demonstration, wherein he is defective, wil undertake shortely to supplie his wantes, and make him more absolute. My selfe dare not hope to hoppe after him, til I see something or other, too, or fro, publickely and autentically established, as it were by a general Counsel, or acte of Parliament: and then peradventure, standing upon firmer grounde, for Companie sake, I may adventure to do as other do.

At the end of his annotated copy (on sig. L4ᵛ) of Sir Thomas Smith's *De Recta & Emendata Linguae Anglicanae* (1568) Harvey alludes to Smith's reformation of orthography, to John Barett's alphabet and orthographic studies in *An Alvearie, or Quadruple Dictionarie* (1580), and to John Hart, Chester Herald, who in 1569 published *An Orthographie* and in 1570 *A Methode or comfortable beginning for all the unlearned*.[24]

and Thomas Blundevill's *Exercises*, printed in London in 1594 and 1597. See Carroll Camden, Jr., 'Some Unnoted Harvey Marginalia' in *PQ*, 13 (2 Apr. 1934), 214–18.

[22] Early examples of his unusual spelling are his letters to John Young (pp. 35–49). See pp. 124–5 for a later example, the 1598 letter to Robert Cecil.

[23] *Three Proper and wittie familiar Letters*, 1580, sig. D4ᵛ.

[24] Harvey's notes read: 'Istius Thomae Smithi reformatae Orthographiae, ecce honorifice

I have recently come across an annotated copy of John Hart's *Orthographie* which undoubtedly belonged to Harvey (see p. 218 below) and it is replete with his comments on spellings and pronunciations. On sig. Cii[r] he notes: 'Two falts in th'Alphabet

> 1. One letter for divers elements.
> 2. Divers letters for one element.

Example of the

> 1. c. for k. & s. / g for gam & jod.
> 2. c. & s. / g' & j. /'

On sig. Diiii[r] where the text states 'a writing is corrupted when any worde or sillable hath more letters, than are used of voyces in the pronunciation'. Harvey cites as examples: 'Comptroller, Bloudde, Adde, Speake'. For the most part, he approves of Hart's statements in the text but occasionally he differs from or clarifies them.

Before adopting his own version of proper spelling, Harvey had evidently studied the work of various sixteenth-century orthographic reformers.[25] But, as he explains to Spenser in the letter mentioned above, until some form of standardization be decreed by governmental fiat, Harvey dares give 'no Preceptes' nor 'set downe any certaine general Arte'. However, he will not hesitate to use 'particular Examples' according to his own orthographic tenets. Harvey's aim seems to be to omit unpronounced letters so as to obtain the simplest possible spelling: instead of 'debt', 'might', and 'ought', he writes 'det', 'miht', 'ouht'.[26] He attempts to spell in such a way that there will be no uncertainties of pronunciation: instead of 'our', 'promise', 'endeavour', 'stomach', 'ready money', 'done', 'they', he writes 'owr', 'prommis', 'endevur', 'stummock', 'reddi munni', 'dun', 'thai'. His spelling also shows his awareness of

saepe mento in Baretti Anglico, Latino, Graeco, Francico. In literis A:H:S: Smitho de orthographia contra Cacographiam secundus, in suo alphabeto Barettus. Nec vero contemnendus de Orthographia Chesterus, ille etiam polyglottus. Adhuc tamen plaerique veterani, cum consuetudine delirant: et illustre vetulum abusum etiamdum observant, qui etsi non est ex usu, in usu tamen est'.

[25] For discussion of sixteenth-century orthographic reforms, see E. J. Dobson, *English Pronunciation 1500–1700*, Oxford, 1968 (second edn.), i, 38 ff.

[26] In Shakespeare's *Love's Labour's Lost* (*c.* 1594?) there is an amusing passage in which Holofernes refers to Armado (a character in some but not all respects suggestive of Harvey): 'I abhor . . . such rackers of orthography as to speak "dout" fine when he should say "doubt"; "det" when he should pronounce "debt"—d,e,b,t, not d,e,t. He clepeth a calf "cauf"; half "hauf"; neighbour vocatur "nebor", neigh abbreviated "ne". This is "abhominable" which he would call "abominable". [1963 Pelican Shakespeare, ed. Alfred Harbage, v. i. 17–24]

syllabification,[27] e.g. 'offendid', 'semid', 'drivn',[28] That he had firm views on the subject of orthography is evident in the frequent corrections he makes of the spelling of words in published texts. He even dares to emend (to '*Anglicae*') the Latin '*Anglicanae*' on the title-page of Sir Thomas Smith's work.

Part of Harvey's desire to standardize spelling would have derived from his interest in versification and his early experiments with classical metres. As Derek Attridge points out, 'The inconsistency of English spelling would have been particularly noticeable when compared with the regularity of Latin, and though it gave the quantitative poet a certain amount of freedom, it added to the difficulty of achieving agreement on rules'.[29] Harvey's doubling of the 'm', for instance, in words like 'stummock' and 'prommis' not only determine the pronunciation of the preceding vowel but also justify an accented (long) syllable.

Harvey is precise about many facets of scholarship. He takes great care in punctuation and customarily underlines the textual phrase to which his annotations refer. In at least one of his later books there is evidence that he has used a system of footnoting not very different from that used today.[30] There is considerable cross-referencing within a text and to passages in other volumes. Sometimes he will comment that a certain statement is incorrect and will cite the authority for his view. He employs a number of languages in his notes. The majority are written in Latin and a fair number in English. Italian and Greek inscriptions are found in texts written in these languages. Occasionally one comes upon a French or Spanish phrase, but there are no lengthy notes in these tongues. Often Harvey takes the trouble to add an index to a volume which has

[27] In the letter to Spenser on reformed versifying (published in *Three proper and wittie familiar letters*, 1580) Harvey writes that he advocates making 'heavn' and 'seavn' monosyllabic in writing as well as speaking. He points out that in *Toxophilus* Roger Ascham has done this with the word 'iron' by writing it 'yrne'. Ridiculing some of the absurdities of current spelling which Harvey terms 'pseudography', he gives the following examples: '*Mooneth* for *Moonthe*, *sithence* for *since*, *whilest* for *whilste*, *phantasie* for *phansie*, *even* for *evn*, *Divel* for *Divl*, *God hys wrathe* for *Goddes wrath*, and a thousande of the same stampe: wherein the corrupte Orthography in the moste hath beene the sole, or principall, cause of corrupte Prosodye in over many'. (sig. F2ʳ)

[28] The above examples of Harvey's spelling are culled from pp. 8–14 of the *Letter-book*.

[29] Attridge, *Well-Weighed Syllables*, Cambridge Univ. Press, 1974 p. 113, n. 1.

[30] In Battista Guarini's *Il Pastor Fido* (1591) Harvey sometimes places a tiny letter symbol above a textual passage and an explanatory note or translation in the margin text to a notation of the same tiny letter. Some pages have a number of such footnotes each designated by a different letter. (See Plate F, between pp. 148–9.) Other writers before him sometimes use a single footnote designated by an asterisk or other mark but I have found no previous instance of the use of a whole series of notes organized on one page in sequence.

none.[31] His comments frequently stress the need for good organization of material and classifying it under proper headings. Sometimes he will dichotomize the subject matter in a Ramus-type of diagrammatic outline.[32]

One of the interesting features of Harvey's marginalia (and of his writings as well) is the number of neologisms he coins. Building on sound etymological principles, he creates English words from Greek or Latin roots (e.g. 'acumen' and 'canicular' which are found in *Pierces Supererogation*), or he logically constructs a Latin term from Greek roots or from a current English word (e.g. 'megalander', see p. 153 n. 13 below, 'stoicheologia' and 'stentorice', which are quoted on p. 170 n. 63 below and 'polypraxia' and 'axiozelis' which are quoted on p. 184 n. 104). Often he coins a Latin adjective or adverb from a Latin noun.

His notes show his wide reading in each field and his careful evaluation of authorities. He is quick to question a statement which he believes false or doubtful. Frequently he cites the views of a respected authority on a similar point. When it comes to a contemporary authority he can often quote from personal conversations with him, as he was acquainted (through Sir Philip Sidney's circle and the entourage of foreign scholars around Leicester) with a great many eminent and erudite men.

The marginalia of Harvey's most copiously annotated books are repositories of original material from which he can draw for his own writings or lectures. Some of his books include original poems in Latin or English, sometimes fairly extensive prose passages,[33] frequently there are memorable precepts, a succinct analysis, or well-turned phrase which he excerpts to incorporate into one of his published works.[34]

[31] At the end of Sacchi de Platina's *Hystoria de Vitis pontificum* (*c.* 1505) Harvey adds his index of Popes and page references; in G. Bruele's *Praxis Medicinae Theorica* (1585) on the front flyleaf is a manuscript index of various medical problems, together with page references.

[32] For example, the final page of J. Hopperus, *In veram Jurisprudentiam Isagoges* (1580). See Plate G, between pp. 148–9.

[33] Poems are found in James VI, *Essayes of a Prentise* (1585), and Peter Whitehorne, *Certaine wayes for the ordering of Soldiours* (1573); long prose passages are found in Titus Livy, John Florio, William Thomas, and at the front or end of many of Harvey's volumes.

[34] Some passages in his printed works derive almost word for word from his marginalia, e.g. in *Gratulationes Valdinenses* (1578) on sig. Dii, the page and a half of verse and introductory prose entitled '*Italorum duorum Xena Encomiastica*' is duplicated (with a few minor variations) almost word for word from notes of Harvey's, presumably written in 1577. They were noticed (inscribed in a copy of Harvey's works) by Thomas Baker and copied out by him in Baker MS. 36, fols. 109–10. There is a passage in *Pierces Supererogation* (1593), sig. G4<sup>r</sup>, which is apparently derived from notes in Harvey's copy of James VI's *Essayes of a prentise* (1585), sig. C4<sup>r</sup> (next to a translation by James VI of Du Bartas' '*Uranie*'). The annotations in Ramus's *Ciceronianus* (1557) and in Quintilian's *Institutiones Oratoriae* (1542) seem to have been used as source material for Harvey's lectures. Many other examples could be cited of passages in the marginalia being used as seminal material for Harvey's printed works.

# 2. Typical Kinds of Marginalia

IN a classification of Harvey's many notes by subject matter, the following extant volumes contain his more interesting and extensive observations in each field:

(1) *Oratory and Eloquent Written Expression*
Demosthenes, Erasmus, Quintilian.

(2) *History and Politics*
Livy, Foorth, Florio.

(3) *Modern Languages*
Florio, Holyband, Corro.

(4) *Gentlemanly Training*
Castiglione and Guazzo (manners and deportment). Blundevill (horsemanship). Frontinus, Machiavelli, Whitehorne (technique of warfare).

(5) *Jurisprudence*
Hopperus, Duarenus, Ramus's *Oikonomia*, Foorth.

(6) *Cosmology and Other Sciences and Pseudosciences*
Sacrobosco and Firminus (cosmography, astronomy, meteorology). Blagrave (navigation). Gaurico (astrology).

(7) *Medicine*
Hugkel, Bruele, Braunschweig.

(8) *Literature and Drama*
Gascoigne, Euripides, Dolce, Chaucer.

(9) *Philosophical Outlook and Personal Observations*
Domenichi, Guicciardini, Demosthenes, Foorth, Ramus's *Oikonomia*.[1]

Following are some characteristic comments in each of the above fields.

## ORATORY AND ELOQUENT WRITTEN EXPRESSION

Harvey's tiny vellum-bound volume of selections from Demosthenes[2] was one of his favourites, and he evidently carried it with him frequently,

---

[1] These and other Harvey-owned volumes are catalogued on pp. 198–241 of Part III which deals with his library.

[2] The diminutive octavo volume contains a number of books in continuous pagination: Demosthenes' *Gnomologiae* and *Similia* which are memorable sayings extracted from his

# PARABO-
## LAE, SIVE SIMI=
### LIA DES. ERASMI
Roterodami.

*Quas Ipse profitetur & se exquisitas Gemmas.*

*cum uocabularum liquot non ita uul-*
*garium explicatione.*

*Gabriel Harveius.*
*mense Ihnuario.*
*1566.*

EPI SCOP.

*A quibus nihil boni spero, quia nolunt:*
*ab ijs nihil mali metuo, quia nõ possunt.*

# BASILEAE, M. D. LXV.

*Vel Arte, uel Marte.*

Title-page of Erasmus, *Parabolae* (1565)

(By courtesy of the Folger Shakespeare Library)

PLATE A

*a.* Early Italic, *c.* 1573 (Sloane MS. 93, fol. 9ʳ)

*b.* Late Italic in 1598 Chaucer (fol. 422ᵛ)

(Both specimens reproduced by courtesy of the British Library Board)

PLATE B

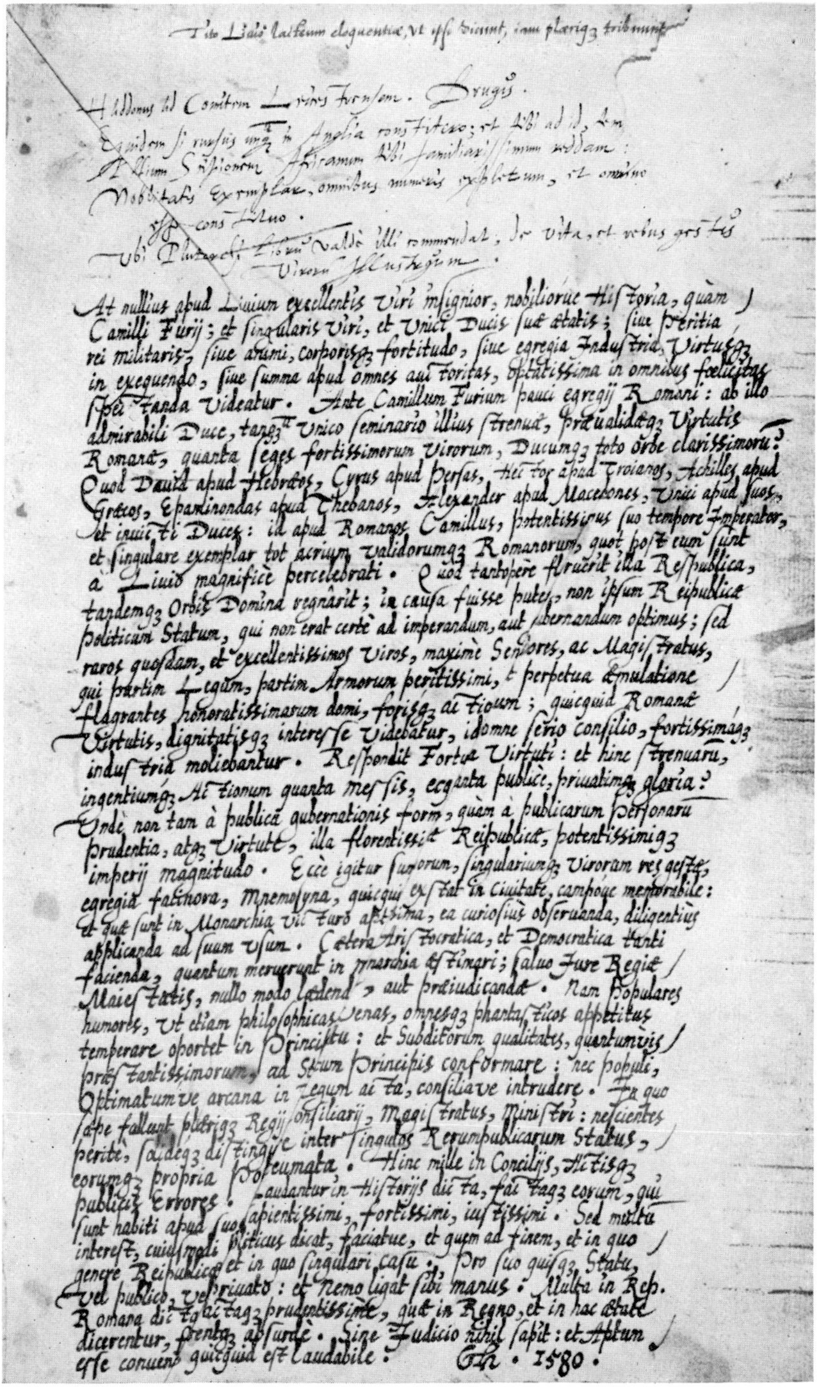

Flyleaf preceding title-page of folio Livy: early Secretary hand followed by later Italic

(By courtesy of a private collector)

PLATE C

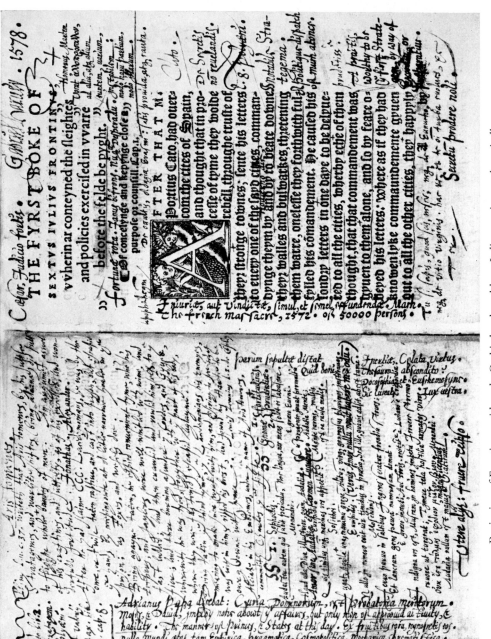

Beginning of Frontinus, with early Gothic hand, and later Italic superimposed vertically

(By courtesy of the Houghton Library, Harvard University)

PLATE D

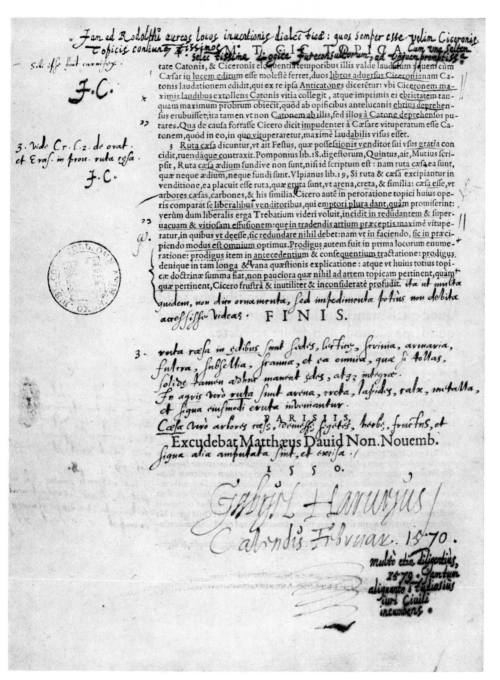

Final page of Cicero, *Topica* (1550), with dated 1570 signature and 1579 added notes
(By courtesy of the Warden and Fellows of All Souls College, Oxford)

PLATE E

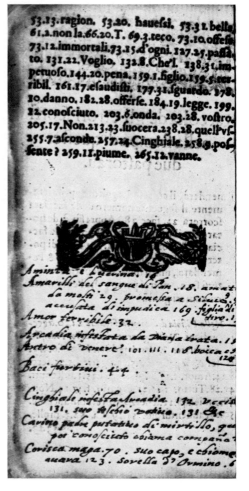

Sig. H9ᵛ of Guarini's
*Il Pastor Fido*

Part of the manuscript index
added by Harvey at the end of
Tasso's *Aminta* (sig. O3ᵛ)

Miniscule annotations in Harvey's hand on the tiny 2½ inch × 5 inch pages of a 1591 volume
(By courtesy of the Columbia University Libraries)

PLATE F

Title-page of Hopperus, showing
various hands and symbols

Final page of Hopperus,
with Ramus-type diagram

(Both specimens reproduced by courtesy of the British Library Board)

PLATE G

Typical page of Domenichi (fols. 5ᵛ–6r), with marginalia covering nearly all blank spaces

(By courtesy of the Folger Shakespeare Library)

PLATE H

for in the margin of sig. C2ʳ he writes of it, 'One of my pockettings, and familiar spirits'. If no other Greek authors were extant, he observes, he would miss none of their books provided he had this one, as it incorporates the complete strength and charm of the Greek language (and about no other writings except Cicero's orations has anything so strong been said, Harvey affirms). He comments that Demosthenes and Cicero, Bembo, and Sturm are parallel orators, each pre-eminent in his native tongue. When Cicero and Demosthenes are compared as orators, Demosthenes is found to be more concise and vigorous. Also cited as outstandingly eloquent are Epictetus, Isocrates, Dion, the 'Hymn to Apollo', Julian's *Misopogon*, Eunapius, Philostratus' *Sophistae*, and the 'exquisite judgement of Dionysius of Halicarnassus on Thucydides and on Demosthenes himself'.³

Within the marginalia in Erasmus' *Parabolae* inscribed in the same Italic hand as the 'Letter-Book' and therefore probably dating from the late 1570s is Harvey's conception of good and poor oratory: 'Nothing more pleasing than a brilliant and facile speech; than an affected and officious one, nothing more disagreeable. The sweetest eloquence flows smoothly but it should have neither too much honey nor too much salt'.⁴

In notes written on sig. T3ʳ of his Quintilian, Harvey comments that [Geoffrey] Chaucer, [Thomas] More, and [John] Jewel are three very vivacious talents and to these he adds three 'natural geniuses': [John] Heywood, [Philip] Sidney, and [Edmund] Spenser. Other illustrious English talents who are dear to him are [Thomas] Smith, [Roger] Ascham, [Thomas] Wilson, [Leonard] Digges, [Thomas] Blundeville, and [Richard] Hakluyt.⁵ Pietro Aretino's hyperbolical style was much

---

speeches and epistles; the spiritual *sententiae* of Bishop Gregory Nazianzen; the *Arithmologia Ethica*; and excerpts from Isocrates and others. Most of the marginalia are inscribed vertically (probably because of the tiny pages).

³ The Latin originals of the non-English Demosthenes marginalia described above are as follows:
'Si alii absunt Graeci auctores; nullis careo libris, cum hunc habeo unum.' (sig. c2ʳ)
'In *Demosthene*, quem crebro habebis in manibus, absoluta est Linguae Graecae vis, ac gratia: non aliter quam de *Ciceronis* orationibus diximus. *Isocrate* simplicius, ac purius cogitari nihil potest.' (verso of title-page)
'Demosthenes, et Cicero; Bembus et Sturmius, Oratores paralleli: singulares suae linguae.' (inside front cover)
'Cicerone multo strictior, nervosiorque Demosthenes.' (sig. a3ᵛ)
'1. Demosthenes, et Arithmologia. 2. Epictetus, et Isocrates. 3. Dion, et Apollinis Hymnus. 4. Juliani Caesares, et Misopogon. 4. Eunapii, et Philostrati Sophistae. Cum exquisito iudicio Halicarnassei de Thucydide, ipsoque Demost.' (sig. G5ᵛ)

⁴ The Latin passage in Erasmus referred to above is: 'Splendida, et facili oratione, nihil gratiosius: affectata, et curiosa, nihil putidius. Dulcissima Eloquentia fluit facillime: nec nimium habet mellis, nec parum salis'. (sig. h2ᵛ)

⁵ Harvey's Quintilian passages quoted above are the following: 'Tria vividissima

admired by Harvey, but he recognizes that it can only be used when suitable to the subject matter. On the last printed page of his Quintilian immediately following brief notes for the July 1578 disputation at Audley End he writes in English:

Unico Aretino[6]—in Italian, singular for rare and hyperbolical amplifications. He is a simple orator that cannot mount as high as the quality or quantity of his matter requireth. Vain and fantastical amplifications argue an idle brain. But when the very majesty and dignity of the matter itself will indeed bear out a stately and haughty style, there is no such trial of a gallant discourser and right orator. Always an especial regard to be had of decorum, as well for orators and all manner of parleys as in other actions.

### HISTORY AND POLITICS

An avid student of classical history, Harvey profusely annotated his copy of Livy's *Romanae Historiae Principis*; his dated signatures give evidence of the volume having been perused in 1568, 1580, and 1590. The most copious manuscript notes are found in the first of the three decades of Livy. This was read with Sir Thomas Smith's son shortly after Harvey's sixteenth birthday at which time he inscribed some few scattered notes in an irregular Secretary-type hand. In 1580, with Thomas Preston at Trinity Hall, Harvey re-read the First Decade along with Machiavelli's *Discorsi* in Italian. The abundant 1580 marginalia are carefully inscribed in a large, neat Italic hand as are the 1590 notes which can usually be distinguished by slightly more rounded letter forms. Marginalia on sig. h5$^r$ at the end of the First Decade's first three books state that these books together with various commentaries on them were 'intimately' discussed by Harvey with Philip Sidney just before the latter's mission as the Queen's representative to the Emperor Rudolph II (in March 1576/7).[7] This provides interesting confirmation of Harvey's more than casual acquaintance with Sidney. Harvey may have been considered somewhat of an authority

Britannorum ingenia, Chaucerus, Morus, Juellus. Quibus addo tres florentissimas indoles; Heiuodum, Sidneium, Spencerum. Qui quaerit illustriora Anglorum ingenia, invenie, obscuriora. Perpaucos excipio; eorumque primos, Smithum, Aschamum, Vilsonumt Diggesium, Blundevilum, Hacluitum, mea corcula'. (sig. T3$^r$)

[6] Harvey frequently uses the adjective 'unico' (unique) to describe Pietro Aretino as he apparently considered it a very apt epithet. It is not known whether or not he was aware that the early Italian lyric poet Bernardo Accolti was often called 'l'Unico Aretino'.

[7] The marginalia read: 'Hos tres Livii libros, Philippus Sidneius aulicus, et ego intime contuleramus, qua potuimus politica analysi ultro, citroque excussos: paulo ante suam Legationem ad Imperatorem, Rodolphum II. Cui profectus est regineo nomine honorifice congratulatum; iam tum creato Imperatori summus noster respectus erat ad rerumpublicarum speties; et personarum conditiones, actionumque qualitates. De Glareani, aliorumque annotationibus parum curabamus'.

on political history, for elsewhere in Livy (sig. aa1ʳ) he notes that Sir Edward Dyer and Sir Edward Dennie thanked him for information he had given them.[8]

This handsome folio edition of Livy printed in 1555 contains frequent memoranda by Harvey on recommended readings in the field of Greek and Roman history. After commenting that knowledge of classical history assembles for the learned man a continuous chain of events in uninterrupted succession (by implication useful in understanding the present), he lists the sources of such a chain, starting with Herodotus who began his history at the end of the prophets, Thucydides at the end of Herodotus, Xenophon at the end of Thucydides. Harvey observes that the early part of Roman history was recorded by Dionysius of Halicarnassus, the next portion by Polybius, and the third and most forceful part by Livy. Tacitus began his *Annals* shortly after the end of Livy's account, Marcellinus proceeded from the end of Tacitus (i.e. from the ascendancy of Nerva) and Herodian from the death of Marcus Antoninus at which Dion stops. Harvey admires Livy's fine style, his knowledge of Roman jurisprudence, and his pious treatment of subject-matter; in Tacitus, also, Harvey finds noteworthy the political and legal, and a depiction of the vigour of Roman courage as well as its prudence, temperance, and justice.[9]

[8] 'Certe non nihil lucis a Lodovici Regii Commentariis in Aristotelis politica; arcanis Bodini libris de Republica, et Methodo historica; equitis Poncetti in Gallica aula Turcicis; politicis Sansovini maximis; novissime Althusii, Lipsiique politicis; paucis aliis et prudentes decet, eo, quod est politicae lucis, enixe uti. Extendere etiam quo ad licet. Pro hac politica historicaque animadversione, magnas mihi gratias egerunt duo praeclari aulici; Eques Eduardus Dierus, et Eques Eduardus Denneius. Sed res ipsa agat gratias, penitus probata: nec quicquam vehementius opto, quam vivam, efficacemque summarum historiarum politicam analysin. Praesertim, cum Annibal, et Scipio; Marius, et Sylla; Pompeius, et Caesar in flore'.

[9] The marginalia on sig. S3ᵛ from which the above is derived read: 'Scire Graecae, Romanaeque historiae continuatem seriem, valde conducit polyhistori. Herodotus historiam suam est orsus a fine prophetarum: Thucydides a fine Herodoti; Xenophon a fine Thucydidis. Romanae historiae primam partem praecipue suscepit Halicarnasseus; professus eam se historiam velle producere usque ad initium primi belli Punici. Secundam partem antea suscepit Polybius. Postea Livius tertiam potissimum partem explicare voluit. Tacitus non procul a fine Livii suos Annales orditur: Marcellinus a fine Taciti, id est, a principatu Nervae: ut Herodianus ab exitu M. Antonini, in quo desiit Dion. Nec est difficile, talia continuae successionis exempla in aliis plaerisque historicis observare.

Duo autem Romanae historiae doctissimi, et diligentissimi scriptores, Dionysius [Dionysius of Halicarnassus who wrote *Roman Antiquities*, referred to above as Halicarnasseus], et Livius. Quod attinet ad res, et tempora, Dionysii antefero diligentiam: verba a Livio malo repetere, et sententias. Neque enim minima historiae, ipsiusque iurisprudentiae Romanae pars, visque sita est in ipsis formulis; quas ubique Livius diserte recitat, atque adeo religiose: Dionysius, et reliqui Graeci sua lingua exprimere non possent, ne si vellent quidem maxime. Unde politicum decet, in primisque iurisconsultum, Livianae phrasis maxime omnium esse observantem: ut postea Taciteae'. (Probably written in 1580.)

No historian has such a fund of civil law and so much material for forensic use as does Livy, and after him Tacitus or experienced Frontinus. Although Harvey is quick to recognize worth, he is impatient with those whom he considers second-rate historians. He notes that among the more recent history writers (Francesco)Guicciardini, (Theodore) Zwinger, (David) Chytraeus, and some few others are of pragmatic value. 'But today there are many asses who dare to compile histories, chronicles, annals, commentaries as, for example, Grafton, Stowe, Holinshed, and a few others like them who are not cognizant of law or politics, nor of the art of depicting character, nor are they in any way learned. How long shall we yearn for a British Livy?' he queries. 'Or when will there emerge a British Tacitus or Frontinus? We abound in petty things but how dry is our record of matters of pragmatic or military import. Almost the only men who approach former art and industry are More and Ascham. After all, even Camden, Hakluyt, . . .' (the passage is unfinished).[10]

Harvey's marginalia evince his interest in the character traits of outstanding men and what caused them to succeed. Those he most admires are carefully studied, presumably in an attempt to mould his own personality. In the 'First Decade' he comments in 1580 in English on Hannibal: 'Annibal a craftie Foxe & even for theise dayes a notable example. The finest Politicians, or Pragmaticians may finde in Livie to serve their turne'. (sig. ooo3ʳ). Eminent men who have failed in any way can teach Harvey other lessons:

Worthie Pescenninus tooke exceptions to the untrained & unexercised youth of Scipio: but might not just exception be taken to his fading and drouping age: when right magnanimitie, & pollicie might have made him admirable? His frend Masinissa, & his adversarie Cato, atchieved great actions, when they were mutch elder, then Scipio was, when he dyed. Which may seeme to die in the melancholie moode of a drouping malcontent. A base end for so brave a captaine, & so worthie a man.   (sig. mmm2ᵛ)

Harvey observes that he who knows Livy knows the world.[11] He further comments that the study of texts like Livy lead him to do more

---

[10] 'Nullus historicorum tantum habet civilis iurisconsulti, aut forensis pragmatici, quam Livius, et post eum Tacitus, callidusque Frontinus. Novissime etiam Guicciardinus, Zwingerus, Chytraeus, alii perpauci nonnihil habent pragmatici. Hodie multi asini ausi sunt componere historias, chronica, annales, commentarios: ut fere Graftonus, Stous, Holinshedus, non pauci tales: nec iurisconsulti, nec politici, nec ethologi, nec ullo modo eruditi. Livium Britannicum quam diu desiderabimus? Ah quando exstabit Tacitus, Frontinusque Britannicus? Nugis affluimus: pragmaticarum, bellicarumve rerum quam ieiuna memoria? Solus fere Morus, et Aschamus nonnihil attigerunt pristinae artis, et industriae. Tandem etiam Cambdenus, Hacluitus, . . .'  (sig. A1ʳ, following 1580 signature)

[11] 'Novit mundum, qui novit Livium'.  (sig. aaa2ᵛ)

and more reading for their practical worldly wisdom and to forgo writing and bend his efforts toward becoming a man of action (not of letters).[12]

Harvey's books and marginalia indicate that he provided himself with a good background in classical and in biblical history and in Renaissance events and politics as well, for it was through such studies and historical biographies that he hoped to gain the insight and resourcefulness to himself become a great man, a 'megalander'. At about this time he had written in his Quintilian: 'Whatever may occur in the course of human affairs, Eutrapelus [i.e. Harvey] always is a great man. . . . Without great knowledge and enormous virtue no one is a great man, for the highest knowledge requires a loftier spirit and is animated by enduring virtue'.[13]

He points out that great men are usually distinguished orators:

Almost all great men were outstanding orators either by nature or by art. What sort under King Henry VIII? Cardinal Wolsey, Prorex Cromwell, Chancellor More, the pragmatic Gardiner: four heroic councillors. Under Prince Edward VI, the Duke of Northumberland, Archbishop Cranmer, Secretary Smith, Cheke the pedagogue. Under Queen Elizabeth, Smith a Cineas, Cecil a Nestor, Bacon a Scaevola, Essex an Achilles. How many courtiers and city dwellers, Ciceronians and Virgilians, Columbuses or Sforzas?[14]

Many of Harvey's notes are related to the study of statesmanship. In Freigius's *Mosaicus* (1583), an account of various prominent biblical figures from Adam to Moses, Harvey writes at the bottom of sig. p8$^r$ that without religion it is impossible either to found a new state or to destroy an old one. Machiavellians and Atheists believe that they can govern republics while allowing neglect and contempt of religion to

---

[12] The same sentiment (to act not to write) is found in many other examples of marginalia. Here it is implicit in the annotation on sig. rrr6$^r$ of Livy: 'Minus scriptionis plus plusque lectionis mihi conducit, expedit actori. Ecce Livius ipse instar omnium notarum scholae, aut observationum mundi'.

[13] 'Quicquid humanitus acciderit, Eutrapelus semper megalander. . . . Sine magna scientia, et ingenti virtute, nemo megalander. Summa enim scientia excelsiorem requirit spiritum: et vivaci virtute animata, invicte corroboratur. Nec tali scientia, tantaque virtute quicquam exstat in ambitioso mundo eminentius'. (sig. T2$^r$ and T3$^r$)

[14] 'Omnes fere Megalandri, egregii erant vel natura, vel arte Oratores. Quales sub rege Henrico 8$^o$: Cardinalis Volsaeus: Prorex Cromvellus: Cancellarius, Morus: pragmaticus, Gardinerus: quatuor heroici Consiliarii. Sub principle Edouardo 6$^{to}$: Dux Northumbrius: archiepiscopus Cranmerus: secretarius Smithus: Checus paedagogus. Sub regina Elizabetha, Smithus Cineas; Cecilius Nestor; Baconus, Scaevola; Essexius, Achilles. Quot aulici urbicique, Cicerones, et Virgilii: Columbi et Sfortiae?' (sig. T3$^v$). Harvey considers Smith like Cineas, a peacemaker and a man of retentive memory, Cecil a wise man, Bacon a wealthy patron, Essex a brave warrior, but perhaps an impetuous one.

take place. How impossible it is politically to sustain for long a public or private state without some divine worship![15]

Like many of his contemporaries, Harvey advocates the active not the contemplative life and on the title-page of his copy of Joannis Foorth's *Synopsis Politica* (1582) writes 'Il pensare non importa, ma il fare'. In the course of his studies he observed that certain men in the stream of history ('Megalandri', as he terms them) were better able than others to make their mark. He became eager to comprehend the inherent qualities and outward actions which were involved. On sig. A1ᵛ of Foorth he notes: 'But fower right politiques of late memory: Wulsey, Crumwell: Gardiner: & Cicill. All the rest, children in comparison. But noovices, & pupills in pollicy. Incipients: not perficients'. One finds a number of similar comments elsewhere on Stephen Gardiner and on William Cecil. The former seems to have been of especial interest to Harvey because Sir Thomas Smith had made so many personal observations about him. As Harvey records (in John Florio's *First Fruites*, 1578, sig. Tt1ᵛ), Smith was 'a great adversary & frend' of Bishop Gardiner's. The last thirteen pages of the Florio volume are almost completely filled with closely written notes inscribed by Harvey perhaps with the intention of putting forth a life of Gardiner, for at the top of sig. Ssiiiiʳ in Harvey's Italic hand is captioned 'The politique history of Doctor Stephen Gardiner, bishop of Winchester, & afterward L. Chancelour of Ingland'. Below this is appended: 'Dʳ Gardiner of manie surnamed the Foxe: Dʳ Wootton [Nicholas Wotton, Secretary of State under Henry VIII] the Ape, Wootton had the text, & glosse of the Lawe bye hart verbatim: Gardiner the matter, & substance. Two pregnant advocats in anie dowtfull or subtile case of whatsoever importance.' Because of craftiness and deceitfulness Gardiner was frequently characterized by his contemporaries as 'the fox'[16] but the linking of a fox and an 'ape' (in this case Wotton) as two shrewd and designing 'advocats' suggests that Harvey may have discussed the fox and ape of 'Mother Hubberds Tale' with Edmund Spenser at least five years in advance of its publication in 1591. The date '1580' is inscribed by Harvey on sig. Ss3ᵛ of the Florio volume at the 'Finis' of the major portion of the

---

[15] 'Impossibile est, aut novam politeian fundare, aut veterem tueri, *sine Religione*: in quo, plaerique Machiavellitae, et Athei, phantastice, ac pueriliter sapiunt, somniantes se posse, non obstante Religionis neglectu, aut contemptu, Respublicas, atque Regna politice gubernare: cum experti omnes, satis superque senserint, quam plane ἀδύνατον sit, sine divino aliquo cultu, vel publicam Maiestatem, vel etiam privatum aliquem statum, diu sustinere'.

[16] On sig. Tt1ᵛ Harvey writes: 'Gardiner, a crafty deceytfull fox, thorowghout all the predicaments as was lately said by a French gentleman, & of the Queen Moother, & County Rhetz'.

text, the section containing chiefly Harvey's comments on language study. The last thirteen pages (at the beginning of which on sig. Ssiiii[r] is the above allusion) are datable after 1585 by certain of their contents.[17]

The closely written notes on Gardiner's life are described by Harvey in a vertically written inscription on sig. Ssiiii[r]: 'Winchester, pro, et contra [the two subsequent words are illegible], Language, Lawe, Industrie, and confidence: temperance, & exercise. His singular perfection, experience, and cunning practis in the affaires of the world'.[18] At the bottom of this page Harvey has written horizontally: 'Exemplarie patterns for imitation, or observation. J.C. [Jurisconsultus]'. The ensuing notes relate details of Gardiner's life, together with some of his apt sayings,[19] quotations from his letters, and judgements about him by some of his contemporaries. Harvey attempts to remain an impartial critic, for he includes both the laudatory and the censurable.[20]

On sig. Tt2[r] is an account of the 'surrendring of the college of Trinityhall in Cambr. wherof the said B[ishop]. was master'. (A portion of the notes on this and some subsequent pages are lost since Harvey has here written to the page edges and these, unfortunately, have been reduced when the volume was rebound.) On this same page inscribed vertically is the provocative assertion: 'Ad mnemosyna Gabrielis Harveii regia, nobilia,

---

[17] One inscription (unconnected with the life of Gardiner) reads: 'Owre brave men now in flore, & owre arch-discoverers at sea and land. Sir Humfrie Gilbert [died 1583 but his reputation would still have been very much "in flore"], Sir John Hawkins, Sir Martin Forbisher [i.e. Frobisher], Sir Francis Drake, Jason Candish [Thomas Cavendish?], Sir Walter Ralegh, Sir Richard Grenvile, Captain Carlell [Christopher Carleill], M. Heriot [Thomas Hariot whose American expedition was in 1585], & such like valorous adventurers, brave commanders of golden Fortune with more golden vertu. (sig. Tt3[r])'. The vertical inscription on sig. Tt2[r] (see pp. 155–6), must have been written in 1585 or after if, as seems likely, it relates to the loss of Harvey's Mastership of Trinity Hall.

[18] At the top of sig. Tt1[r] Harvey again summarizes: 'The most memorable pointes in the *famous Historie* of *Stephen Gardiner*. . . . A man reputed singularly wise, politique, & learned: especially in *Lawe*, and matters of *state*. A man much emploied in greatest *Councells*, *Ambassages, Judgments*, and all Occurrences *of state*: . . . next his master, *Cardinall Wolsey*: even aboove his fellow, the famous Lord Cromwell: an other rare politician in his kinde, & Experimenter of Fortune, by a singular Industry, & Audacity. Two notable & rare men in their desines, & practices'.

[19] Some of Gardiner's sayings are found elsewhere in the marginalia, as, for instance, his amusing remarks to Cromwell (sig. S2[r] of this volume): 'Every cuntry hath his peculiar inclination to nawghtines: Ingland, & Germany to the Belly; the on in liquor, th'other in meate: France a lyttle beneath the Belly: Italy to vanity, & pleasures devised. Lett an Inglish Belly have a further advauncement, & nothing can stay it'.

[20] Re the imprisonment in the Tower of the Lady Elizabeth (the future Queen): 'A writt cam down from certain of the counsell for her execution, it is owt of controversy but that wily Winchester was the only Dacdalus, & framer of that ingin. Who no dowt in that on day had browght this whole Realme in wofull ruine, had not M[r] Brydges, then Lieutenant . . . preventid Achitophel's Bluddy devises'. (sig. Uui[v])

generosa, popularia, omnique modo pragmatica. Multa popularia: nihil vulgare.'[21] Did the ousting of the belligerent Gardiner from Trinity Hall recall to Harvey his own milder, more gentlemanly behaviour when by the devious manipulations of the 'Heads' he was prevented from becoming Master?

### MODERN LANGUAGES

Florio's teaching manual for Italian conversation *First Fruites* (1578) is, despite the previously discussed extensive marginalia dealing with Stephen Gardiner, primarily annotated with Harvey's references to the study of modern languages. On sig. Eei[v] he has written: 'Florio, & Eliot[22] mie new London companions for Italian, & French. Two of the best for both'. Though in youth Harvey had become well versed in Latin and in Greek and seems to have used both languages with ease in his marginalia,[23] about 1578 he decided to undertake the study of Italian and French, as knowledge of these tongues was an important attribute for life at court; he admired the fluent Italian of Leicester, Hatton, and Sidney and at this time worked hard to become proficient himself. On sig. Aiii[r] of his Florio, his marginalia bemoan the fact that he ('Axiophilus') does not have the same dexterity as these outstanding English courtiers. He has learned the principles of law, he says, in three days, why is it not possible in two or three days to learn the Italian language which is half akin to the Latin which he knows so well? He who has the face of an Italian, as the Queen recently remarked, why should he not also have the mouth and tongue of an Italian? He now addresses Florio: 'How often have you instantaneously created blossoming Italians?' (Harvey uses the word '*florentes*', a pun on Florio's name.) He continues: 'Florio and [William] Thomas in close connection will intensely inspire me with their language. This I shall learn, for where love exists there does the eye fasten itself. . . . Repeat, repeat as the ardent trainer of a gladiator would do!'[24]

---

[21] 'As memorials of Gabriel Harvey, regal, noble, gentlemanly ones, of the people, pragmatic in every way, very much of the people but not vulgar'.

[22] John Eliot, a teacher of French, was author of *Ortho-epia Gallica* (1593) which Harvey later owned. Eliot was one of the reader-translators who in the 1590s worked in John Wolfe's printing house.

[23] The greater part of Harvey's extant marginalia are in Latin, but Greek words and phrases are used freely throughout and his copy of Aristotle's *Rhetoric* is annotated almost completely in Greek.

[24] The passage is as follows: 'Quomodo Comes Leicestrensis, Dominus Hattonus, Eques Sidneius, multique praeclari Aulici nostrates fluentissime loquuntur Linguam Italicam. Cur non Axiophilus eadem iam iamque dexteritate? Triduo ille J.C. Cur non ego biduo, aut triduo Italus semilatinus? Qui vultum habet Itali, ut aiebat nuper Regina: cur non etiam

Despite Harvey's dissatisfaction with his progress in Italian, in 1580 he managed to read the 'First Decade' of Machiavelli's *Discorsi* and in numerous Italian texts subsequently inscribed marginalia in Italian.[25] During his period of assiduous language study he had at hand two Italian grammars (William Thomas's, 1550, and that of Scipio Lentulo, 1575, purchased in 1579) and Claudius Holyband's *Arnalt & Lucenda with certen rules and dialogues for the learner of th'Italian tong* (1575), annotated by Harvey in 1582. At this period he also acquired several books for the study of French: Claudius Holyband's *French Littleton* (1576), a book of grammar and dialogues; Pierre Du Ploiche's *Treatise in English and French* (1578), a gift from the author to Harvey and annotated by him in 1580; and *Images of the Old Testament* (1549), a little volume of twenty-four woodcuts with captions in French and English. This popular book helped to teach simple French and was acquired by Harvey in 1580. Although he read Rabelais and several other French authors in the original, it is unlikely that Harvey's mastery of this language approached that of Italian, for one finds very few marginalia or phrases in French. In the early 1590s, he studied Spanish with the aid of Antonio de Corro's *Spanish Grammer* (1590) and Richard Perceval's *Bibliotheca Hispanica* (1591), an English, Spanish, Latin dictionary printed as an adjunct to Corro's grammar. On sig. S4[v] of his Corro, Harvey commends a number of Spanish books some of which would have been available to him only in the original language. However, one finds no annotations in Spanish other than an occasional phrase like '*Poco, y bueno* [slowly and well]' which he frequently uses as advice in studying, especially in the case of languages.

In Pierre Du Ploiche on sig. A2[v] he writes about the proper way to master languages: 'A paradox in lerning: Quo plus, eo minus. [By whatever more, by this the less.] Beginners must not leap over hastely, lest they overleape all. Apt & reddy pronunciation of the Alphabet, one weeks exercise'. On sig. H3[r] he records his progress in French: 'The eight Chapter is of verbes impersonall, & personall. This with the first [the a, b, c in French] will serve for good part of the grammer. Pronunciation & the verbs perfetly lernid: little other grammer needed. My homogeneal Dictionary,[26] with daily reading & speaking will soone supply the rest'. Underlinings of the following text of Corro's *Spanish Grammer*

---

os, et linguam? Florio quot fecit ex tempore florentes Italos? Me Florio et Tomaso contesti inspirabunt, nobis linguis flagrantem. Hoc age. Ubi amor, ibi oculus. . . . Repete, repete; ut fervidus lanista!'

[25] This is true of Guicciardini, Domenichi, Porcacchi, Guazzo, etc.

[26] Apparently no longer extant, unless perhaps he is referring to Perceval's *Bibliotheca*.

(sig. D1ʳ) suggest that Harvey may have found the pronunciation of French difficult. His underlinings are printed in italics:

any *man may easily learne to read the Spanish toong*. But as for the *French toong*, I would exhort them that are willing and desirous to learne the French toong, to *take the lectures and helpe* of *some naturall borne French man*: And then they may of him learne how *distinctly to read and pronounce* the French toong. Which I take to *be so difficult* for many reasons, that I thinke a man shall or *may hardly* (nay and scarce so too) learne the pronunciation thereof, unlesse *he heare a French man pronounce* it.

Harvey was much impressed with the Queen's multilingual proficiency. Next to the printed text of *First Fruites* (on sig. Ciiiᵛ) where Florio refers to the Queen's speaking eight languages including Scottish and Flemish, Harvey notes: 'The Queen commendid by Utenhovius,[27] and M. Ascham, not only for her Latin, & Greek, but also for her French, Italian, & Spanish. Tam ad loquendum quam ad intelligendum. [As accomplished in speaking as in comprehending.]'

### TRAINING IN GENTLEMANLY MANNERS

His ambition to shape himself for a milieu at court led Harvey not only to seek skill in languages but courtly manners and arts as well. In Hoby's translation of Castiglione's *Courtier* on sig. E2ᵛ he notes in about 1580: 'A Courtier must do, & speak everie thing aswell, as possibly he can: yet with such a dexteritie, & such a negligent diligence, that all may think, he might do much better, if he woold. Summa summarum. [The highest of the highest.]' However, this negligent diligence or *sprezzatura* seems to have been alien to Harvey's nature. Whatever he had learned to do well generated in him a natural pride which is evident throughout his marginalia and probably was made apparent to those around him.

On sig. Yy3ᵛ he sets down aphorisms for the courtier:

Aphorismi Aulici.
 { No excellent grace, or fine cumlie behaviour without three cunning pro-
 { perties; a sound judgment to informe; an apt dexteritie to conforme; & an
 { earnest intention to performe.
The rarest men extend their utterest possibilitie, with a fine (as it were) familiar sleight: & they that do not enforce themselves to display their best, cum ever short of their reckoning.

[27] Charles Utenhove, the multilingual poet, was a Flemish resident of England. For a brief biographical sketch, see Van Dorsten, *Poets, Patrons, and Professors*, p. 16 n. 2. Also see p. 46 above.

At the end of the translation Harvey sums up Castiglione's portrayal of the characteristics to be desired in a courtier:

Above all things it importeth a Courtier, to be gracefull & lovelie in countenance & behaviour; fine & discreet in discourse, & interteinment; skilfull & expert in Letters, & Armes; active & gallant in everie Courtlie Exercise; nimble & speedie of boddie & mind; resolute, industrious & valorous in action as profound & invincible in execution, as is possible: & withall ever generously bould, wittily pleasant, & full of life in all his sayings & doings. His apparrel must be like himself, cumlie & handsom; fine & clenlie to avoid contempt, but not gorgeous or statelie to incurr envie, or suspicion of pride, vanitie, selfloove, or other imperfection. Both inside, & outside, must be a faire paterne of worthie, fine, & Loovelie Vertu.

GH. 1580.

Nimbleness of body is further commented upon on sig. Zz1 next to Castiglione's listing of 'marciall feats both on horsebacke and a foote' and other physical activities in which the courtier should be skilful. Castiglione puts first 'To play well at fense upon all kinde of weapons', and upon this Harvey comments, 'Ars Grassi', an allusion to Giacomo di Grassi's *Ragione di adoprar sicuramente l'Arme si da offesa come da difesa* (1570). Castiglione's listing of gymnastic activities for the courtier 'privilye with himselfe alone, or among hys friendes and familiars' includes 'to be nimble and quicke at the play of tenise, to hunt and hawke, to ride and manege wel his horse, to be a good horsman for every saddle'. Next to the last two items Harvey notes 'Ars Blundevili'. This refers to his much annotated copy of Thomas Blundevill's *The fower chiefest Offices belonging to Horsemanship. . . . The office of the Breeder; of the Rider; of the Keeper; and of the Herrer* (1580). Its marginalia indicate that it was carefully read and studied.[28] On sig. Zz1ᵛ of *The Courtier* Harvey observes next to the continued listing of exercises in which the courtier should be proficient: 'All such exercises, honorable for a Gentleman: necessary for every right active man'. The printed text continues: 'to swimmen wel:' and here Harvey notes 'Ars Digbei', a reference to Everard Digby's *De arte natandi* (1587). Next follows: 'to leape wel, to renn wel, to vaute wel, to wrastle wel, to cast the stone, to cast the barr'.

---

[28] On sig. A1ᵛ of Blundevill, Harvey inscribes: 'I use Mʳ Astley [John Astley's *The Art of Riding* (1584)] for the compendious, & fine Art: and Mʳ Blundevil for the larger & fuller Discourses upon the Art: & commend them both, for two right-proffitable, & gallant Writers, in the excellent veine of Xenophon. A necessary, & singular preparative to the brave Art of Warfare: & one of the noblest exercises in the world. It importeth a Courtier, to be a perfect horseman. sᵃ [see above]. Chivalrie, & Armes, his principal, & singular profession, above all other qualities.'

Harvey comments only on the last item about which he observes, 'Heroica Gymnastica apud Homerum'. The ensuing list of seven 'thinges in open syght to delyte the commune peple withall' includes tilting, tourneying, fighting at barriers, flinging a spear or dart, etc. and is without comment. Although probably an active man, he apparently did not indulge in any of the courtly displays of prowess.

A good part of Book Two of *The Courtier* is devoted to courtly jesting and this seems to have been of particular interest to Harvey. At the top of sig. R3<sup>r</sup> he inscribes the caption: 'The Art of Jesting: pleasurable, & gratious'. At the top of sig. T3<sup>r</sup> he observes that there is the greatest and most ingenious use in the world for salty sayings. Outstanding in this respect, he believes, are those of Aristippus and Diogenes.[29] On sig. Y2<sup>r</sup> Harvey praises Pietro Aretino's use of jests: 'In derriving mens opinions, and frustrating the most probable expectation; Unico [i.e. Pietro] Aretino superexcellent. . . . Without any offence, & with many delights.' On sig. K3<sup>r</sup> he has quoted Lucian the Rhetor's opinion that in jesting Plato is cold and Isocrates is silly whereas Demosthenes is skilled in the art of pleasing.

The importance which Harvey attached to skill in jesting is confirmed by the quantity of his marginalia on this subject. In Thomas Wilson's *Arte of Rhetorike* (1567) on sig. 15<sup>r</sup> he especially praises Wilson and notes other texts which have been helpful to him in learning this skill: 'One of mie best for jesting: next Tullie, Quintilian, the Courtier in Italian, the fourth of mensa philosoph.[30] Of all, the shortest, & most familiar, owr Wilson.' And on sig. 15<sup>v</sup> below the printed marginal caption, 'Mirthe how many waies it is moved', Harvey notes some experts in the art: 'Few delitiae Atticae. Wilson of Cambridg: & Jewel of Oxford then, since Clark of Cambridg: & Tobie Matthew of Oxford. Before, More, & Heywood of London. The first, & last, Chaucer, & Sidney. Fine, & sweet men: almost like Tullie, & Cesar'.[31]

In Lodovico Guicciardini's *Detti et Fatti* (1571) on fol. 82<sup>v</sup>, Harvey writes of the pragmatic metamorphosis of Eutrapelus (i.e. of Harvey):

He likes to bring into jest the salt of the earth and the light of the world both very savory and very splendid. Other men present serious discourse; only Eutrapelus pursues serious matters and effects distinguished ones. Eutrapelus

---

[29] 'Salsorum dictorum, usus in mundo maximus, et ingeniosissimus. In quo genere excellent arguta dicta Aristippi, et Diogenis.'

[30] A reference to the fourth book of *Mensa Philosophica* of which Th. Anguilbert is thought to have been the author. See p. 199, below.

[31] In addition to Wilson, the men to whom Harvey is referring are: John Jewel, Bartholomew Clerke, Sir Tobie Matthew, Sir Thomas More, and John Heywood.

changes great matters into small ones, small into great ones. This is Eutrapelus's secret metamorphosis: serious matters of others must be converted into jests....[32]

On fol. 83ʳ he advocates the combination of urbanity and political skill: Many are urbane, not skilled in civil affairs: many skilled in civil affairs but not urbane. Give me the urbane man skilled in civil affairs, as much politically oriented as elegant: the only combination of cunning wit and stratagems, one who is capable of being of best use at that moment. A hook of favour and fit occasion should always be hanging down to you.[33]

In the 'Tavola' of Guazzo's *La Civil Conversatione* (1581) Harvey aptly summarizes his ideal of urbane behaviour by quoting and commenting on the proverb:

Play with me & hurt me not: ⎫    A notable rule
Jest with me & shame me not. ⎭    of Civilitie.

Aware that courtesy and civility preclude any display of anger, he advises on sig. M2ʳ of Hoby's *Courtier*: 'That man whom reason, not anger, rules is closest to the gods'.[34] In Foorth's *Synopsis Politica* (1582) on sig. C8ᵛ Harvey makes his point even more strongly: 'To be angry [is] rude, bestial; I have said too little; surely, indeed, it is barbarous and profane to be angry in any way, unless by chance in simulation or ironically'.[35]

A very different branch of Harvey's marginalia, also oriented toward possible service at court, has to do with his study of the technique of warfare. Extensive notes in this area are found in his copies of Frontinus, Machiavelli (Peter Whitehorne's 1573 translation of the *Arte of Warre*), and Whitehorne's *Certain wayes for the ordering of Soldiours* (1574).

In 1578 Harvey purchases *The strategemes, sleyghtes, and policies of warre gathered together, by S. Julius Frontinus, and translated into Englysshe, by*

[32] The complete original of the partial and free translation above is as follows: 'Eutrapeli Pragmatica Metamorphosis. Iuvat in iocis haurire salem terrae, et lucem mundi, utrunque sapidissimum, et splendidissimum. Alii seria praetendunt: solus Eutrapelus seria exequitur perpetrat egregia. Magna in parva mutat Eutrapelus: parva in magna. Arcana metamorphosis Eutrapeli. Aliorum seria, in iocos convertenda: tui ipsius Ioci in seria. Alienis addatur hyperbolicos, aut ironicos tuis detrahatur: sed praeter tua, todos es nada [everything is nothing]. (Musa magis floriat, quam vir.) Malim eos Auctores, in quibus nihil deest, quam eos, in quibus multa redundant'.

[33] 'Multi urbani, non pragmatici: multi pragmatici, non urbani. Da mihi urbanum pragmaticum: tam politicum quam facetum: solum combinatorem salium, et stratagematum: ad optimum qui esse potest, praesentem usum. Semper tibi pendeat Hamus Favoris, et Commoditatis'. (i.e. someone's aid and a timely opportunity should always be accessible to you.)

[34] 'Diis proximus ille est, quem Ratio, non Ira reget'.

[35] 'Irasci, rude, et ferinum: parum dixi: certe quidem barbarum est, et prophanum, ullo modo irasci; nisi forte simulato, et Ironice'.

*Richard Morysine* (1539). On sig. A3ʳ of this copiously annotated volume Harvey notes:

It is almost unpossible in conveying intelligence, to escape al casualties of miscarriage. Therefore it neerly, & deeply concernith every Intelligentiar, fitt to be imployed in secret affaires, to provide with all possible circumspection & cautele, even in every respect, & circumstance, subject to any jeoperdy. The only surist way, to prevent al circumvention, and to avoide all incident daunger. So procede, so guarde.

In 1580 Harvey purchases the English translation by Whitehorne of Niccolò Machiavelli's *Arte of Warre* (1573) and on the title-page verso writes:

The Art of Defence, newly published in Italian.³⁶ A necessary preamble to the Art of Warr. The Art of Defence: The Art of Riding: the Art of Navigation:³⁷ three necessary artes for every publique, or private actor in the world; that intendeth to travail, or to do any thing abroade. And if he purposeth to sturr in greater actions; he must not want the Arte of Warfare: with the bravest new Inventions of Pyrotechny, Fortification, & such like. And also with the new Aggiunta of Gandino to the Strategemes of Frontine and Polyen [Frontinus and Polyaenus].

At the bottom of sig. A2ʳ Harvey sets down:

Machiavel, a fine master of a brave Art: after the terrible Invention of gunnes unknown to Vegetius.³⁸ . . . That Courage is well grownded, that is grownded upon cunning, & practise; strength, & nimblenes. The Roman foundation of magnanimitie. The lesse strength; the more cunning, practis, & agilitie necessarie: with extraordinarie courage.

Having examined the bungled 'campaign' of the Earl of Leicester in the Netherlands with Sir John Norreys as military chief, Harvey compares the English morale with the better discipline of the Spanish as reported by Sir Roger Williams in his *Brief Discourse of War* (1590). On sig. Ee1ʳ of Machiavelli's *Arte of Warre* Harvey writes:

Owr Inglish militar Discipline under General Norris, in the Dialogue intituled the Castle of pollicy: under the Earle of Leicester. In his own lawes, & ordinances The Spanish Discipline under the Duke d'Alva, & the Prince of Parma, the best Disciplin now in esse, newly discouvred by Sir Roger Williams.

³⁶ Grassi's text (see p. 159, above), published in Venice in 1570.

³⁷ On sig. S4ᵛ of Corro's *Spanish Grammer*, Harvey refers to 'Cortesii, et Medinae Libri Hispanici, *de arte Navigandi*'. Harvey may have owned these books either in the original Spanish or in English translations. Martin Cortes's *The Arte of Navigation*, translated by Richard Eden, was published in 1561; Pedro de Medina's *Arte of Navigation*, translated by J. Frampton, was published in 1581.

³⁸ Flavius Renatus Vegetius (*c*. A.D. 380) was author of *Epitome Rei Militaris*. In 1572 it was translated by J. Sadler and published as *The foure bookes of martiall policy*.

At the bottom of the page Harvey lists his preferred military authors:

Mie principal Autors for Warr, after much reading, & long consideration: Caesar, & Vegetius: Machiavel, & Gandino; Ranzovius, & Tetti: with our Sutcliff, Sir Roger Williams, & Digges Stratioticos: all sharp, & sound masters of Warr. For the Art, Vegetius, Machiavel, & Sutcliff: for Stratagems, Gandino, & Ranzovius: for Fortification, Pyrotechnie, & engins, Tetti, & Digges: for the old Roman most worthie Discipline & Action, Caesar: for the new Spanish, & Inglish Excellent Discipline & Action, Sir Ro: Williams. Autors enowgh; with the most cunning, & valorous practis in esse.[39]

Harvey must have inscribed the above marginalia after 1595, for Heinrich Rantzau (Ranzovius) published his *Commentarius Bellicus . . . praecepta, consilia et stratagemata* in Frankfurt in that year. On sig. Cc2[r] of the Machiavelli translation where the text discusses the carrying of messages and the danger of their seizure, Harvey comments: 'Ciphring, & deciphring morr cunningly used now, then in Machiavel's time'.

In Whitehorne's *Certaine wayes for the ordering of Soldiours in battelray* (1573) Harvey has inscribed a number of Latin poems and poetic prose passages, apparently his own. The first three are entitled 'Polemica', 'Parasitica', and 'Oeconomica'. The text of the volume is also supplemented by a number of annotations of Harvey's, for instance on sig. D3[v]:

The bravest use of Vitruvius Art [i.e. architecture], in fortification; & in building of Shippes. In warres, it is not Engins of wood, but of iron, that must strike the stroke: & therefore Vulcan more serviceable, then Daedalus: and Peter of Navar, then Archimedes, himself: Unles Archimedes leave his wood-workes, and betake himself wholly to iron-workes, fire-workes, & such terribilities.[40]

On D4[r] he writes: 'The Spanish fortification, newly discoverid by Sir Roger Williams: in his new discourse of war'. On E3[v] Harvey notes: 'The circular forme with many corners, much preferrid, before any other form of fortresse'. On F3[v] he complains: 'The pyrotechnie of M. Digges, promised in his Pantometria [1571], and his Stratioticos [1579]: but not yet extant'. At the top of I1[r] he praises, 'Bournes Art of Gunning

---

[39] Carlo Theti (Tetti) was author of *Discorsi di Fortificationi*, originally printed at Rome in 1569 and later in a folio edition at Venice in 1589. Marco Antonio Gandini was Italian translator of Sextus Julius Frontinus' *Strategematicon* to which he added an important 'aggiunta' treating of modern historians. The Italian title is *Stratagemi militari di S.G.F.* (1574).

[40] References in this paragraph are to the following: Marcus Vitruvius Pollio who was renowned for his books on architecture, and Peter I (king of Aragon and Navarre (1094–1104) who fought against the Moors and won the battle of Alcoraz in 1096.

[William Bourne, *The art of shooting in great Ordnance*, 1587]; necessarie next Grassies Art of Defence'.

## JURISPRUDENCE

Harvey's esteem for the law as a profession is evident in much of his marginalia. In Latin inscriptions in Joachim Hopperus's *In veram Jurisprudentiam Isagoges* (1580) Harvey asserts on sig. A1ʳ and A2ʳ that Jurisprudence is the master builder, the governor of the republic, protectress of remains, queen of arts, and the kingly and imperial profession.[41] Nevertheless, he is well aware of the areas in which law can be abused. On sig. V3ʳ he writes: 'The natural use of Testimonies is, to proove, where dowbt is, not to accloy, where all is cleare'. On sig. Ii4ᵛ he warns:

Law is a woman, as is Fortune: not a man, nor God. Who will expect perpetual constancy from a woman regardless of how uncorrupted she may be? Astraea may be a chaste and perfect virgin; Themis may be a goddess; but they are of feminine sex. This will be understood; a word to the wise is sufficient.[42]

In Florio on sig. I2ᵛ he quotes a sardonic saying by Aeneas Sylvius (Pope Pius II) on the dangerous attraction of the law court: 'Suyters in Law, be as byrdes: the Court, is the bayte: The Judges, ar the netts: and Lawyers, the fowlers'. At the end Harvey adds his symbol for subject matter relating to law, 'J.C.'.

Harvey emphasizes that it is important to keep law alive and healthy. On the title-page of Foorth's *Synopsis Politica* (1582) he writes: 'Use Legges, & have Legges: use Law and have Law. Use nether and have nether'. He finds that lawyers have neglected some very strategic areas and he remarks in Foorth on sig. C6ʳ: 'Our lawyers know how to use the forum but not the court, the Consistorial Court but not the royal court. They involve themselves in services to the citizens but not in offices of the palace'.[43] It was because Harvey hoped to function in governmental circles that he specialized in civil rather than common law.

He quotes the advice of the Doctor of International Law and (Regius) Professor of (Civil) Law at Cambridge, Sir Thomas Smith, to become

---

[41] 'Jurisprud. Architectonica. Reipublicae gubernatrix. Jurisprudentia, reliquarum patrona, et Regina Artium'. 'Jurisprudentia, Regia, et Imperatoria professio'.

[42] 'Lex foemina est: ut Fortuna: non Vir, non Deus. Quis expectet a Muliere, quantumvis incorrupta, perpetuam constantiam? Sit Astraea, virgo illibata, et integra: sit Themis Dea: at foeminei sexus sunt. Verbum intelligenti sat'.

[43] 'Nostri Jurisconsulti norunt uti foro, sed non Aula: Curia Consistoriali, sed non Curia Regali: officiis civilibus, sed non honoribus palatinis'.

well acquainted with the various laws just as one does with the people one meets:

Just as one learns the names of citizens, one must know the principles of many laws and their meaning which is their worth and order. The strength and rationale of a law is like an individual's face and the appearance of his mouth. It is to be inspected and recognized as is the law which is neighbouring to it. One must learn to know that law which is related and closely akin to it, and that which is inimical and contrary.[44] (Hopperus, title-page verso.)

At the end of Book Four of Hopperus, he discusses those he considers outstanding jurists and the law books he esteems:

Not many skilled in the law, few skilled in discriminating law; there are very few like Hopperus and Vigelius prudent in law. No one has become an acute lawyer without Nicasius upon the *Institutions*, Deicus and Dynus on the Rules, Bartholus on the *Digest*. Baldus on the *Codex*, Vigelius in disputed points of law, Speculator [Durandus] on the court.[45]

At least twelve law texts which Harvey owned are extant, most relating to canon or civil law.

## COSMOLOGY AND OTHER SCIENTIFIC INVESTIGATIONS

In the satire on the three academic gardeners in *The Thirde Parte of the Countesse of Pembrokes Yvychurche* (1592) Gabriel Harvey is depicted as being an avid student of cosmology and astronomy.[46] The books in his library and his marginalia amply confirm this description.

Early examples of marginalia in these fields are found in the large 1527 folio containing Sacrobosco's *Textus de Sphaera*, Bonetus's *Annuli . . . super astrologiam* (described on p. 203, below), and Euclid's first book of geometry translated into Latin by Boethius. On the title-page is Harvey's signature and the date '1580', which seems to have been the period for many of his annotations in this volume.

Although he is interested in the presentation of material, he does not

---

[44] Harvey's marginalia read: 'Th. Smithus, Legum Doctor transmarinus et professor Cantabrigiensis: paulo etiam post Eques Auratus, Tenenda sunt, inquit, Multarum Legum *principia* tanquam Civium nomina: non est ignorandus *sensus*, quasi eorundem dignitas, et ordo: *vis et ratio* Legis tanquam vultus et habitus oris; inspicienda, atque agnoscenda est. Neque est illus ignorandum, quae cuique *vicina* sit Lex: quae cognata, et tanquam affinis; quae quasi inimica et contraria'.

[45] 'Iuris periti non multi: iuris subtiles pauci: juris prudentes. Ut Hopperus, et Vigelius, paucissimi. Nemo argutus J.C. sine Nicasio in Institutiones: Decio, et Dyno in Regulas, Bartholo in Digesta: Baldo in Codicem: Vigelio in ius controversum: Speculatore in forum'.

[46] The satire states that the 'thistle' is 'wholly addicted to contemplation' and has his head 'full of Cosmographicall Proclamations'. See also pp. 90–1, above.

subscribe to all of Sacrobosco's statements, for he has read other authors who question some of these views. On sig. a3ᵛ Harvey notes:

Concerning today's corrected astronomy and geography, behold the subtle and exquisite observations of Bodin in chapter 7 of his *Methodus Historicus* and in chapter 2, book 4, of his *de Republica*. The same and other questionings of Ptolemy are to be found in Regiomontanus, Cardanus, Copernicus, Reinhold, Apian, Joachim Rheticus, Gemma Frisius, Jofranc Offusius, Statius, Maestlinus, Tycho Brahe, and many others. Until now, however, Ptolemy (who flourished in Hadrian's empire) has been considered chief of astronomers and a mathematician worthy of immortal honour. My views are that one should use what has been most truthfully proven by just demonstration and observation. It is prudent to use the principles of Ptolemy, namely phenomena and proofs.[47]

Harvey was soon, however, to reject Ptolemy's geocentric principles for the newer Copernican heliocentric view of the planetary system. He had become aware that the Copernican system was substantiated by recent more accurate measurements of the heavens. In a later hand[48] he points out on sig. B6ᵛ:

Firminus in his prologue and chapter 1, part 1 of the Reportorium de Aeris mutatione:[49] teaches that all experiments apply to our times and climates and the judgements of Ptolemy ought to be changed now that the locations of the fixed stars have been changed. A necessary fundamental is the observation of the heavens.[50]

In the text of Firminus itself Harvey's annotations at the bottom of sig. b3ʳ, probably made in the early 1590s, further clarify his changed views:

Changed judgements: since the fixed stars have changed places. The old tables of Ptolemy and of others are not congruent with today's observed phenomena:

[47] This is a somewhat abbreviated translation of the Latin (and Greek) marginalia which read: 'De hodierna Astronomia, et Geographia correcta, ecce Bodini subtiles, et exquisitae observationes cap. 7, Methodi Historicae: item cap. 2. libri 4. de Republica. Et eadem, et plures animadversiones Ptolemaicae apud Regiomontanum, Cardanum, Copernicum, Reinholdum, Apianum, Joachim Rheticum, Gemmam Frisium, Jofrancum Offusium, Statium, Maestlinum, Tichonem Braheum, complures alios. Adhuc tamen Ptolemaeus Astronomorum suo merito princeps, et dignus immortali honore Mathematicus: Ut erat Adriani, quo floruit, imperio maxime omnium admirabilis. Mihi optima, quae iusta demonstratione, et observatione probantus verissima, et efficaciss. prudentis est, uti Ptolemaei principiis, Τοῖς φαινομένοις, καὶ ἀποδείξεσι'.

[48] This hand is considerably more slanted with different shaped 'f's and comma-like dots over the 'i's.

[49] This is a meteorological text (published in 1539) which Harvey owned but he has not specified the date when he acquired it.

[50] 'Firminus in prologo, et capite I, partis I. Reportorii de Aeris mutatione: docet Experimentia omnia, temporibus, et climatibus nostris aptare et Judicia Ptolemaei mutanda, mutatis iam Locis Stellarum fixarum. Necessarium fundamentum omnium observationum coelestium.'

it is very clearly evident from the Copernican revolutions and the Prutenic tables of Reinhold,[51] where the errors of the old tables ought to be reformed.[52]

A pre-Baconian, Harvey becomes more and more an advocate of experimental science and a respecter of observed evidence. On the first page of John Blagrave's *The Mathematical Jewel* (1585), a practical manual containing 'instruments' (diagrams with rotatable dials) which can be used to study the heavens and for navigational purposes,[53] Harvey writes:

All sciences are founded upon perception and reason. . . . Experience [is] the firmest demonstration and an irrefutable criterion. Give me ocular and rooted demonstration of every principle, experiment, geometric instrument, astronomical, cosmographic, horologiographic, geographic, hydrographic, or mathematical in any way.[54]

In Blagrave, Harvey refers to Sacrobosco's *Sphaera* as his 'once pregnant introduction' to things cosmological and concludes, 'An empirical world cares only for those matters proven by experience'.[55] Scientific instruments are valuable for empirical proof, and Harvey has great respect for those who make such instruments. Next to Blagrave's modest prefatory poem 'The Authour in his own defence', Harvey comments: 'An Youth, & no University-man. The more shame for sum Doctors of Universities, that may learn of him'. And in similar vein, he observes on sig. C2$^r$: 'Schollars have the bookes: & practitioners the Learning'.

On sig. a1$^v$ of Sacrobosco, he bewails the shocking mathematical ignorance of today's academic men and points out the importance of mathematics to workers in many fields: opticians, architects, craftsmen in wood, metal, stone or in mechanical arts, merchants, surveyors, navigators, painters (presumably in the use of perspective), and many others. On sig. a4$^r$ he observes that 'it expedites mathematics to know

---

[51] Erasmus Reinhold (1511–53), professor of astronomy at Wittenberg, refined the rough tables of Copernicus in his *De Revolutionibus* (1543) by more accurate measurements and called them the Prutenic (i.e. Prussian) Tables.

[52] '*Judicia mutanda*: quia stellae fixae *mutaverunt* loca. Veteres Tabulas Ptolemaei, et aliorum, *non congruere hodie cum Phaenomenis*: patet maxime, ex *Revolutionibus Copernici*; et *Tabulis Prutenicis* Rheinoldi: unde antiquarum tabularum errores reformandi'.

[53] The instruments are here made of heavy paper but could be made in more permanent form. A later version of the same instruments, as Harvey notes on the title-page, was made and sold in brass by 'M$^r$ Kynvin' of London.

[54] 'Omnes Artes fundatae super Sensu, et Ratione, plane constant Ratione, et Sensu. Ratio, anima cuiusque principii. Experientia, anima animae, firmissima demonstratio, et irrefutabile κριτήριον. Da mihi ocularem, et radicalem demonstrationem cuiusque principii, experimenti, instrumenti Geometrici, Astronomici, Cosmographici, Horologiographici; Geographici, Hydrographici; et omnino cuiusvis Mathematici'.

[55] 'Empiricus Mundus sola curat Empirica'.

Faber and Regiomontanus, Reinhold and Apian'.[56] He praises Robert Recorde's *Pathwaie to Knowledg* [1551] and Euclid's Geometry. On sig. e8[v] at the end of the Boethius Latin translation of Euclid's *Liber Primus Geometrie*, Harvey writes, 'No Euclide more profitable, then that lately published [1570] in Inglish; with a large and notable preface of M. Dee, one of the profoundest Mathematicians now living'.[57]

In Luca Gaurico's *Tractatus Astrologicus* (1552) on sig. Yi[r], Dee is again praised by Harvey in an interesting list of contemporary mathematicians and technicians whom he rates highly:

M. [Richard] Benese, M. [Thomas] Digges, M. [John] Blagrave, M. [Cyprian] Lucar, & Valentine Leigh, artificial & expert Surveiours. But most of theise fine Geometricians & greater artists: especially Digges, Blagrave, Lucar. As notable mathematicall practitioners, & polymechanists, as the most commended beyound sea. But for cunning points, profound conclusions, & subtill experiments in Geometrie, Astronomie, Perspective, Geographie, Navigation, & all finest mathematical operations, I knowe none like unto M. [John] Dee of Murclake [sic], M. [Thomas] Hariot at Durrham howse, & M. [Edward] Wright of Caius Colledg in Cambridg. For Arithmetique, & Geometrie M. Alderman Billingesly, as singular as the best. For artificial navigation, M. William Borrough, controller of her Majesties navie, esteemed exquisite. Now for the mathematiques, divers other begin to carrie credit, M. Christopher Heydon, M. [Thomas] Blundevil, M. [Thomas] Hood, M. [Robert] Norton, M. [Giles??] Fletcher,                                   [*The list is unfinished.*]

Luca Gaurico's *Tractatus Astrologicus* is a work giving horoscopes and brief descriptions of noted persons and of city-states and towns, each being illustrated by an astrological diagram. Harvey read and annotated the volume in 1580 and his notes indicate that he is impressed with the teachings of the more learned and conscientious astrologers among whom he places Gaurico. He is aware that there has been much opposition to and questioning of astrological principles [he, too, has been a sceptic], but, probably because of the prodding of his brother Richard,[58]

[56] Harvey is referring to Jacques Lefèvre d'Étaples (*c.* 1455–1537) known as Jacobus Faber Stapulensis, to Johann Regiomontanus (1436–76), Erasmus Reinhold (1511–53), and Peter Apian (1495–1552).

[57] Euclid's *Elements of Geometrie* translated into English by Sir Henry Billingsley contains a long and eloquent preface by Dr. John Dee of Mortlake on the importance of mathematics and its value as a study.

[58] Harvey's 1576 notes on fol. 6[r] of Dionysius Periegetes are critical of 'owr vulgar Astrologers, especially such, as are commonly termed Cunning men or Artsmen'. He writes:

'Theise be theire great masters: & this in a manner, theire whole librarie: with sum old parchment-roules, tables, & instruments.

Erra Pater, their Hornebooke.

decides to study and thus evaluate the subject[59] for himself. He indicates at the beginning of the text that he has considerable esteem for the twin Gaurico brothers (Luca and Pomponio) and concludes that Luca's authority is especially to be respected because of his experience at the Papal Court [of Paul III],[60] because he is discriminating in the making of horoscopes, and because he is critical of the careless practices that he has, for instance, found in the work of Girolamo Cardano and Johann Schöner. On sig. F1[v] he describes Luca Gaurico, a knight of St. Peter's and Bishop of Civitate, as a truthful seer and observes, 'Certainly one ought not to despise so much experience and authority joined to so much science, and it in the Court of a prudent pontiff'.[61]

Harvey's 1580 comments (on sig. E4[r]) reveal that he then believed that the stars have an influence on human physiognomy and actions and that by their study in relation to other factors one could to a certain extent foretell the future. After an appraisal of the more careful and scientifically minded astrologers, he writes:

Hence, my (method of) physical prediction carefully corrected consists as much in stoicheology[62] and especially in physiognomy as in planetology or horoscopy. The junction of these yields me credible and in no way fallacious natural prediction. This becomes complete when, from the other part of philosophy, ethics, also education and the conduct of the rest of life are closely examined. Particularly, if one likewise has considered the kind of civil causes which prevails in each political government. Hence, he is ignorant and unjust who spurns astrological judgements in so far as they proceed artfully from natural causes

The Shepherds Kalender, their primer.
The Compost of Ptolomeus, their Bible.
Arcandum, their newe Testament.
The rest, with Albertus secrets, & Aristotles problems Inglished, their great Doctours, & wonderful Secreta secretorum'.

In the Astrological Discourse, which Richard Harvey addresses to Gabriel, he states on sig. a2[v] that his brother had originally objected to the study of judicial astrology as fruitless but Richard insists that it is 'neither any vaine and idle studie, nor forbidden and unlawful Arte', as Gabriel 'having long since taken some reasonable paines therein . . . will readily (I thinke) testifie'. The tract (which was published in 1583) is dated by Richard 'from my fathers in Walden the 6 of December 1582'.

[59] While relaxing at Trinity Hall during the study of civil law, Harvey amuses himself by cataloguing astrological data in Oldendorp's Loci Communes Juris Civilis (1551). These annotations were probably made in the late 1570s. See p. 68, above.

[60] Astrology and theology were considered irreconcilable. That Gaurico was highly esteemed by Pope Paul III and became Knight of St. Peter's and Bishop of Civitate is unusual.

[61] 'Certe non contemnanda tanta experientia, et auctoritas, tantae scientiae c[   ]idque in Aula prudentis pontificis'.

[62] 'Stoicheology' is the physics of elements.

to natural effects, from moral causes to moral effects, and finally from political causes to political effects.

Gabriel Harvey, no matter what others may think to the contrary.[63]

Harvey's credence in astrology at this time may seem puzzling in view of his stress on the importance of experimental evidence (see p. 166, above), but one must remember that the Renaissance was an age when astronomy and astrology were closely interwoven and many of the foremost scientific men of the age were advocates of astrology. According to George Sarton, there were few astronomers, physicians, or mathematicians who did not believe in the subject, 'and many were engaged in astrological pursuits of one kind or another'. Agrippa, Paracelsus, Cardano, Dee, Bodin, Cornelius Gemma, and Della Porta were all believers. The superstitious notions of these men were combined with admirable achievements in many diverse fields. Although men like Pico, Reuchlin, d'Étaples, Trithemius, and Erastus did not hesitate to attack astrology, 'their attacks were almost always qualified'.[64]

One must also take into account how readily Harvey is impressed by a learned man of outstanding position like Gaurico. Some of the appeal of astrology for Harvey undoubtedly was its highly organized structure and myriad colourful details. Later in life he probably became less credulous and not so easily awed by astrological 'erudition'. One recalls that after the earthquake of 6 April 1580 Harvey wrote his 'Earthquake Letter' exposing the false scholarship of pseudo-scientists at Cambridge. Perhaps it was not too long thereafter that this attitude of his may have broadened enough to puncture the cant of contemporary astrologers. In Abraham Fraunce's 1592 satire on the three academic gardeners (see pp. 89–91, above), although Thistle (Gabriel Harvey) is described as a 'doctor of Astrologie', he is depicted as a specialist in astronomical and cosmological knowledge and rather sceptical of astrological predictions. For instance, when he finally reached the 'first heaven', 'he began to observe and marke, whether *Strabo*, *Ptolomaeus*, and other measurers of the world, had made a good survey thereof'.(sig. P2ʳ) Again, when the three Harveys are

---

[63] Mea tandem Physica divinatio exquisite rectificata, tam Stoicheologia, in primisque Physiognomia constat, quam Planetologiam, aut Horoscopographiam. Quorum accurata coniunctio credibilem mihi exhibet, minimeque fallacem Physicam divinationem. Quae tum demum fit, absoluta, cum ex altera philosophiæ parte, Ethica etiam educatio reliquaeque vitae consuetudo intime examinatur. Praesertim, si civilium etiam causarum ea habeatur ~io, quae in quaque politica gubernatione invalescit. Unde ignarus, improbusque est, qui ~logica iudicia spernit, quatenus artificiose procedunt a physicis causis ad physicos ~s, ab ethicis ad ethicos; a politicis denique ad politicos.

　　　　　　　　　　　　　　　　　　gabriel harveius: quicquid alii contra.

~ton, *Six Wings: Men of Science in the Renaissance* (Bloomington, Ind., 1957),

asked to account for their presence at such elevated altitudes, they reply that they are 'Academikes' who 'troubled and perplexed with the repugnant conceipts of Astrologers, and menaced and threatened with their unhappy predictions, have travelled hither of purpose to understand whether their divination be true or not'. (sig. P3ʳ) Thistle elaborates: 'myself am a doctor of Astrologie, & can yeeld you an accompt of the opinions of the *Chaldees, Aegyptians, Indians. Mores, Arabians, Jewes, Grecians, Romaynes,* modernes & ancients whatsoever: al whose conceipts I finde as variable as the moone, & themselves altogether Lunatike'.

## MEDICINE

From astrology Harvey turned to the study of medicine when he became a Medical Fellow at Pembroke in 1584 (see pp. 74–5, above). According to marginalia in Jacob Hugkel's *De Semeiotice Medicinae* (1560), although he is cognizant of astrologically oriented medicine, he is evidently aware of its inadequacies, for he finds Bruele's theories and empirical practice of medicine[65] preferable. On sig. A1ᵛ Harvey writes:

To knowe everie *Complexion* according to the *signes,* & *planets* whereof they cum, bie judgments of *sores*. Other of Mʳ Leas rules, or aphorismes.[66]

1. If the sore be of colour red, & watrish, it is ingendred of corrupt blud. If it be hott, & moist, his signes ar *Gemini, Libra,* & Aquarius: his planets ♃, & ☿ *[Jupiter and Mercury]*.

2. If the sore be red, & hard, it is ingendred of red choller, & is hott, & drie. His signes are *Aries, Leo,* & Sagitarius; his planets ☉ [Sun] & ♂ [Mars]. Whiles the moone is in anie of theise signes, do no medicin, or plaster to it.

3. If the sore be grey or browne, blackish or white, & hard withall, it is of melancholie, & is cold & drie. His signes, *Taurus,* Virgo, & Capricorne: his planets ♄ [Saturn].

4. If the sore be white & pale, bright or soft in feeling, it is of fleame: & is cold & moist. His signes, *Cancer,* Scorpio, & Pisces: his planets ♀ [Venus] & ☽ [Moon].

If the disease be hott, & drie, use *remedies cold,* & moist.
If hott & moist, use cold & drie.
If cold & moist, use hott, & drie.
If cold & drie, use hott & moist. The best instructions in Mʳ Le

[65] The text by Gualterus Bruele which Harvey owned is *Praxis medicinae theo empirica familiarissima* (1585). It was annotated in 1589.
[66] Preceding this passage Harvey describes Lea's analysis of urine by a sim John Lea was the Earl of Leicester's physician and surgeon (according to p. 139 of Bruele).

Paperbooke which he commonly called his boosum-booke: sumtime his vade mecum. But nothing comparable to Bruels theorique, & empirique practis of physique.

Harvey also owned Hieronymus von Braunschweig's *Apothecarye*, and its marginalia include evaluations of early physicians and medical writers. His preferred medical text, however, seems to have been Bruele, for it was copiously annotated by his brother John, who had owned the book previously, and then by Gabriel himself. The notes include aids to diagnosis together with various and sundry remedies, many of them herbal concoctions or common sense physiological or psychological measures, although a few in Gabriel's hand marked 'prooved by M. Lea' sound quite fanciful in terms of modern medicine.[67] The volume contains notes on the treatment of melancholia, vertigo, epilepsy, spasm, lethargy, cataracts, angina, asthma, plague, dysentery, colic, kidney stones, diabetes, haemorrhoids, etc.

Near the end of the Bruele volume Harvey has copied a brief letter which he evidently sent to Dr. (Nicholas) Wathe when the doctor himself was seriously ill. This warm and witty note suggests a close relationship with the Saffron Walden doctor who may well have been the Harveys' family physician.[68]

One also finds medical marginalia in some of Harvey's non-medical books. For instance, on the final leaf of Olaus Magnus's *Historia de gentibus Septentrionalibus* (1555), a history of the Goths, is a page devoted to the discussion of the gout and cures (both fanciful and practical) advocated for it. At the bottom of a lengthy paragraph headed 'Remedia aliquot podagrae: Podagra vitrum' (some remedies for gout: a gout woad plant) Harvey queries, 'Were Bacon, the Keeper of the Great Seal and Cecil, Lord High Treasurer of England—men of exceptional prudence & intelligence—aware of this remedy?' Apparently Nicholas Bacon and William Cecil were among the more famous sufferers from gout.

---

[67] One wonders whether M. Lea was involved in the fiasco of medical treatment which led to the wounded Sir Philip Sidney's death in the Netherlands in 1586.

[68] The Saffron Walden Parish Register for 1601/2 bears the burial entry: 'Mr Nicholas Wathe Doctor of Visicke'. He may have been the father of Nicholas Wathe of Clare Hall, B.A. 1569/70, M.A. 1573. There are a number of references to Dr. Wathe in Harvey's marginalia. The letter alluded to above is the following: 'Doctori Watho Salutem. Quid ibi vis, Ignave, quod te tam graviter aegrotare pateris, et tam diu? Non enim possum, ⸱erite, dicere, cui tot ad unguem peritissimi. Sed tamen ut aliis iampride[m] ita tibi ipsi ⸱n teipsum proba Medicum: et ante omnia Medice cura teipsum. Frustra studi tot ⸱e, philosophiae, Chymiae secretis, qui nescit curare semetipsum. Nam nolle aut sane esset insaniae. Quamprimum igitur vale, et salve; salve et vale. Sat bene,

Tuus G.

Sympathy for the mentally ill and an understanding of insanity beyond that of most of his contemporaries is shown in an original poem of Harvey's inscribed on sig. O3ʳ of James VI's *Essayes of a Prentise* (1585):

A charme for a mad wooman

Ô heavenlie Medcin, Panacea high,
Restore this raging Wooman to her health,
More Worth then hugest Summes of worldlie Wealth
Exceedingly more worth then anie Wealth.

Ô light of Grace, & Reason from the Skie,
Illuminate her madd-conceipted minde,
And Melancholie cease her wittes to blinde.
Cease *fearful* Melancholie her wittes to blinde.

Axiophilus.[69]


## LITERATURE AND DRAMA

Perhaps the best known of Harvey's marginalia are those relating to literature and drama. His literary appraisals in his copy of Chaucer have been mentioned on pp. 125–8 above and marginalia in Gascoigne's *Certaine Notes of Instruction* have been reprinted by Gregory Smith.[70] The latter give evidence of a sound approach to literary criticism. For instance, of Gascoigne's presentation Harvey writes on sig. Y2ᵛ:

His aptest partition had bene, into precepts of Invention} Elocution} And the several rules of both, to be sorted & marshialled in their proper places. He doth prettily well: but might easely have dun much better, both in the one, & in the other: especially by the direction of Horaces, & Aristotles Ars Poetica.

At the top of T3ʳ he observes, 'The difference of the last verse from the rest in everie Stanza, a grace in the Faerie Queen'. At the bottom of the same page he indicates his dislike of words unnaturally accented in poetry: 'The reason of manie a good verse, marred in Sir Philip Sidney, M. Spenser, M. Fraunce, & in a manner all owr excellentest poets: in such words, as heaven, evil, divel, & the like; made dysyllables, contrarie to their natural pronunciation'. Less well known but interesting notes are found in Harvey's Euripides. On flyleaf 7ʳ, which is filled with marginalia in Italian, he discusses comedy and tragedy and expresses his especial

---

[69] 'Axiophilus', as has previously been noted, is Harvey's name for himself as poet and lover of poetry. The marginalia in this volume were for the most part written about 1592–3, as pointed out by Eleanor Relle in 'Some New Marginalia and Poems of Gabriel Harvey', *RES*, 23, no. 92 (Nov. 1972), 401–16.

[70] In *Elizabethan Critical Essays*, Oxford, 1904, vol. I.

admiration for Aretino's 'Il Filosofo', Ariosto's 'Suppositi', and Machiavelli's 'Mandragola' and 'Clitia'. Of the last two he writes that they are very incisive and elegant and that to read them nourishes one's cleverness. Changing to Latin he then observes on 7ᵛ that as with comedies so with tragedies: he who knows three or four intimately knows almost all of them. He praises this little 'golden' volume of Euripides' 'Hecuba' and 'Iphigenia in Aulis' (translated into Latin by Erasmus) and terms Euripides the wisest of poets, even more sinewed and profound than the Attican Sophocles or the Latin Seneca. On sig. 8ᵛ he observes that it is useless to read tragedies unless one knows how to distinguish philosophical from tyrannous concepts.[71]

In Lodovico Dolce's *Medea Tragedia* (1566) on sig. Aiᵛ Harvey comments on recent translations from the great Greek playwrights. The 'Giocasta' of Dolce, and Gascoigne, the 'Thebais' of Seneca and Statius, and the 'Oedipus' of Seneca, he considers, are almost synopses of all tragedies. The 'Hecuba' and 'Iphigenia' of Euripides he finds especially pleasing, both because of the very skilled strength of the author and the singular discrimination of Dolce, the translator. A similar feeling is to be found in Sophocles' 'Antigone' translated by Bishop Watson,[72] both on account of the very grave style of the original author and of the careful

---

[71] The marginalia referred to are the following:

'Quattro comedie del divino Aretino. Cioè, Il Marescalco, o pedante, La Cortegiana. La Talanta, Lo Hipocritico. Habeo, et legi: sed nondum comparare potui Il Philosofo: quae tamen ipsius Comoedia dicitur etiam exstare. Memorantur etiam *duae illius Tragoediae* L'Hortensia. Tragedia di Christo.

Comedie, Dialoghi capricciosi, Le Lettere, e capitoli dell' Unico. Historie del suo tempo· La quinta essenza del suo unico ingegno; et Lo Specchio di tutte L'Arti cortegiane. Due Comedie agutissime et facetiss. di Machiavelli politico. La Mandragola: La Clitia. A legere nutrica lo ingegno.  (sig. 7ʳ)

Ut comoedias, sic tragoedias; qui tres aut quatuor intime, novit, novit fere omnes. Tanti valet hic aureus libellus. Meo tandem iudicio, poetarum sapientissimus, Euripides: vel ipso Sophocle magis Attice nervosus, et profundus. Ut Seneca Latine.  (sig. 7ᵛ)

Inutiliter Tragoediae legit, qui nescit philosophiae sententias, a Tyrannicis distinguere'. (sig. 8ᵛ)

[72] This is apparently a slip of Harvey's pen, for it was Thomas Watson, the poet and law student, not the Bishop of Lincoln, who translated 'Antigone'; Harvey elsewhere refers to the translator of 'Antigone' as merely 'Thomas Watson' or 'Watson', e.g. on fol. 5ᵛ of Dolce: 'Gascoigni Jocasta, magnifice acta solenni ritu, et vere tragico apparatu. Ut etiam Vatsoni Antigone: cuius pompae seriae, et exquisitae'.
And on sig. 11ᵛ of Euripides: 'Vatsoni Antigone meruit annecti Erasmi Hecubae, et Iphigeniae: eademque potius superior, quam inferior ullo modo'.
Although Harvey is as a rule remarkably accurate, occasionally there are slips of his pen, as for example, at the end of F. Duarenus, *Pro Libertate Ecclesiae Gallicae* (1564) where he refers to 'J. Fox. Actes, & Monuments, Anno 1523.', instead of 'Anno 1563'. For a discussion of 'Gabriel Harvey and the two Thomas Watsons', see the paper by Louise George Chubb, *?N*, 9 (Summer 1965), 113–17.

judgement of the translator. Harvey considers the tragic dramatists superior to even the great epic poets such as Homer or Virgil.[73]

In Lodovico Domenichi's *Facetie, motti, et burle* on fol. 1[r] Harvey notes his literary preferences among recent authors: 'To the classical authors I add Du Bartas on account of his divine frenzy, Rabelais on account of his humanity, Chaucer for a mixture of the two. I prefer Chaucer to Petrarch, and Boccaccio, Ariosto, and Rabelais to Aretino'.[74] On fol. 37[r] he observes: 'Of all writers, Homer, & Livie excell for lively portraiture of persons, places, things, & acts'; and on fol. 64[v] vertically: 'Virgil refined Ennius: a second Pindar [Du Bartas] refined Virgil. That Pindar may still be refined'; and on fol. 53[v]: 'Rabelais alone will putt downe all the Italian, & Spanish poets, either in words, or in deeds. In jest, Rabelais: in earnest, Bartas singular'.

### Philosophical Outlook and Personal Observations

Having briefly surveyed Harvey's typical marginalia in a number of fields, I shall now turn to his annotations of a personal nature: declarations of his aspirations, aids to self improvement, reports of praise from some he has esteemed, and other 'morale boosters'. Curiously, the above are expressed not by the personal pronoun 'I' but in terms of various Greek and Latin epithets which represent Harvey in different aspects and at various phases of his life.

'Axiophilus' (to whom allusion was made on pp. 126–8, above) seems to have been one of the first *personae* to be used by Harvey, appearing in his marginalia at least as early as 1572.[75] It is a name by which he refers to himself as a poet in the broadest sense and one who has sensitivity for

---

[73] The marginalia in Dolce referred to read *in toto*: 'La Giocasta d'Euripide, Dolce, et Gascoigno. Senecae, et Statii Thebais. Item Senecae Oedipus. *Quasi Synopsis Tragediarum omnium*. Non gioco, ma Giocasta. Omne genus scripti, gravitate Tragoedia vincit. Hae quatuor Tragoediae, instar omnium Tragoediarum pro tempore: praesertim reliquarum non suppetit copia. Duae Euripidis placent in primis, et propter auctoris prudentissimam venam; et propter interpretis singularem delectum. Eadem in Sophoclis Antigonen affectio, ab Episcopo Vatsono tralatam: cum propter inventoris gravissimum stilum; tum propter interpretis accuratum iudicium. Qui tanti fecit optimos tragicos, ut eosdem soleret cum Checo, et Aschamo, omnibus aliis poetis anteferre; etiam Homero, et Virgilio'.

[74] 'Classicis auctoribus addo Bartasium, ob divinum furorem: Rabelaisium ob humanum: Chaucerum ob mixtum. Malo Chaucerum quam Petrarcham: Boccatium, aut Ariostum: Rabelaisium, quam Aretinum'.

[75] It is found on one of the blank flyleaves (3[v]) preceding the title-page of Thomas Twine's 1572 English translation of the *Surveye of the World* by the third century B.C. Greek poet Dionysius Alexandrinus and refers to Axiophilus's knowing by heart certain works of Bartas and of Buchanan.

the finer and nobler values of life. Most of the original poems in his marginalia are signed in this way and it is 'Axiophilus' (from the Greek, meaning 'lover of the worthwhile') who acts as literary and poetic critic, notably in the Chaucer annotations where he discusses contemporary writers and the future of English literature and poetry.

Other *personae* found frequently in the eight books which contain most of Harvey's personal marginalia[76] are 'Angelus Furius', 'Eudromus', 'Eutrapelus', 'Chrysotechnus', and 'Euscopus'.[77]

'Angelus Furius', used by Harvey in the early 1580s, is described by him as angelic in speech and a fury in action. This *persona* seems to relate to experiences in Italy, whether anticipated or actual ones is difficult to ascertain. In Joannis Foorth's *Synopsis Politica*, 1582, on sig. D1$^v$ Harvey inscribes:

Angelus Furius, the most eloquent Discourser, & most active Courser, not in this on Towne, or in that on Citty; but in all Italy, yea in all Christendome, yea even in the whole Universal Worlde. No on so persuasively eloquent; or so incessantly industrious.

On sig. A5$^r$ of *Oikonomia* he reveals his grandiose conception of the role he would like to assume: 'In one Caesar many Mariuses and Sullas; in one Angelus Furius many Caesars'.[78] It is conceivable that by the end of 1580 (after he left the English court) Harvey may have gone to Italy, for in October 1579 he was writing to Spenser of his expectations of being in Italy for a year or two.[79] But, unless one credits as first-hand accounts Harvey's descriptions of Florentine customs as inscribed in Florio (see p. 252 n. 12 below), I have found no clear reference to such a sojourn. Perhaps, therefore, as in many other Harvey notes where verbs are omitted, these 'Angelus Furius' marginalia indicate wishful thinking rather than actuality.[80]

The *persona* one meets most frequently in Harvey's marginalia is

[76] Domenichi, Guicciardini, Demosthenes, Foorth, Joannis Ramus, Quintilian, James VI, and Chaucer.

[77] *Personae* only fleetingly mentioned are 'Eutuchus' (the well-walled) who substitutes love for envy and attempts to restore the golden age and the heroic world (e.g. Domenichi, fol. 9$^r$) and 'Euhecticus', a jovial and sunny man (Domenichi, fol. 19$^r$); further study of the marginalia may disclose still others.

[78] 'In uno Caesare multi Marii, et Syllae: In uno Angelo furio, multi Caesares'.

[79] Mention is made in Harvey's 1578 'De Vultu Itali' of the Earl of Leicester's plans to send him abroad to France and Italy but he apparently left the English Court in the autumn of 1579 and at least temporarily returned to Cambridge. Nashe's version is that Harvey was summarily dismissed as Leicester's secretary. See pp. 42, 68, above and Nashe, *Works*, III. 79.

[80] For other references to 'Angelus Furius', see Ramus's *Oikonomia*, sigs. L8$^v$, M1$^r$, N2$^r$, N3$^r$, and Add. MS. 32,494, fols. 14$^v$ and 51$^r$.

'Eutrapelus', one who is clever and ingenious with words. He is the eloquent orator, teacher of rhetoric, persuasive man in speech (and in writing) and one who engages in witty jesting and very often in irony. His name derives from the Greek word Εὐτραπελία meaning 'turning well' and was used by Aristotle for 'pleasantness in conversation', one of the seven moral virtues enumerated by him. The almost obsolete English word 'eutrapely' was used similarly to mean, according to the *OED*, 'courteous, civil conversation, or urbanity'. Unlike 'Angelus Furius' who is alluded to only during a brief portion of Harvey's career, 'Eutrapelus' continues to be referred to throughout Harvey's lifetime.

In terms of 'Eutrapelus' Harvey says of himself on fol. 5ᵛ of Domenichi's *Facetie*, 1571:

Eutrapelus scornes himself, till he teaches all other, to pronounce more sensibly; to expresse more lively; to speake more effectually; to resolve & persuade more powrefully, then anie heretofore. Lett Eutrapelus excell all other in speaking, designing and doing: or lett Eutrapelus be accounted a meacock, & a base fellow.

On fol. 82ᵛ of Guicciardini's *Detti et Fatti*, 1571, he inscribes in Latin the following which reveals some of his methods of discourse:

The pragmatic metamorphosis of Eutrapelus. It is helpful to draw the salt of the earth and the light of the world into jests either very savoury or very splendid. Others present serious discourses but only Eutrapelus works at serious matters and effects serious results. Eutrapelus is able to convert great matters into small ones, small into great ones. This is Eutrapelus's secret metamorphosis: serious matters of others being converted into jests, your own jests into serious discussions. Hyperbole is added to the remarks of others and removed from yours. But next to yours *todos es nada* [i.e. 'everything is nothing'].[81]

'Eudromus' (one who runs well as in a race) is the pragmatic, hurrying competitive man of the world. He is constantly striving to excel, for he is frequently asked: 'What are you able to do more than others? What do you have of greater excellence than others? The great question and continuous judgement of the world'.[82] And he replies:

So many of the secret matters, the smallest worthwhile points or momentous

---

[81] For the original Latin see p. 161 n. 32.

In S. Stefano Guazzo's *La Civil Conversatione* (1581) which Harvey read and annotated about 1582, he notes in Italian on sig. Ff3ʳ that one's speech should be full of variety, acuteness, and learned judgement. It should be also concise yet very pleasing ('*discorsi pieno di varietà, acutezza, e sententiosa . . . brevità, con molta piacevolezza*').

[82] 'Quid potes, plus quam alii? Quid habes excellentius aliis? Magna mundi quaestio, et perpetua censura'. (Domenichi, fol. 42ʳ.)

discussions and negotiations. Writing of their exploits and subsequently reading about them can be considered a kind of self-therapy for Harvey to help him avoid melancholy or bitterness and maintain the cheerful demeanour for which he strove.

Of the profusely annotated volumes the one that contains most notes of a personal nature and one of those that makes frequent reference to *personae* is Lodovico Domenichi's *Facetie, motti, et burle, di diversi signori et persone private*, Venice, 1571. It was printed in seven books with an eighth added by Thomaso Porcacchi in 1574. Harvey's text contains the Porcacchi addition and is bound with Lodovico Guicciardini's *Detti et Fatti*, Venice, 1571. The little octavo volume is printed in Italian with wide margins. Both works are collections of wise and witty sayings and anecdotes. On fol. 37$^v$ of the Domenichi, Harvey refers to 'Meus Rabelaesius: meus Martialis: Domenicus meissimus. Ad unguem totus. [My Rabelais, my Martial, my very own Domenichi—totally at my fingertips.]' and on fol. 39$^r$ to 'Mei proprii auctores, Gandinus et Weccherus: Speculator et Domenicus. Spiritus meissimi. [My very own authors, Gandini and Wecker, Speculator[90] and Domenichi—the spirits most mine.]' Of these cherished works only a portion of Harvey's copy of Domenichi (the final quarter of it, about one hundred pages) is extant, and the margins have been considerably trimmed. On most pages Harvey has written in all four margins and often interlinearly as well. If the missing leaves were similarly annotated, the notes in this volume must have been very abundant. Marginalia in the extant portion probably date from the late 1570s to at least 1608.[91]

A good many of the extant annotations are declarations of Harvey's aims and achievements, usually expressed in terms of *personae*.

On fol. 4$^v$ Harvey writes of himself (underlinings are his): 'The *most sensible voice living*: & most emphatic stile that ever was: both, as

[90] Marco Antonio Gandini seems to have been especially cherished by Harvey for his writing on military history (see p. 163, above). Hans Joseph Wecker, the eminent physician, wrote *Practicae Medicinae Generalis*, 1585, and *Antidotarum Speciale ex opt. authorum*, 1601. Gulielmus Durandus, known as 'Speculator', (d. 1296) was author of many legal works (see p. 165, above).

[91] Since the title-page is missing, it is impossible to ascertain when Harvey acquired the volume. However, there are a number of notes relating to courtiers and the court so it is possible that the early annotations relate to the period when Harvey was in Leicester's service or anticipating a post at court. Although the volume is no longer in its original binding, it seems likely that the Domenichi and Guicciardini were originally bound together by Harvey since the subject-matter and method of annotating is similar for both. The Guicciardini work has a title-page signed and dated '1580'. Annotations in Domenichi continued until at least October 1608, for on fol. 47$^v$ is the inscription: 'Its a madd world mie masters. The title of a new booke'. The play was entered at Stationers' Hall on 4 October 1608.

expressive, as is possible, & most effectually significant. A voice most clere, & sounding: as sweet & thundring, as maie be of vocal instruments'.

The above is written vertically in the outside and inside margins; the following is written horizontally at the top of and in any available spaces through the text: 'Voice, & stile. Knowledg, & action. An extent, & emproovement in all deeds. Eutrapelus: his onlie braverie in both; hand, & toung. Discourse, & dispatch attonce. Still extent, & multiplication: but refined, & cumlie'. One also finds the following inscriptions relating to Eutrapelus: 'The most industrious, & impetuous man of all other [52ᵛ]'. 'Sum do all bie halves: as Castor lived halfe a life: & Pollux dyed half a death. Tota vita, Eutrapeli vita [56ᵛ].' The marginalia on fol. 3ᵛ record Eutrapelus's reading:

What kinds of unique authors does Eutrapelus read daily? Eunapius, with Tacitus, Philostratus with Julian, Zwinger's *Theatre* with Gandino, Bartas with Rabelais, Theocritus's 'Idyll I' with the epitaphs of Bion and Adonis. Three heroic shields (Homer, Hesiod, Virgil) with the 'seventh day' of Bartas, Solomon's 'Song of Songs' with the Behemoth of Job and the Leviathan.[92]

and on fol. 40ᵛ some of his studies:

Eutrapelus always earnest and absolutely steel-like in his most worthy studies. In chemistry, in politics, or pragmatic matters, in polemics, or stratagems, finally in the most powerful secrets of the world or new practices which are dug up from the depths and very greatly amplified. He examines all the commonplaces and plucks out only those especially useful in terms of his target. All the endless nonsense of authors he relinquishes to the idle, as much to professors as to rhetoricians..[93]

He follows a similar process with classical authors and attempts to assimilate all the worthwhile knowledge of the past. On fol. 21ᵛ he inscribes: 'All deade lines of antiquitie, in one living centre of this age'.

Such seeming outbursts of self-adulation are primarily self-exhortation,[94]

---

[92] 'Quales Eutrapelus quotidie legit singulares auctores: Eunapium, cum Tacito: Philostratum, cum Juliano: Zwingeri theatrum, cum Gandino: Bartasium, cum Rabelaesio. Theocriti Idyllion I, cum Bionis, et Adonidis Epitaphio. Tria scuta heroica (Homeri, Hesiodi, Virgilii) cum septimo die Bartasii. Canticum Canticorum Salomonis: cum Jobi Behemeth, et Leviathan'.

[93] 'Semper serius, et prorsus adamantinus in suis dignissimis studiis Eutrapelus. In chymicis: in politicis, aut pragmaticis: in polemicis, aut stratagematicis: in mundi denique potentissimis arcanis, aut neopracticis; quae sunt de profundis eruta fortissimeque multiplicata. Omnes excuti texcellentissimos Locos Comm[unes] et sola decerpit suo scopo maxime conducentia. Ociosis [sic] tam professoribus quam scholasticis infinitas tot auctorum nugas relinquit'.

[94] He also writes: 'Miseret me mei, si Fortius violentior: aut Ramus illustrior: aut Lipsius securior: aut Tycho peritior: aut Draco validior: aut Bartasius divinior ullo modo. [Alas for me if Fortius is more vehement, or Ramus more distinguished, or Lipsius more

but Harvey does quote some of the men who have praised him highly.
On fol. 35$^r$ he notes:

Doctor Daro the Frenchman told Monsieur Bodin that besides mie learning,
I was all spirit. Antonie de Corro the Spaniard was woont to salute me, Mag-
nanime Domine. Summa animi corporisque contentio nimis adhuc est remissa.
[High minded scholar, the highest striving of mind and body is still too small.]
    Sir John Skinner when he speakes of mee, calles me the great scholler.
Sir Jon Woodes opinion of mee, greater & higher, then Sir Jon Skinners.$^{95}$

At the end of these marginalia Harvey quotes Tycho [Brahe]: 'Non
haberi, sed esse: [Not to be considered but to be.]'
    The inscribing of these comments must have served as partial solace for
Harvey's lack of tangible success in the outside world. On fol. 5$^r$ he even
refers to his marginalia as his 'divine notes'.$^{96}$ Other attempts to console
himself are comments such as the following:

A great world abrode, is not worth a little world at home.   [fol. 49$^v$]
Witt turneth feare into hope: griefe into joy: hate into loove. No miserie in
a hart of felicitie.   [fol. 50$^v$]
A great world of trubble, not equivalent to a little world of content.   [fol. 50$^v$]
Everie place a paradise, thowgh a purgatorie: everie occurrent a restorative,
thowgh a corsive. Everie crosse a blessing: & everie losse a gaine.   [fol. 51$^r$]
Animus est, qui facit foelicem, aut miserum. Vilis animus est, qui non dominatur
corpori. [It is the mind which makes happiness or misery. It is a worthless
mind that does not rule the body.]   [fol. 33$^r$]
Malo vivere laetus, et pauper, quam tristis, et dives. [I prefer to live happy
and poor rather than sad and rich.]   [fol. 69$^r$]

    Eutrapelus has a predilection for irony and finds it a useful tool. On
fol. 52$^v$ Harvey writes (in Latin): 'Eutrapelus is he who laughs at solemn
matters and very deftly accomplishes serious things everywhere', and on
50$^v$: 'Irony within, elegance without, entelechy round about'. On 48$^v$ he

fearless, or Tycho more skilled, or Drake more powerful, or Du Bartas more divine in any
way.]'
    $^{95}$ I have been unable to identify 'Doctor Daro'. Perhaps Harvey meant 'Doctor Dalo'
about whom there are annotations in the folio Livy on sigs. Z5$^r$ and YYY5$^r$ which
indicate that he was ambassador to France and to the Low Countries and was an acquaintance whom Harvey admired. Jean Bodin, the French political scientist was a friend
to whom Harvey frequently refers in his marginalia. He probably owned copies of Bodin's
Methodus Historicus and De Republica. Corro (discussed on p. 157 above) was associated
with John Wolfe's printing house. Sir John Skinner was son of a Lord Mayor of London;
Sir Jon Woodes, as previously mentioned, was a nephew of Sir Thomas Smyth.
    $^{96}$ He writes: 'No more scribling: but enjoy the excellent, & divine notes, which you have
alreddi written.'

comments: 'The irony of Eutrapelus is impregnable',[97] and then alludes to two occasions when irony was formidably used by others:

The Baker of Bononia [i.e. Bologna], & owr Martin Martprelate, two good od fellows, made it a pollicie, to jest at reverend Fathers, & all solemn ceremonies as matters of show, not of bonde, or valour—the stratagem of sum greater politicians in flore.

Harvey continues on fol. 49[r], this time in Latin:

To speak the truth in a laughing manner, what does forbid it? [Harvey here inscribes his symbol for eloquence.] The letters of a certain baker of Bologna to the Pope, faithfully turned from the Italian original printed at Florence. By cutting to the quick, certain matters are thus able to be touched upon, disdain having been deeply fastened by the stings of a jest. Whence it is prudent to arm oneself with an efficacious art, or with a serious skill, or with any cogent force which depends upon oneself and spurns all men of the world.[98]

On fol. 23[r] he writes: 'Irony, the remedy at hand for all major things. The device at hand for all good things, Entelechy. Worldly vanity even demands it'.[99] On fol. 43[v] he clarifies further: 'Eutrapelus is all spirit and pure industry and nevertheless sport and jest precede accomplishment itself'.[100] In his youth Harvey may have frequently been told that he was too serious and that he should indulge more in play and jest. Spenser's prescription of the reading of *Howleglas*, Scoggin, Skelton, and *Lazarillo*[101] certainly suggest this as does the following note on fol. 15[r]:

P. Cevalerius, citizen of Geneva, and likewise a great Professor of religion and the Hebraic language,[102] often was in the habit of saying to me: Come Sir,

---

[97] The Latin originals of the three annotations quoted above are: 'Ridet graves iste Eutrapelus: et levissimus peragit longe gravissima'. 'Intra Ironia: extra euschemosyne: circa entelechia'. 'Eutrapeli, ironia, inexpugnabilis'.

[98] 'Ridentem dicere verum, quid vetat? Litera cuiusdam Pistoris Bononiensis ad Papam. Fideliter versa ex Italico exemplari, impresso Florentiae. Sic gravissima quaeque possunt ad vivum perstringi, alte defixis aculeis faceti contemptus. Unde prudentis est, seipsum armare efficaci arte; aut seria facultate; aut aliqua cogenti vi, quae a seipso dependeat, et omnes spernat mundi'. I have been unable to identify the 'baker of Bologna'.

[99] 'Maiorum omnium praesens remedium, Ironia. Bonorum omnium, praesens machina. Entelechia. Mundi etiam vanitas exigit'.
    On fol. 172[v] of Guicciardini he writes: 'Ironia, a present restorative at all oppositions, or afflictions'.

[100] 'Totus est spiritus, et mera industria Eutrapelus; et tamen ludus iocusque prae ipso Enteleche'.

[101] See p. 49, above.

[102] P.[rofessor] Cevalerius (Anthony Rodolph Chevallier) was at Cambridge as Hebrew lecturer from 1568–72. In C. H. and T. Cooper's *Athenae Cantabrigienses*, Cambridge, 1858–61, I, 306–8, he is described as a remarkable scholar and teacher and a very modest and pious man.

let us go into the fields for laughing. Moreover, when first we came into the fields, instantly without any cause he used to laugh in a stentorian tone. Nor did he stop laughing with the loudest voice, as though insane, until his breath failed. Ha-ha-ha. Hi-hi-hi. Ho-ho-ho. Ha-hi-ho. Nor did he cease at all until he had almost burst open. Never have I seen such a small man, so gigantic in laughing.[103]

However, whatever the attempts to stimulate humour and playfulness in Harvey, his personality seems to have taken a sardonic turn, possibly because of his embittering experiences, and he developed a rather acid wittiness and irony, especially evident in the *persona* of Eutrapelus.

The competitive Eudromus, on the other hand, is an extrovert who prefers 'to be a critic of the world rather than a critic of words' and 'to be contemporary rather than antiquarian' (fol. 40$^r$). Of him Harvey writes:

The life of Eudromus: multitechnical and multipragmatic. O how many insignificant studies there are! Alas, distinguished studies how few! To study the highest matters, to search out the highest, to unravel the highest, to teach the highest, to follow, to expand, it is my life. [fol. 10$^r$]

With facility Eudromus skims through all books for his worthwhile passages, which neither are many nor individually lengthy but are the choicest memories of authors and the only treasury of all libraries. Hidden are the secrets of his ownership and the unique practice of Eudromus. The common or the insignificant he bypasses: matters of grammar, philology, imaginary and superfluous things. He cares only for energetic, distinguished matters of superior forcefulness. Nor, indeed, does he deem worthy of his serious study the rare matters themselves except to whatever extent they are very dynamic. The pure choice of the greatest. [fol. 41$^v$][104]

---

[103] 'P. Cevalerius, Civis Geneuensis, idemque magnus religionis, et Hebraicae linguae professor, mihi saepe solebat dicere: Veni Domine: Eamus in campos ad ridendum. Cum primum autem in campis essemus, statim sine ulla causa ridebat stentorice. Nec cessabat altissima exclamatione ridere, ut insanus; donec spiritus deficeret. Ha-ha-ha. Hi-hi-hi. Ho-ho-ho. Ha-hi-ho. Nec finem ullam faciebat, donec fere rumperetur. Nunquam vidi talem pigmaeum, tantum agere in ridendo gygantem'.

[104] 'Malo esse Cosmocriticus quam Logocriticus. Et mavult Eudromus esse hodiernarius quam antiquarius'. [40$^r$]

'Vita Eudromi (polytechnia et polypraxia). O quot minuta studia? Heu egregia quam pauca? Discere summa; exquirere summa; retexere summa; summa docere; sequi, extendere, vita mea est'. [10$^r$]

'Facile omnes libros pervolutat Eudromus pro suis unius axiozelis locis: qui nec multi sunt, nec longi: sed exquisitissima tot auctorum mnemosyna, et soli omnium bibliothecarum Thesauri. Suae occulta proprietatis arcana: et unica Eudromi autopraxia vulgaria, aut minuta obiter praetermittit: grammatica, philologica, phantastica, ociosa omnia: sola curat energetica, egregia, praepotentia. Ne ipsa quidem rara suo serio dignatur, nisi quatenus praevalida. Merus maximorum delectus'. [41$^v$]

A good many of the manuscript notes in Domenichi are Harvey's exhortations to himself to try harder, for instance:

Everie day better, & better. . . . Continual improovement of everi part. [fol. 8ᵛ]

Whatsoever you have to speake, or do; do ever the Best you can. He often forgettes himself, & becummes vulgar, that doth not all wais his Best. The Best woold be refined with habitual perfection. The sublimation of art, & vertu multiplied. An extent of secrets: and emproovement of stratagems. Continual multiplication of Art, Vertu, & Fortune.   [fol. 32ᵛ]

The best is but well: none well enowgh.   [fol. 16ʳ]

Harvey expresses firm ideas on proper health measures. On fol. 62ᵛ he specifies: 'Diaeta ingenii, et exercitium agilitatis; quotidianus. [A natural diet and exercise of agility daily.]' Throughout, he advocates a simple diet, regular exercise, and abstention from excesses of all kinds.[105] On fol. 24ʳ he describes Julius Caesar's habits as an admirable example since he was sparing of meat, wine and sleep, daily walked vigorously up to the point of a 'hygienic sweat', and abstained from 'agitation of the senses'.[106] Harvey himself seems to have followed similar procedures: Exercise, a minimum of sleep, moderation in eating, and near-abstention from drinking, and as he grows older cautions himself against indulgence in venery (see p. 189 below). Excessive wine-drinking and banqueting evidently affected him adversely, for on fol. 1ᵛ he reminds himself: 'Eutrapelus, [who] always is quite animated, is never duller than among the wine goblets and at banquets, a table companion of little value who becomes dull in wine and meat'.[107]

He is obviously embarrassed by this 'disability', for on the same page he adds: 'He who is not a man of all hours is an ass of exceptional hours.'[108] On fol. 23ʳ he again alludes to the subject of wine-drinking, querying: 'What language speakes this wine, pure Spanish, or nett French, or right Italian? Is it brave or fine wine?' Below this he comments: 'It is a pretty thing but does not have bearing on my present course. Don't touch me. All things in due time'.[109]

105 In Guicciardini on sig. Ff3ᵛ he observes: 'Repletion, & venerie, cheife causes of diseases: Exercises, used with discretion, causes of health. Noman sturres everie part enowgh'.

106 'Caesar cibi, vini, somni parcissimus. Maximeque Caesaris *ambulatio, plus* quam, Asclepiadis Exercitationes. Nec eget alia abstinentia, frictione, gestatione, aut articulorum, sensuumque agitatione; cui Caesaris diaeta, ambulatio, quotidianus sudor; generosissima mundi disciplina. Caesaris diaeta, ambulatio, sudor unica Hygiena'.

107 'Interpocula, aut epulas nunquam hebetior Eutrapelus, sed semper vividior. Vilis conviva, qui hebescit vino, aut cibo'.

108 'Qui non est homo omnium horarum, asinus est exceptarum horarum'.

109 'Pulchrum est; sed non pertinet ad meum iam cursum. Noli me tangere. Omnia tempus habent'.

In later notes in this volume (on fol. 43ʳ) he observes, 'Vino corrigitur: vino corrumpitur aetas. [Age is corrected by wine and corrupted by wine.]' and 'Moderatus Bacchus corroborat: immoderatus enervat. [Moderate drinking strengthens, immoderate enervates.]'

Some of the Domenichi annotations relate to service at court. On fol. 27ᵛ Harvey depicts what he aims to be: 'The most seemly of courtiers, the most discriminating of those skilled in civil affairs, of all the most highly esteemed, indisputably as preeminent as honest'.[110] The inscription on fol. 53ʳ probably reflects a disillusioning experience:

His duobus, La Corte, la sorte.
Illis duobus millibus, La Corte, la morte.
[For these two men: the court and fortune.
For those two thousand men: the court and death.]

A note on fol. 50ᵛ is perhaps related to this period: 'Aparrel alwais clenlie, & hansom, & sumtime fine: but never vaine, or glorious'.

The Domenichi marginalia also include a few contemporary allusions. There are several references to the Sherley brothers, one on fol. 18ʳ pointing out that worthy service at home is as important as more spectacular actions abroad: 'O brave Sir Antonie Sherlie at larg: but O worthie Earle of Salisburie at hand'. The Sherley brothers became renowned for their brave exploits and breathtaking adventures in foreign lands, but Harvey apparently realizes that more was accomplished by the conscientious and quiet administrative work of Robert Cecil on the home front.

On fol. 50ʳ Harvey alludes to the death of Jean Bodin in 1596: 'Politique Monsieur Bodine (ô simple Bodine) died in the streets for want'.[111]

After praising Roger Ascham's Discourse of . . . Germany [1570?], John Heywood's Proverbes and Epigrams [1562] and Sir Thomas Chaloner's [1549] translation of Erasmus's Encomium Moriae on fol. 59ᵛ, Harvey notes on 60ʳ that it was customary for Chancellor [Nicholas] Bacon and Treasurer [William] Cecil to have these three works at their fingertips.[112]

On fol. 28ʳ is a cryptic inscription: 'Ad omnia opposita: non querela,

[110] 'Aulicorum decentiss. pragmaticorum subtilissimus, industriorum diligentiss. omnium facile tam praestantis. quam candidissimus'.

[111] I am grateful to P. L. Rose for the following comment: 'I think that Harvey may be wrong about "Politique Monsieur Bodine" dying in the streets of want. He seems to have had a decent legal business at Laon in his last years and some good patrons. He also owned a little property. He actually died at home in 1596 of plague and was buried in church. (See Roger Chauviré, Jean Bodin, La Flèche, 1941.)'

[112] 'Unde oportet illas etiam tres habere ad unguem: additis necessaria addendis. Sic qui maxime dudum excellebat, Bacon Cancellarius; non multa sed multum. Sic adhuc Caecilius Thesaurarius sensi saepius'.

sed medela. [At all interventions: not quarrels but healings.] With Sir
Francis Veres Tutt-tutt a figg: quoth A.O'.

Sir Francis Vere, commander of the English army in the Netherlands,
was noted for his successful sieges. 'A. O' may be the Lady Alpha Omega
(of John Grange's *Golden Aphroditis*, 1577) who at one point advises her
discouraged suitor, Sir N. O., on shrewd 'military' strategy.

On fol. 61ʳ below a paragraph in the text about the disciplining of a
presumptuous and unscrupulous peasant, Harvey writes: 'A posse ad
esse [From possibility to being], who but HEE?' This is undoubtedly a
reference to the unscrupulous behaviour and statements of Thomas
Nashe. But Harvey is not a snob, intellectually at least, because one finds
many notes showing his appreciation of uneducated wisdom, for instance:
'I have heard more witt from sum apt Mercurial boyes, then from manie
learned & grave men'. [50ᵛ].[113]

A recurrent theme of the manuscript notes in Domenichi is the necessity
of depending solely upon oneself. The following (on fol. 31ᵛ) is a typical
example:

He who is not dependent upon himself is not a man. From which it follows that
as much fortune as art or virtue is to be based in yourself; and unworthy sub-
jugation or contemptible need are not due to anyone else. To be sufficient unto
oneself is magnificent; to beg for the help of another is abject. May he who is
able to be his own not be another's.[114]

Harvey likewise believes in self-sufficiency in the field of learning,
repeatedly stating that the only real knowledge is what one has absorbed
and is able to use extemporaneously, for example:

Non est doctus, qui non est sua bibliotheca.[He who is not his own library is
not learned.] [Domenichi, fol. 63ᵛ]

Nondum viri sunt, qui egent libris ad dicendum, aut agendum peritissime.
Clericus in libro, non valet obolo. [Not yet men are those who need books for
speaking or performing skilfully. A clergyman in a book is not worth an obol.]
Bookewormes: & Scriblers: pen & inkhorn men: paperbook men, men in
their bookes or papers: not in their heds, or harts.   [Guicciardini, sig. F5ʳ.]

Lodovico Guicciardini's *Detti et Fatti* (1571), from which the last

---

[113] Nor is Harvey a pedant. In his Erasmus, *Parabolae* (1565) on sig. h3ᵛ, he finds commendable a passage because it is 'Against those, that go abowte to make shewe of all their
lerninge atonce'. In Ramus's *Oikonomia* on sig. K1ᵛ is the following manuscript note:
'The cunningest in schoole, May learne of many a foole'.

[114] 'Vir non est, qui a seipso non dependet. Unde in teipso tam est fundanda Fortuna,
quam ars, aut virtus. Ne cuique servias vilis: aut cuiusque indigeas contemptibilis. Sibi ipsi
sufficere, magnificum est: alterius opem implorare, abiectum est. Alterius non sit, qui suus
esse potest'.

marginalia come, is, as previously explained, bound following the Domenichi work and is profusely annotated in a similar manner. A number of
the Guicciardini marginalia have already been mentioned; of the others
the following are noteworthy.

On fol. 147ᵛ Harvey refers to advice given to him by John Young when
Bishop of Rochester (1578–1605):

Leave scribbling: quoth Rochester: & now in deed to the purpose. Aut me
decepisti: aut tarditatem compensabis gravitate. Prudens aliquando sero sapit;
nunquam nimis sero. [Either you have deceived me or you will compensate
for tardiness by authority. The skilled one sometimes is wise at a late hour (but)
never at too excessively late an hour.] So that sharp Bishop to miself, & sum
other whome he thowght as sufficiently qualified, as Dʳ Lewen, Dʳ Clark, or
other fine pragmaticians in the sun.

Young seems to have believed firmly in Harvey's outstanding abilities and
that he would eventually achieve eminent status as had William Lewin
and Bartholomew Clerke. Harvey himself seems to have bolstered up his
hopes with comments on late achievers such as the following:

The natures of men ar not ripe in few yeares: & what wittes soone mellow, to
consider, do, & suffer; with sound discretion, & fitt distinction? Homerus,
Hesiodus, Pindarus, maioribus nostris optimi poetae: quia nescierunt meliores.
Sic optimi oratores, Demosthenes, Isocrates, Cicero. [Homer, Hesiod, Pindar,
the best poets to our ancestors since they did not know better ones. Likewise
the best orators: Demosthenes, Isocrates, Cicero.]    [Guicciardini, fol. 85ᵛ.]

On fols. 91ᵛ–92ʳ Harvey quotes a flattering appraisal of him by William
Bird (1561–1624), a member of the influential Walden family (later Sir
William Bird, M.P. and Judge of the Prerogative Court of Canterbury).[115]

William Bird spoke thus about Axiophilus: I never saw that man angry, never
sad, never fearful, never irresolute, never lazy or inactive, never unmindful of
himself or of others, never thrown into confusion on any occasion. I have not
known a man more mindful of kindnesses or more forgetful of injuries, I do
not know anyone whose principles and actions emanate so much from himself.
These things that level-headed gentleman frankly [said] about Axiophilus.[116]

Several humorously urbane leaves of the Guicciardini text treat of

[115] There is a large tract of land on Little Walden Road which is still known as 'Byrd's
Farm'. Christopher Bird of Walden (d. 25 October 1603) whom Harvey addresses in the
second of his *Foure Letters* (1592) was probably a member of the same family.

[116] 'Sic Gulielmi Birdi dictum de Axiophilo. Illum hominem nunquam vidi iratum;
nunquam tristem, nunquam timidum; nunquam irresolutum; nunquam ignavum, aut
languidum; nunquam sui, aut aliorum immemorem; nunquam perturbatum ullo casu.
Non novi hominem, beneficiorum magis memorem; aut iniuriam magis immemorem.
Nec scio aliquem cuius rationes, et actiones ita dependent a seipso. Haec ille sobrius genero-
sus de Axiophilo ingenue'.

women's wiles and the foolishness of old men in love. Harvey in his later years jots down his own reflections on these pages. On fol. 169$^v$ he writes that it behoves a courtier to speak in an honourable way about women but that they are often scheming politicians and one should follow Julius Caesar's advice to be cautious about involvement with them.[117] On fol. 170$^r$ Harvey reminds himself:

Read often to yourself and your dear ones.
No sexual love ought to be allowed in the decline of life.
After fifty all coitus is pernicious.[118]

On fol. 170$^v$ he cautions:

Earnestly being prudent after fifty-five, altogether no sexual love or any wasting titillation of the spirits of life. The life of Julian, golden practice, as prudent as strong.[119] Not 'to live life' but 'to be vigorous in life'. In an ideal diet, exercise, chastity, and joy there is a certain immortality. Many mortals perished before their time in dissipation or concern.[120]

Although Nashe writes that in Harvey's youth he was always in love,[121] his interest in women apparently waned or was suppressed in later years as his prime consideration became the preservation of his health, strength, and tranquillity. That he reached the age of eighty is perhaps a tribute to his preoccupation with physical and mental well-being.

In later years he evidently was able to maintain a *joie de vivre* and to avoid whatever earlier tendencies he may have had toward melancholy, for on fol. 79$^r$ of Guicciardini he writes: 'Eutrapelus is always joyous,

[117] 'Aulici est, loqui honorifice de foeminis. Viri serii noli me tangere: nisi torte politice; ut arcanorum causa, acerrimus in omni genere pragmaticus, Caesar. Alioqui spretor foeminarum, quoties gerenda aliqua res virilis, parcissimus cibi, vini, somni; idemque laboriosissimus. Quae probant veneris rarissimum usum; nec nisi politicum'.

[118] '⎰Lege saepe et tibi, et tuis carissimis.
⎱Nulla Venus permittenda declinanti aetati.

Post quinquagesimum, omnis coitus perniciosus'.

[119] Harvey was a great admirer of the Roman Emperor Julian ('the Apostate') and perhaps in some respects such as his great learning, ascetic life, and failure to realize his dreams, Harvey identified with him. On fol. 178$^r$ of Guicciardini are the following notes: 'Juliani vita, a Martinio concinnata, synopsis virtutum, et historiarum omnium. [The life of Julian gathered together by Martinius, a synopsis of virtues and of all histories.]' The book to which reference is made is probably the Greek–Latin edition of Julian's *Misopogon* and his letters, edited by Petrus Martinius and prefaced by Martinius's life of Julian. It was published in Paris in 1566.

[120] 'Serio prudens, post quinquagesimum quintum, nulli omnino utitur Venere, aut titillanti aliquo spirituum consumptione. Juliani vita; aurea praxis t[am] prudentium quam fortium. Non est vivere, sed valere, vita. In aurea diaeta, et exercitatione; castitate, et laetitia, quaedam immortalitas. Plaerique mortales sua luxuria, aut cura ante diem perierunt'.

[121] *Works*, III, 81.

more joyous, most joyous, of all living men easily the most joyous . . .
Happy is he who continues unvanquished up to the end'.[122]

This study can offer only a sampling of the variety and interest to be
found in Harvey's marginalia. There is really no adequate substitute
for examining the volumes themselves. Some notes are, of course, cor-
related with the text, but most are relatively extraneous (some even being
written upside down thus emphasizing the fact) and most are supplemented
by the succinct marginal symbols to which reference has previously
been made.

The inscribing of Harvey's notes was only part of the pleasure and
benefit he derived from them; he spent much time rereading, mulling
over them, and repeating to himself those points which he wanted to
imprint on his memory. On fol. 36[r] of Domenichi he counsels himself:
'Enjoy the soverain repetition of your most excellent notes. Quotidie lege,
lege: sed quotidie repete, repete, repete. [Daily read, read; but also daily
repeat, repeat.]'

Harvey had undoubtedly read and studied a number of memory
treatises; he makes several references to Ravenna.[123] In Guicciardini
on fol. 97[v] he inserts the following: 'It is not bookes, that make the skill-
fulman, but the knowledg of bookes: & the memorie of knowledg, &
the practis of memorie, both in words, & in deeds. He deserves to be
esteemed the most cunning man, that can best negotiate his lerning,
*viva voce* & *vivo opere*'. The pragmatic use and negotiation of his learning
was Harvey's aim—to be involved in noble and useful action. But he
achieved no spectacular success as the 'active man' although his dreams
of such achievement continued strong. In his later years he seems to
have mellowed and to have turned more to his family[124] and to his Walden
neighbours to whom he probably gave medical and legal counsel, gaining
increasing reputation as a local sage. Perhaps this helped to ease the pangs
of frustration he must have felt. Certainly he found considerable solace
in his magnificent library and marginalia of a lifetime.

---

[122] 'Eutrapelus semper laetus, laetior, laetissimus, omnium vivorum facile laetissimus.
Tamen prudentior quam laetior . . . . Foelix, qui ad finem usque perseverat indomitus.'

[123] On sig. A6 of Franciscus Duarenus's *De Sacris Ecclesiae* (1564) he refers to the 'memorative compendium of Petrus Ravenna in 4°'. Petrus Tommai (Peter of Ravenna) was author
of *Phoenix, sive artificiosa memoria*, first printed in Venice in 1491, translated by R. Coplande
and published *c.* 1548 as *The art of memory*. This was one of the best-known memory text-
books. (See Frances A. Yates, *Art of Memory*, 1966, (Peregrine reprint), pp. 119–20.)

[124] The reference on fol. 170[r] of Guicciardini to reading to his 'dear ones' suggests this
as do the disfiguring childish scrawls (possibly the work of Harvey's young nephews or
nieces) found on the title-pages and beginnings of some of his books, e.g. in Florio on
sig. ✶✶✶3[r].

# PART III

# HARVEY'S LIBRARY

# 1. General description of the library

THE titles in an individual's library are usually an excellent indicator of his interests. In Harvey's case this is certainly true. Not surprisingly, what is extant of his library reflects most of the intellectual and practical interests of his times since he was a man of whom it can truly be said, 'all learning was his province'. His books, more than half of which are in the vernacular, are, as we have seen, the working library of a learned man of all-encompassing interests: history, government, law, languages, medicine, science, and literature, both in their classical foundations and in their topical relevance. His marginalia in these volumes disclose the value which he attached to them.[1]

Several other large personal libraries of the period have been investigated in some detail. In 1956, Sears Jayne and Francis R. Johnson edited the 1609 catalogue of the famous Lumley library. Of Lord John Lumley's 2,675 volumes about 88 per cent were in Latin, and the majority of his books dealt with theological subject-matter. The emphasis on theology and Latin texts was typical of the larger collections of the period, as shown by Sears Jayne's listing in his *Library Catalogues of the English Renaissance* (Berkeley, 1956), a tabulation of booklists found in wills, inventories, and bequests. An extensive, largely theological library which has not to my knowledge yet been recorded in print is that of Harvey's nemesis Dr. Andrew Perne. A parchment roll in the Cambridge University archives[2] contains a 1589 inventory which includes a listing (with prices) of his approximately two thousand books and manuscripts in the Master's Lodge at Peterhouse at the time of his death. An interesting shelflist of very different subject-matter is that in Sir Walter Ralegh's hand briefly itemizing about five hundred books which he owned: chemical, medical, geographical, cosmological, and many historical ones probably used in the composition of his 1614 *History of the World*.[3]

One of the most famous English libraries of the period was that of the great mathematician, astronomer, and magus Dr. John Dee at Mortlake. An excellent scientific library covering other fields as well, it was consulted by many of his contemporaries and even drew a visit from Queen Elizabeth. As Peter J. French notes in his chapter on Dee's books: 'Any library of more than a few hundred volumes must be considered remarkably large for the sixteenth century. . . . In

---

[1] The material for much of Part III, including the major part of the catalogue of Harvey's library, originally appeared in *Renaissance Quarterly*, 25, 1 (Spring 1972, pp. 1–62). I am indebted to the Renaissance Society of America for permission to include this material here.

[2] Box D11, 191, 10 May 1589.

[3] See Walter Oakeshott's 'Sir Walter Ralegh's Library', in *The Library*, Fifth Series, 23, 4 (December 1968), 285–327.

1582, the University Library at Cambridge only had about 451 books and manuscripts.'[4] On 6 September 1583 just before his departure for the continent, Dee
made a library catalogue in his own hand,[5] listing about 170 manuscripts and
several thousand printed works. Dee estimated the total number of titles at
'neere 4000'.

A large secular collection of a slightly later period was recently investigated
by Robert H. MacDonald. Early in 1971 he edited an informative study of the
1,745 titles in the library of the poet William Drummond of Hawthornden
(1585–1649), many of whose books reflected his appetite for drama and for
literature in the vernacular, only about half of his volumes being in Latin.[6]

Unlike Lumley, Perne, Ralegh, Dee, or Drummond, neither Harvey nor any
of his heirs or executors seems to have left a contemporary catalogue of his
holdings. The earliest listing that I have found is that of W. Carew Hazlitt in
his chapter 'Gabriel Harvey' in Bernard Quaritch's compilation, *Contributions
toward a Dictionary of Book Collectors*, Part xiii (London, 1899). Hazlitt enumerated twenty-five printed books and one manuscript commonplace book. In 1913
G. C. Moore Smith published his important *Gabriel Harvey's Marginalia* (Stratford-upon-Avon, 1913) in which he drew from Hazlitt's material and his own
findings to reconstruct a list of fifty-three printed books plus nine manuscripts.
Frank Marcham, Moore Smith in later years, William A. Jackson, Harold S.
Wilson, and D. M. Rogers have each located a number of additional marginalia
volumes,[7] and readers' responses to my paper in the spring 1972 issue of *Renaissance Quarterly* have elicited a few further titles.[8]

The catalogue of the present study comprises one hundred and eighty printed

[4] *John Dee: The World of an Elizabethan Magus*, London, 1972, Chapter 3, p. 44.

[5] There are extant two copies of Dee's catalogue in his own hand: Harleian MS. 1879,
arts. 5–6 and MS. 0.420 at Trinity College, Cambridge.

[6] *The Library of Drummond of Hawthornden*, ed. with an introduction by Robert H.
MacDonald, Edinburgh University Press, 1971.

[7] Frank Marcham, *Lopez the Jew, An Opinion by Gabriel Harvey*, with some notes of
other books annotated by Harvey (Middlesex, 1927); Moore Smith, *MLR*, 28 (1933),
78–81; *MLR*, 29 (1934), 68–70, a report by Moore Smith of the findings of William A.
Jackson; *MLR*, 30 (1935), 209, additional discoveries by Moore Smith and Jackson; Harold
S. Wilson, *HLB*, 2 (1948), 344–61; and in Wilson's manuscript notes at the Centre for
Reformation and Renaissance Studies, Victoria University, Toronto; and David M. Rogers,
of the Department of Printed Books at the Bodleian Library, who has kindly written to
me of his findings at Oxford.

[8] I am grateful to the Revd. Walter Ong for calling to my attention the copy of Cicero's
*Topica* in the Codrington Library at All Souls, Oxford. Emmanuel College, Cambridge,
advised me of their Justinian's *Institutes*; Durham University Library apprised me of the
Richard Cosin volume. Paul Morgan of the Bodleian wrote me of a number of his findings
at Oxford: at Balliol, *Epistolae obscurorum virorum* and *De generibus ebriosorum*; at All Souls,
the N. Barnaud and Gui Du Faur as well as the Omer Talon, *Academia* volume which also
contains Talon's commentary on Cicero's *Lucullus* and on his *Topica*. The three last have
been included in Walter Ong's definitive *Ramus and Talon Inventory* (Folcroft, Pa., 1958).
I am also grateful to Clifford Huffman for identifying one of the books referred to in
Harvey's *New Letter* (1593) as *A True Discourse . . . the wonderful mercy of God shewed toward
the Christians against the Turke before Sysek*.

books, a broadsheet, a folio sheet, and ten manuscripts.[9] It has been compiled by a variety of methods. At the outset I assembled in one listing titles recorded in any of the above sources and endeavoured to track down to their present whereabouts any of Harvey's books mentioned in old auction-sale catalogues. By writing to and visiting libraries and a few private collections throughout this country and England, I have unearthed a considerable number of additional titles. By personally inspecting nearly all of Harvey's already identified books I have gathered further information about them; for instance, I have discovered a previously unrecorded folio leaf bound at the end of Hieronymus von Braun-schweig's *Apothecarye* and ascertained that Harvey owned two English editions of Machiavelli's *Arte of Warre*, whereas only one had been recorded previously. In leafing through a Scolar Press facsimile of John Hart's *Orthographie* I recog-nized as Harvey's previously unidentified marginalia and subsequently located the original volume from which the facsimile was made. Quite by chance I have discovered two of Harvey's Italian titles in the Columbia Library (Battista Guarini's *Il Pastor Fido* and Torquato Tasso's *Aminta*), and while visiting the home of a bibliophile friend one evening, I turned up an Italian edition of Ptolemy with notes unmistakably in Harvey's hand. Sometimes, of course, I have found a book tentatively classified as Harvey's, which upon inspection does not seem to warrant such identification, and I have therefore omitted it from the catalogue.[10] An important source for identifying further titles has been Harvey's own writings. In his marginalia, his letters, and in his individual works he refers to certain books which he indicates he owns. There are others to which he alludes so frequently and so specifically (with chapter and line references) that it seems certain that they were also in his possession even though he fails to state that they belong to him. However, unless Harvey's particular copy has been found, I have placed an asterisk in the subsequent catalogue at the left of the title to signify that it is still unlocated. Sometimes a title referred to by Harvey can be found (in a work of the proper date and size, for this information is sometimes provided by him) entered in the 'British Museum Catalogue of Printed Books' or other library catalogue together with a notation that it con-tains unidentified manuscript notes. Occasionally when I have inspected such a volume, I have had the good fortune to come upon one which has the earmarks

---

[9] There is also appended on p. 264 below a list of twenty-six additional titles, probably owned by Harvey, for which there is as yet insufficient evidence to warrant a place in the catalogue proper.

[10] I find it questionable that the marginalia in any of the following are Harvey's: *A particular declaration or testimony, of the undutifull and traiterous affection borne against her Majestie by Edmond Campion Jesuite, and other condemned Priestes* (1582) (Pforzheimer Library); Abraham Fraunce, *Insignium Armorum* (1588) (Huntington Library); Gabriel Harvey, *Pierces Supererogation* (1593) (Houghton Library 14465.63.18*); Philip Sidney, *Arcadia* (1913) (Houghton 14457.23.8.7F*); William Warner, *Albions England* (1612) (Brit. Lib. C.71. c.11); and Edmund Spenser, *Faerie Queene* (1590) several pages reproduced and the copy described by Edward Almack in his *Fine Old Bindings* (London, 1913), present where-abouts unknown.

of Harvey ownership and so have been able to add it to the list of library books which he possessed.

There are a number of ways by which a Harvey-owned volume can be identified. It can be a simple matter when the title-page has not been cropped or mutilated in any way, for it is Harvey's frequent (although not invariable) practice to place his signature or monogram, often both, on the title-pages of his books. He may enter below this the date of purchase and sometimes the price paid or the name of the person who has given him the book. Harvey tends to include on the title-page of a cherished volume a comment on its contents and on its value to him.

In addition to signatures other criteria for assigning an existing volume to Harvey's library are penmanship, type, and content of manuscript notes and their placement on the page and in the volume, and chalk-like coloured markings (which are present to some degree in nearly half of Harvey's books);[11] with a few titles that I have been unable to locate I have accepted as dependable the earlier identifications of W. C. Hazlitt or of G. C. Moore Smith. In the case of a book that has never been found, my criterion has been a statement or an obvious implication of Harvey's that he owned it. Such an indication of ownership is inherent in the volumes enumerated under the manuscript heading 'Propria Eulogi bibliotheca ... 1590' (Guicciardini, fol. 209$^r$), at the beginning of *A New Letter of Notable Contents*, 1593 (sig. A2), where he expresses gratitude to John Wolfe for the various books just received from him, and in the manuscript reference to Pietro Aretino's *Comedie* (at the beginning of Euripides' *Hecuba*) after which Harvey adds 'Habeo et Legi'. In listing such not-yet-found volumes as part of Harvey's library, I have in each case indicated the evidence attesting to his ownership; in the case of extant volumes with signatures or marginalia which have already been identified, I have endeavoured to give their present whereabouts or last known locations.

In the following pages is a partial catalogue of Harvey's library, including as many Harvey-owned volumes as can be identified with reasonable certainty. Marginalia quoted are from title-pages unless otherwise specified. Although some titles may now be bound together, each has been entered separately. In cases where it seems evident that Harvey himself had a group of books bound together, this fact is noted. The catalogue is arranged alphabetically by author; a short-title index classified according to publication date has been appended. All books listed have been examined by me with the exception of those accompanied by the following notations:

[micr]                     means that I have examined a microfilm but not the volume itself.

[desc]                     designates that the description of this volume has been derived from an article listed in the Appendix E (pp. 272–3) or from a sale catalogue.

---

[11] Each of these criteria has been discussed in Part II (pp. 138–47 above).

[tpx]                          indicates that I have seen a xerox or other reproduc-
                               tion of the title and of perhaps a few other pages, but
                               have not seen the actual volume.

[whereabouts unknown] refers to a supposedly extant volume, described by
                               Hazlitt, but whose whereabouts are no longer known.

*                              at left margin, signifies a book not yet found but for
                               which there is evidence of Harvey's ownership.

In quoting Harvey's marginalia, his underlinings have been represented in
print by italics.

# 2. Catalogue of Harvey's books

*AGRICOLA, Rodolphus the Elder. *De inventione dialectica libri . . . scholijs illustrati, Ioannis Phrissemij, Alardi Aemstelredami, Reinardi Hadamarij. Quorum scholia . . . contulit ac congessit Ioannes Nouiomagus.* Cologne [G. Fabritius], 1557. 8°

Harvey evidently owned the above or one of the earlier editions of this book, for on sig. Y5ʳ of his Quintilian he refers to his copy of 'Rodolphus de inventione dialectica'.[1] In various other notes in Quintilian, Harvey commends Agricola as a dialectician and in Simlerus's bibliography underlines the above approximate title and refers to its author as 'Lux altera Germaniae'.

*AGRIPPA, H. Cornelius ab Netesheim. *Henrici Cornelii Agrippae de beatissime Annae monogamia . . . iuxta disceptationem Iacobi Fabri Stapulensis in libro de tribus & una, initulato. Eiusdem Agrippae defensio propositionum praenarratarum contra quendam Dominicastrum earundem impugnatorem . . . Quaedam Epistolae super eadem materia atque super lite contra ejusdem ordinis haereticorum magistro habita.* [Cologne] 1534 (or possibly other edition of letters). 8°

On sig. m6ʳ of the Simlerus bibliography next to the entry for H. C. Agrippa, Harvey writes: 'Et haec omnia legi, et plura Agrippae habeo; septemque in primis Epistolarum libros, et nonnulla Epigrammata.'

*—— *Henrici Cornelii Agrippae . . . Orationes X . . . Ejusdem de duplici coronatione Caroli V Caesaris, apud Bononiam historiola. Ejusdem ac aliorum doctorum virorum epigrammata.* Cologne, J. Soter, 1535. 8°

See Harvey's comment in the Simlerus bibliography, as quoted in the Agrippa entry above.

ALCIATO, D. Andrea. *De verborum significatione libri quatuor. Eiusdem, in tractatum eius argumenti veterum JureconsultoR, Commentaria.* Lyons, Seb. Gryphius, 1530.
Folio      Bodleian Douce A subt. 75(2)

'GabrielHarvejus.' Below title: 'His adjunge praetermissorum Libros duos.' Bound with two other Alciato legal tracts all of which contain brief marginal notes. On first title is: 'Gabrielis Harveij' 'pretium Xˢ' and at end of third

---

[1] For Harvey's use of this work, see L. Jardine, 'Humanism and Dialectic in Sixteenth-Century Cambridge', pp. 145–54, in *Classical Influences on European Culture A.D. 1500–1700*, ed., R. R. Bolgar, Cambridge Univ. Press, 1976.

tract: 'gabrielisharveij, et amicorum, 1578.' On second tract above on title-page Harvey has inscribed: 'GabrielHarvejus.' 'His adjunge praetermissorum Libros duos.'

——*Paradoxorum, ad Pratum, lib. VI. Dispunctionum, lib. IIII. In treis libros Cod. lib. III. De eo quod interest, liber unus. Praetermissorum, lib. II. Declamatio una. De stip. divisionib. Commentariolus.* Basle, A. Cratandrus, 1531.
 Folio  Bodleian Douce A subt. 75(1)
 'GabrielisHarveij'. Below title: 'De verborum significatione, &c. De pactis. De transactionib. &c. Parerga alio tomo. &c.' At bottom of page: 'pretium X$^s$'. The three folios of Alciato's legal discourses are bound together in what appears to be a sixteenth-century binding, this volume being placed first, although not published first. Probably ten shillings was the price for all three, as on the title-page of the 1530 volume (listed above) Harvey notes that the three books are joined.

—— *Ad rescripta principum commentarii, de summa trinitate. Sacrosanct. eccl. aedendo. In ius vocando. Pactis. Transactionibus. His accessit eiusdem de quinque pedum praescript. lib. I.* Lyons, Sebastianus Gryphius Germanus, 1532.
 Folio  Bodleian Douce A subt. 75(3)
 Elaborate monogram 'GH.' At end of volume: 'gabrielis harveij, et amicorum, 1578'. All three volumes have some manuscript notes and some underlinings.

ALKINDUS, (Jacobus) [Yakub ibn Ishak]. *Alkindus De Temporum Mutationibus, sive de imbribus, nunquam antea excussus. Nunc vero, per D. Io. Hieronymum a Scalingiis, emissus.* Paris, Jacob Kerver, 1540.
 Folio  Brit. Lib. c.60.0.8
 'iiij$^d$.' 'A: 1299.' 'pret. 4$^d$.' At bottom of the title-page is a quotation in Harvey's hand from Tycho Brahe: 'O quam mira et magna potentia Coeli est? Quo sine, nil pararet Tellus, nil gigneret Aequor'. At end of volume: 'gabrielis harveij et amicorum. 1579'.

*[ANGUILBERT, Th.] (supposed author). *Mensa Philosophica, seu enchiridion in quo de questionibus mensalibus, rerum naturis, statuum diversitate, variis & jucundis congressibus hominum philosophice agitur.* Various 15th- and early 16th-cent. eds.
 4°
 Harvey's marginalia contain many references to this book, e.g. on fol. 69 of Porcacchi to the '4$^{to}$ Mensae Philosophicae' and on fol. 69 of Wilson's *Art of Rhetoric* where he implies ownership of the *Mensa Philosophica* by observing that Wilson is 'One of my best for the art of jesting: next Tullie, Quintilian, the Courtier in Italian, the fourth of mensa philosoph.' On title-page verso of Frontinus, Harvey writes: 'My Barlandi Joci and Mensa

Philosophica: two continuall & inseparable companions of Frontin, & Polyen'.

APHTHONIUS, the Sophist. Ἀφθονίου Σοφιστοῦ Προγυμνάσματα. *Aphthonii Sophistae Praeludia. Cum interpretatione Rodolphi Agricolae Phrysii.* [Paris?], 1543.
   8°.     St. John's College, Cambridge Aa.3.30
'Gabrielis Harveij'. No MS. notes. Introductory textbook of rhetoric comprising Greek text followed by Latin commentary.

*ARETINO, Pietro. *Quattro Comedie. Cioè Il Marescalco, La Cortegiana, La Talanta, L'Hipocrito.* [London, John Wolfe] 1588.
   8°     STC 19911
On fol. 7 (a flyleaf) of his Euripides, Harvey writes: 'Quattro Comedie del divino Aretino.
   Cioè, Il Marescalco, o pedante.
   La Cortegiana.
   La Talanta.
   Lo Hipocrito.
Habeo et legi.'

ARISTOTLE, Ἀριστοτέλους Τέχνης ῥητορικῆς. *Aristotelis de Arte Dicendi Libri Tres . . . a Petro Victorio correcti & emendati. IIdem Latinate donati per Hermolaum Barbarum.* Paris, Vascosanus, 1549.
   8°     Earl of Leicester, Holkham, Norfolk.
   Harvey has written: 'Dedit Gabrieli Harvejo. 1572'. below is the following inscription in another hand: 'Sum Thomae Brouni'. Thomas Browne was translator of John Sturmius' *Nobilitas literate,* which Harvey may have owned. At top of the Aristotle title-page Harvey has written in Greek: 'The decisions of the best men are best'. In the first of the three books are Harvey's marginalia, almost completely in Greek.                                        [micr]

—— Ἀριστοτέλους ὄργανον. Πορφυρίου εἰσαγωγή. [Aristotle's *Organon* & Porphyry's *Isagoge.*] 2 pt., Paris, Morelius, 1562.
   4°     [whereabouts unknown]
   As described in the Samuel Weller Singer sale, Sotheby, 3 Aug. 1858: '[Harvey's] autograph signature, initials, and copious MS. notes, in his very neat hand' including 'Hactenus in Aula Pembrochiani publici ad huc Artium Baccalareus et illis Collegii Socius'; and 'Hactenus legi in Collegio Christi in Publica Aula jam tum Artis Baccalareus'.

*ASCHAM, Roger. *Report and Discourse of the Affaires and State of Germany, and the Emperour Charles, his Court.* London, John Day [1570?].
   4°     STC 830
   In Guicciardini, *Detti, et Fatti* (fol. 209ʳ), Harvey has inscribed the heading

'Propria Eulogi bibliotheca. Poco y bueno: 1590. pro Eulogo'. Below it he has listed a group of titles, one of which is 'Aschami discursus politicarum Caroli 5 negociationum'.

*ASTLEY, John. *The art of riding, set foorth in a breefe treatise.* London, H. Denham, 1584.

    4°    STC 884

Numerous references to this book in marginalia in Blundevill's *The foure chiefest Offices.* On verso of the title-page Harvey writes: 'I use M$^r$ Astley for the compendious, & fine Art: & M$^r$ Blundevil for the larger & fuller Discourses upon the Art: & commend them both, for two right-proffitable, & gallant Writers, in the excellent veine of Xenophon'.

*BALDWIN, William (ed.). *A Myrroure for Magistrates.* London, T. Marshe, edition of 1559, 1563, or 1571.

    4°    STC 1247–49

Mentioned in letter to Arthur Capel *c.* 1573 as one of the books Harvey has recently lent him (see Sloane MS. 93, fol. 90$^v$ or *Letter-Book,* p. 167) and is referred to in Gascoigne's 'Jocasta' on sig. A2$^r$ in the *Posies* volume.

*BARLANDUS, Hadrianus Cornelius. *Iocorum veterum ac recentium duae centuriae; cum scholiis per H. Barlandum: Iovani Pontani . . . de Grammaticarum contentione dialogus, cum eiusdem scholiis. Primae aeditioni nunc adiecti sunt libri duo.* Antwerp, M. Hillenius, 1529.

    8°

On title-page verso of Frontinus, Harvey writes: 'My Barlandi Joci and Mensa Philosophica: two continuall & inseparable companions of Frontin, & Polyen'.

BARNAUD, Nicholaus. *Dialogus quo multa exponuntur quae Lutheranis et Hugonotis Gallis acciderunt. Orange,* Adam de Monte, 1573.

    8°    All Souls, Oxford S.R. 63 c.2(1)

On title-page is the signature 'Gabrielis Harveij' scored through. The tract has no inscriptions or underlinings except on title-page. This anti-Lutheran dialogue is bound with the Du Faur pamphlet.     [tpx]

*[BARNES, Barnabe]. *Parthenophil and Parthenophe. Sonnettes, madrigals, elegies and odes.* [London, John Wolfe, 1593.]

    4°    STC 1469

On sig. A2 of *A New Letter of Notable Contents* (1593) Harvey refers to this as one of the recently-read books received from Wolfe. It is a collection of verses whose publication was (according to Nashe, *Have with you to Saffronwalden,* sig. O2) advocated by Harvey. In the British Library is a copy of

this work (6132.i.50) with notes in several hands. That found at the top of
the title-page resembles Harvey's, but not conclusively so.

[BILLERBEGE, Frauncis]. *Most Rare and straunge Discourses, of Amurathe the
Turkish Emperor that nowe is: Of his personne, and howe hee is governed: with
the warres betweene him and the* Persians: *and also of the* Tartars *and the* Musco-
vites: *Of the peace concluded betweene King Phillip and the great Turke: the Turkish
triumph, lately had at Constantinople: with the confession of the Patriarke of
Constantinople, exhibited to the great Turke.* London, Thomas Hackett [1584?].
    4°    STC 3060 Houghton *70–83
    'Gabrielis Harveij'. 'An other Epistle de rebus *Turcicis*, et *Persicis*, et
*Syriacis* . . . to be annexid unto my other Discourses of States, & Governements'. Bound between Thomas, *Historie of Italie*, and Jovius, *Libellus*, by
the antiquary Maurice Johnson (1688–1755). Copious marginalia.

BLAGRAVE, John. *The Mathematical Jewel, Shewing the making, and most excellent
use of a singuler Instrument so called: in that it performeth with wonderfull dexteritie, whatsoever is to be done, either by Quadrant, Ship, Circle, Cylinder, Ring,
Dyall, Horoscope, Astrolabe, Sphere, Globe. . . . The use of which Jewel . . .
leadeth . . . through the whole Artes of Astronomy, Cosmography, Geography,
Topography, Navigation, Longitudes. . . .* London, Walter Venge, 1585.
    Folio    STC 3119    Brit. Lib. C.60.0.7
    'gabrielharvey. 1585.' 'The Instrument itself, made & solde by M. Kynvin,
of London, neere Powles. A fine workman, & mie kinde frend'. Copious
marginalia. At the end of the volume: 'Gabriel Harvey. 1590'.

BLUNDEVILL, Thomas. *The foure chiefest Offices belonging to Horsemanship,
That is to saie, The office of the Breeder, of the Rider; of the Keeper; and of the
Herrer.* . . . London, Henrie Denham, being the assigne of William Seres,
1580.
    8°    STC 3154    Brit. Lib. C.175.i.4
    'Confer Xenophontis 'Ιππαρχικόν.' 'Gabriel Harvey'. At top right of the
title-page is Richard Harvey's signature 'Ricardi Harveij', followed by the
inscription: 'Ex dono Gabrielis H.' (in G.H.'s hand). The volume itself is
filled with Gabriel's manuscript notes. On verso of the title-page he has
written: 'Blundevile Art of Riding, the most compendious, sensible, & perfect
of any other in this subject: more briefe, & plaine, then Grison in Italian:
more full, & perfect, then Mʳ Astley: albeit Mʳ Astleys Art of Riding, for
the obedience & service of the warriour Horse, be a fine, & excellent traine;
& a very pregnant Method for every brave man on horse back. . . . It importeth a Courtier, to be a perfect horseman'. Richard Harvey has made
few, if any, notes within the book but has signed the final page 'Ricardi
Harveij'.

Boccatius, Joannes [Decembrio, Pier Candido]. *Compendium Romanae historiae, oppido quam succinctum, & jam primum in lucem editum.* Strassburg, J. Jucundus, 1535.

    8°     Pembroke College, Cambridge     I.C.II.102

    'Gabrielis Harveij'. 'Huic adde Petrarchae virorum illustrium vitas, et Augustalia'. A few MS. notes and underlinings. Although this title was printed in the sixteenth century under the name of Boccaccio, the true author has been identified as Pier Candido Decembrio.[2]

Bonetus de Lates. *Hebrei medici Provenzalis Annuli per eum compositi super astrologiam utilitates incipiunt.* Paris, 1527.

    Folio     Brit. Lib. 533.k.1(2)

    On sig. d7$^v$ of this subtitle in the Sacrobosco folio (pagination is continuous), Harvey writes: 'Annuli istrius compositio, multo facilior compositione Astrolabii, et aliorum instrumentorum astrologicorum'. Bonetus was astrologer to Pope Alexander VI and wrote this short tract for him. It describes an astrolabe in miniature fashioned as a ring to be worn on the finger. The 'annulus' is marked with zodiacal signs, degrees, and months and with it one can find the position and altitude of the sun and stars, the sign which is on the ascendant, the hour, the position of the planets, latitude of cities, and height of towers.[3] Harvey notes that Gemma Frisius also describes such a ring.

★Boorde, Andrew (?). *Geystes of Skoggon.* [London, Colwell, *c.* 1566?] This jest book collection of anecdotes was licensed to T. Colwell in 1565–6, but the oldest known edition is 1613. In Harvey's copy of Murner's *A merye jeste of a man called Howleglas* (Bodleian 4° Z.3.Art.Seld.) is a note in Harvey's hand referring to 'Skoggin' as a gift from Edmund Spenser, 20 December 1578. (See p. 49 above and Murner, p. 228 below.)

Borne [Bourne], William. *A Regiment for the Sea, Containing verie necessarie matters for all sorts of men and travailers; whereunto is added an Hydrographicall discourse touching the five severall passages to Cattay; written by William Borne. Newlie corrected and amended by Thomas Hood; who hath added a new Regiment, and Table of declination.* London, T. Est, for Thomas Wight [1592].

    4°     STC 3427     Brit. Lib. C.60.f.8

    'gabrielharvey'. 'His treasure for Travellours'. No other notes except on final page which remarks on strategy in sea warfare. Within the text, brown crayon-like horizontal markings emphasize various topics, e.g. Zenith, Latitude, Longitude, Declination, Parallax, etc. This is a technical navigation manual written in plain style with much practical advice. At the end of this

---

    [2] See L. Bertalot in *Zentralblatt für Bibliothekswesen*, 28 [1911], 73–6.

    [3] See Lynn Thorndike, *History of Magic and Experimental Science*, N.Y., 1934, IV, 465.

work is the subtitle, Hood's *Marriners Guide*, containing Harvey's signature, a few MS. notes, and his inscription of the date, 1594.

BRAUNSCHWEIG, Hieronymus von. *A most excellent and perfecte homish apothecarye, or homely physick booke for all the grefes and diseases of the bodye.* Translated from the German by Jhon Hollybusch. Cologne, [heirs of] Arnold Birckman, 1561.
    Folio     STC 13433     Brit. Lib. C.60.0.10(3)
    'Gabrielis Harveij. 1590'. Signature of 'W. Strachy', probably a former owner, is on the final page. On the blank leaf at the beginning of the volume Harvey writes: 'Ausus est etiam Hieronymus Brunsvig librum scribere de destillationibus, herbarum, radicum, florum, seminum, fructuum, et animalium: excusum Francoforti, 1551. Ut est apud Gesnerum in Bibliotheca. Vel rarus destillandi artifex, vel post Paracelsum satis audax. Nam post eum Matthiolus, Veccherus, Floravantus, tot recentiores pragmatici, et empirici pharmacopoei'. Copious MS. notes. On the final leaf: 'Gabrielis Harveij, et amicorum, 1590'. Bound at end of this book (possibly by Harvey) is a folio sheet (with Harvey annotations) entitled 'An excellent, perfect, and an approved medicine and waie to helpe and cure the stone in the raines'. The folio sheet is in black letter and was published in 1582 but designates no printer or place of publication. Bound with Hugkel and Bruele.

BRUELE, Gualterus. *Praxis medicinae theorica, et empirica familiarissima.* . . . Antwerp, Chr. Plantin, 1585.
    Folio     Brit. Lib. C.60.0.11(1)
    Autographs of both Gabriel and John Harvey and manuscript notes of both. Front flyleaf has index in G.H.'s hand. On final page he writes: 'Emi a Joanne fratre, Aprilis 15°. 1589. cum ille alium sibi Bruelem, totidem suis chartis auctum, compingi curasset, precium Xˢ'.

[BUCHANAN, George]. *Ane admonition direct to the trew Lordis mantenaris of the Kingis Graces Authoritie M.G.B. . . . accordyng to the Scottish copie Printed at Strivilyng by Robert Lekpreuik.* London, John Daye, 1571.
    8°     STC 3967     Brit. Lib. G.5443(1)
    'Gabriel Harvey'. 'A fine Discourse of Buchanan, but bitter in his Invective veine, for elegant stile, none nearer owre Ascham'. Many underlinings and MS. notes.

—— *Ane Detectioun of the duinges of Marie Queene of Scottes, touchand the murder of hir husband, and hir conspiracie, adulterie, and pretensed mariage with the Erle Bothwell. And ane defence of the trew Lordis, mainteineris of the Kingis Graces actioun and authoritie. Translatit out of the Latin quhilke was written by G.B.* [London, John Day, 1571.]
    8°     STC 3981     Collection of Robert H. Taylor, Princeton

'gabrielharvejus'. Some marginalia and underlinings. Text is in Scots. On verso of final page of the text (sig. y iii$^v$) is a full page of MS. notes quoting two passages from James VI, *Basilikon Doron* (Edinburgh, 1603), pages 81 and 93. In the first passage James places blame for Mary's predicament on her father's adultery. In the second passage James gives advice to his son Henry to make himself well versed in authentic histories and chronicles but not 'such infamous Invectives, as Buchanans, or Knoxes Chronicles: & if anie of theise infamous Libels remaine untill your dayes, use the Lawes upon the keepers thereof'.

—— *De Maria Scotorum Regina, totaque eius contra Regem coniuratione; foedo cum Bothuelio adulterio; nefaria in maritum crudelitate & rabie, horrendo insuper & deterrimo eiusdem parricidio: plena, & tragica plane Historia.* [London, John Day, 1571.]

4° 8° is STC 3978 Collection of Robert H. Taylor, Princeton

'Gabrielis Harvey. 1571'. Elaborate and beautifully drawn monogram 'GH'. 'Confer [i.e. Compare] the Queenes Moothers Legendary in French'. 'Georgii Buchanani Invectiva'. On verso of the final page (sig. Qiii$^v$): 'Ecce principum etiam fortuna, obnoxia declamationibus, et supplicijs, ne honoratem ineant, tutamque vivendi rationem'. Many underlinings and a few brief marginalia. Three portraits of Buchanan are added: one on front flyleaf, an engraved frontispiece portrait of the author at age seventy-six, and a final portrait with a signature, possibly in Buchanan's holograph, inscribed 'G Buchanan, August 1577', and written in a rather shaky (aged?) hand. Text of this volume is in Latin.

CASTIGLIONE, Baldessare. *Il Libro del Cortegiano.* Venice, Aldus, 1541.

8° [whereabouts unknown]

'Gabriel Arvejo'. Manuscript notes and other autographs of Harvey's, according to Moore Smith in 1913.

—— *The Courtyer of Count Baldessar Castilio; divided into foure bookes. Very necessary, and profitable for yonge Gentilmen, and Gentilwomen, abiding in Court, Palace, or Place; done into Englyshe by Thomas Hoby.* London, Wyllyam Seres, 1561.

4° STC 4778 Newberry ICN Case Y712 C27495

'X . gabrielharvey. 1572'. Motto on title-page in Harvey's hand: 'Unus quisque Fortunae faber'. At end of the letter he writes: 'Aphorismi Aulici. No excellent grace, or fine cumlie behaviour without three cunning properties: a sound judgment to informe; an apt dexteritie to conforme; & an earnest intention to performe'. Numerous signatures of Harvey's, underlinings, and annotations throughout the volume. [micr]

*—— Balthasaris Castilionis Comitis de Curiali sive Aulico Libri quatuor, ex Italico*

*sermone in Latinum conversi, Bartholomeo Clerke Anglo Cantabrigiensi interprete.*
London, John Day, 1571.
    8⁰    STC 4782

Mentioned in a letter *c.* 1573 to Arthur Capel (Sloane MS. 93, fol. 90ᵛ) as one of the books Harvey will gladly lend him, presumably from his library. Also referred to in marginalia of the Hoby translation (see above) on sig. Aiiᵛ.

CHAUCER, Geoffrey. *The woorkes of our antient and lerned English Poet, Geffrey Chaucer, newly printed* (ed. Thomas Speght). London, George Bishop, 1598.
    Folio    STC 5077    Brit. Lib. Add. MS. 42518

'gabriel harvey. 1598'. Annotations throughout. At end is a lengthy appraisal of contemporary poets; on fol. 422ᵛ is the earliest known reference to *Hamlet* (before Essex's execution early in 1601). The volume comprises Chaucer's life and various works, including his astronomical discourse on 'The Astrolabe' (annotated by Harvey) and is followed by John Lydgate's *Story of Thebes*.

*CHEKE, Sir J[ohn]. *The Hurt of Sedition, how grievous it is to a common welth.*
London, Willyam Seres, 1576.
    8⁰    STC 5111

In a letter to Arthur Capel (Sloane MS. 93, fol. 90ᵛ) Harvey mentions 'that pretti elegant treatis of M. Chek against sedition', presumably as one of the books he has recently lent Capel. The edition owned by Harvey could have been the above or one of the earlier ones of 1549 or 1569.

*CHUTE, Anthony. *Beawtie dishonoured, written under the title of Shores wife.*
London, John Wolfe, 1593.
    4⁰    STC 5262

Referred to at the beginning of *A New Letter of Notable Contents,* 1593, (sig. A2) as received from Wolfe.

CHYTRAEUS, David. *De tribus nostrae aetatis Caesaribus Augustis, Carolo V, Ferdinando I, Maximiliano II, orationes.* Wittenberg, 1583.
    [whereabouts unknown]

Contains the 'autograph of Gabriel Harvey', according to a 'Catalogue of a collection of books relating especially to Italian history and literature . . . sold at auction . . . by F. O. Beggi . . . March 17th . . . 1864' (lot 357).

CICERO, Marcus Tullius. *M. Tul. Ciceronis ad C. Trebatium Iurisconsultum Topica; Audomari Talaei praelectionibus explicata, ad Carolum Borbonium Cardinalem Vindocinum.* Paris, Matthew David, 1550.
    4⁰    All Souls, Oxford a-11-4(3)

'Gabrielis Harveij'. Ornamental initials 'GH'. 'Nullus Romanae linguae liber, magis opportunus, aut [two words illegible] dialectico, vel diserto oratori, vel etiam perito Iurisconsulto'. At bottom of title-page is signature of a previous owner 'Tho. Hatcherus' (see Talon, Omer). Many of Harvey's annotations, some in 1570 and some in 1579. At bottom of final page (sig. E5$^v$) he noted the earlier reading when he wrote in a large Gothic hand: 'Gabriel Harvejus. Calendis Februar. 1570'. Below this in a smaller and heavier Italian hand he wrote: 'Multo etiam diligentius, 1579 iamtum aliquanto studiosius iuri civili incumbens'.

—— *M. Tullii Ciceronis Epistolae Ad Atticum. Ad M. Brutum, Ad Quinctum Fratrem, Cum correctionibus Pauli Manutii.* Venice, Aldus, 1563.
    8° Brit. Lib. C.60.f.9
    'GabrielisHarveij'. 'Arte, et virtute'. 'Epistolae hae pleraeque, omnes maximam partem politicae; et quotidianae vitae communibusque hominum consiliis, atque verum eventibus. . . .' On sig. CC3$^v$: 'Relegi has politices, pragmaticasque epistolas in Aula Trinitatis . . . Mense Julio, sole in Leonis corde flagranti. 1582'. The glossary is divided into eight portions for daily reading, e.g. on CC7$^v$ at top Harvey notes: '2$^{da}$ Latina, Graecaeque lectiunculae. dies ○→ [Tuesday]'; on DD4$^r$ at top: '4$^{to}$ lectiuncula Valdinensis. dies ♃ [Thursday]'. A number of Harvey's signatures and copious annotations throughout the volume.

CORRO, Antonio de. *The Spanish Grammer; with certeine Rules teaching both the Spanish & French tongues . . . with a Dictionarie adjoyned unto it of all the Spanish words cited in this Booke . . . by John Thorius, Graduate in Oxenford.* London, John Wolfe, 1590.
    4° STC 5790 Huntington
    'Gabrielisharveij' and monogram. Manuscript notes including a list of Spanish books. On sig. S4$^v$: 'gabrielharvejus: 1590'. and the admonition 'Poco, y bueno', which seems to be Harvey's motto for language study, oratory, and certain other kinds of learning. This volume is a translation by John Thorius of Corro's Spanish and French grammar.                    [desc]

[COSIN, Richard]. *An Answer to the two fyrst and principall Treatises of a certaine factious Libell, put foorth latelie, without name of Author or Printer, and without approbation by authoritie, under the title of An Abstract of certeine Acts of Parlemement: of certaine hir Majesties Injunction, of certeine Canons, &c.* London, Thomas Chard, 1584.
    4° STC 5815 Durham Univ. Library G.III.33 (3)
    On verso preceding the title-page, Harvey writes: 'Iliacos intra muros peccatur, et extra'. Bound, probably by Harvey, with the anonymous *Abstract* and with Cosin's *An Apologie for Sundrie Proceedings*, 1593.

[——]. *An Apologie for Sundrie Proceedings by Jurisdiction Ecclesiasticall, of late*

*times by some chalenged, and also diversly by them impugned . . . against proceedings*
*ex Officio, and against Oathe ministred to parties in causes criminall.* London,
Christopher Barker, 1593.

    4°    STC 5821    Durham Univ. Library G.III.33 (1)

'gabrielharvey. 1593'. Also an ornamental monogram 'GH'. A few notes
and underlinings throughout the three parts of this more than 500-page
work. Cosin was Dean of Arches at the time that Harvey served in that
Court.

D A V I E S, Richard. *A funerall sermon preached the xxvi. day of november . . . at the*
*buriall of the right honourable walter earle of essex and ewe.* London, Henry
Denham, 1577.

    4°    STC 6364    St. John's College, Oxford, a.1.25

Title-page has Harvey's autograph scored through (evidently by a later
owner) and at the bottom the inscription in Harvey's hand: 'Ex dono
nobilissimi Domini, Robert Devereuxii, Comitis Essexii'. At end of pamphlet
'R:E.' is inscribed. Walter, first Earl of Essex, died in 1576. His son Robert
attended Trinity College, Cambridge, from 1579 to 1581, receiving his
M.A. in the latter year. Robert's sister was Penelope Devereux (later Lady
Rich). Within the volume are Harvey's expressions of esteem and mourning
for a virtuous man.

D E M O S T H E N E S. *Gnomologiae, sive sententiae collectaneae et similia ex Demosthenis*
*orationibus et epistolis, in certa virtutum ac vitiorum capita, collectae per J. Loinum.*
*Divi Gregorii sententiarum spiritualium libri tres J. Lango interprete. Arithmologia*
*ethica, sententiae morales certis numeris comprehensa . . . a J. Camerario conversae.*
Basle, L. Lucius [1552].

    8°    Brit. Lib. C.45.a.9

'Gabrielis Harveij'. 'X$^{li}$' 'G.H. 1571'. At the end of the volume: 'Gabrielis
Harveij et Amicorum, mense Aprili, 1578'. Harvey enters the dates '1573',
1578', and '1580' within the text (evidently signifying various rereadings).
This thick but tiny Renaissance quotation book was previously owned by
Thomas Hatcher whose name, initials, and cipher are on the title-page with
the date '1560'. The Greek text is on the left-hand pages, Latin on the right.
Of this cherished little work Harvey writes on sig. C2: 'Si alii absunt Graeci
auctores; nullis careo libris, cum hunc habeo unum. One of mie pockettings;
and familiar spirits'. It is crammed full of Harvey's marginalia. See pp. 148–9
above.

D E S P R E Z [Des Pres], François. *Recueil de la diversité des habits, qui sont de present*
*en usage tant es [dans des] pays d'Europe, Asie, Affrique & Illes Sauvages. Le tout*
*fait après le naturel.* Paris, François Desprez, 1567.

    8°    Collection of Arthur Vershbow, Newton Centre, Mass.

On title-page: 'gabriel harvey'. In final leaf: 'Gabrielis Harveij' with flourish beneath the signature. No MS. notes. This book on costume (the 1562 edition was the first such work known, according to Colas) contained 121 beautifully executed woodcuts of male and female figures in costume (Russians, Spaniards, Turks, Portuguese, Swiss, Italians, Arabs, Persians, a Brazilian woman, a physician, grotesques, etc.). Below the woodcuts are descriptive verses, probably by Desprez. [desc. in letter from A. Vershbow]

[DIONYSIUS PERIEGETES]. *The surveye of the world, or situation of the earth, so much as is inhabited. . . . First written in Greeke by Dionise Alexandrine, and now englished by Thomas Twine.* London, Henrie Bynneman, 1572.
    8º    STC 6901    Rosenbach Foundation
    '1573. Gabrielis Harveij. GH.' 'poco y bueno'. 'Dionisius Alexandrinus'. At the end (sig. Fiii) Harvey says of this book: 'Synopsis mundi: brevissima, et facillima. Mea tandem mnemonica Geographica; eademque pragmatica Neographia. Qualis etiam in Freigii paedagogo, adhuc brevior, atque facilior'. The tract contains copious MS. notes. It is bound (possibly by Harvey) with Rowlands, Grafton, Turler, and Lhuyd. All title-pages bear Harvey's symbol of a circle with dot in centre, with which he apparently signified geographical matter.

DOLCE, Lodovico. *Medea Tragedia.* Venice, Domenico Farri, 1566.
    8º    Folger PQ4621 D3 M4 1566a Cage
    'GH. 1576'. 'Gabrielis Harveij'. 'Occasio alata'. On sig. A1ᵛ: 'Quattro tragedie. Medea. Thieste. Hecuba. Iphigenia'. Only *Medea* and *Thieste* now remain in the volume. On the title-page of *Thieste*: 'di gabriello harvejo. 1579'. Various annotations. On the verso of first title-page (sig. A1ᵛ): 'Hae quatuor Tragoediae, instar omnium Tragoediarum pro tempore: praesertim cum reliquarum non suppetit copia'. The text is Italian.

DOMENICHI, Lodovico. *Facetie, motti, et burle, di diversi signori et persone private.* Venice, Andrea Muschio, 1571.
    8º    Folger H.a.2(1)
    Title-page and first three-quarters of this volume are missing. Pages 321–431 are bound with Porcacchi (with which the Domenichi was published) and with Guicciardini. The Domenichi is scored in red on a number of pages and contains copious notes in Harvey's hand. On final page: 'Id Domenichus in Facetiis; quod Gandinus in Stratagematis. Uterque dignus, qui ad unguem habeatur. . . . Dulci narratore nihil dulcius'. (See pp. 175–90 above.)

DUARENUS, Franciscus. *De Sacris Ecclesiae Ministeriis Ac Beneficiis Libri VIII. In quibus quicquid ad plenam Iuris Pontificij cognitionem necessarium est, breviter ac dilucide explicatum continetur.*
    *Item, Pro Libertate Ecclesiae Gallicae adversus Romanam aulam Defensio*

*Parisiensis curiae, Ludovico XI. Gallorum Regi quondam oblata.* Paris, Andreas Wechelus, 1564.

    4°     Trinity College, Cambridge L.12.112(1)

'Gabrielis Harveij'. 'pret. ii$^s$. vi$^d$'. 'Sleidanus, de quatuor summis Imperiis; multa habet consentanea'. At end of index: 'Hodie multi sacerdotes, non pascunt oves, sed pascunt boves, id est seipsos. Pastores, a pascendo passive, non active'. End of the second tract (sig. Ziiii$^v$): 'Gabriel Harvejus. Mense Februario. 1580'. 'Prima cursoria, et perfunctoria lectio'. Copious MS. notes and many red markings throughout the book. The text deals with canon law and church history. Bound (possibly by Harvey) with the two following Duarenus titles.

—— *F. Duareni Jurisconsulti Clarissimi Praelectiones In Tit. Ad Leg. Falc. D. in celebri Biturig. academia habitae anno 1555. Opera Leontii Beriacii I.C.* Paris, Andreas Wechelus, 1561.

    4°     Trinity College, Cambridge L.12,112(2)

'Gabriel Harvejus'. No notes within volume.

—— *Fr. Duareni Jurisconsulti Clarissimi in Tit. de Jureiur, lib. xij. dig. Commentarius.* Paris, Andreas Wechelus, 1562.

    8°     Trinity College, Cambridge L.12.112(3)

'Gabrielis Harvejus'. No notes.

D u B a r t a s, Guillaume de Saluste. *See* Saluste du Bartas.

[D u F a u r, Gui (Seigneur de Pibrac)]. *Ornatissimi cuiusdam viri, de rebus Gallicis, ad Stanislaum Eluidium Epistola. Et Ad Hanc De Iisdem Rebus Gallicis Responsio.* [Paris] 1573.

    8°     All Souls, Oxford S.R. 63 c.2(2)

'Gabriel Harvejus'. Volume probably was acquired and annotated shortly after publication since marginalia are in Harvey's early Humanist hand. The first part of the pamphlet tells of a purported conspiracy against the King (Charles IX), his two brothers, and the Queen Mother (Catherine de Medici); the second part of the pamphlet by 'Stanislaus Elvidius' (a pseudonym for Joannis Camerarius) denies that any such conspiracy existed. Harvey apparently agreed with the latter. The accusation of conspiracy seems to have been an attempt to justify the St. Bartholomew's Day Massacre. This tract is bound with that of N. Barnaud.     [tpx]

D u P l o i c h e, Pierre. *A treatise in Englishe and Frenche, right necessarie and profitable for all Young Children.* . . . London, Jhon Kingston for Gerard Dewes, 1578.

    4°     STC 7364     Huntington 53922

'gabrielharvey, ex dono Autoris, Monsieur du Ploiche'. Verso of final page: 'gabrielharvey. 1580'. On sig. A2$^v$: 'A necessary Introduction. petit à petit'. 'A paradox in lerning: quo plus, eo minus. Beginners must not leap over hastely, lest they overleape all. Apt & reddy pronunciation of the Alphabet, one weeks exercise'. Text comprises religious instruction, conversation exercises, and French rules, and is set up in two columns, English at left and French at right. [desc]

[ELIOT, John]. *The survay; or topographical Description of France; with a new Mappe, helping greatly for the Survaying of every particular Country, Cittye, Fortresse, River, Mountaine, and Forrest therein; collected out of sundry approved authors.* . . . London, John Wolfe, 1592.

4° STC 7575 Huntington 56973

'GH. gabriel harvey, 1592'. On verso of last page of text: 'gabrielis harveij, et amicorum, 1592'. Underscorings and a few brief annotations. Included is a folio leaf of MS. in Harvey's hand headed 'Galliae Universalis tabula' which lists the Duchies, Provinces, Peers, Archbishoprics, and Academies of France. [desc]

ELIOT, John. *Ortho-epia Gallica; Eliots First Fruits for the French: Enterlaced with a double new invention which teacheth to speake truely, speedily and volubly the French-tongue.* . . . London, John Wolfe, 1593.

4° STC 7574 Huntington 60231

'G.H.' 'gabrielharvey. 1593'. 'For the French, & Spanish'. Some MS. notes and underlinings. Text is set up in parallel columns (French and English) and consists of three dialogues: (1) satiric picture of contemporary language teaching in England; (2) relative merits of all world languages other than English, discussion of great writers and their works; (3) rapid tour of France, Italy, Spain, and England. [desc]

ERASMUS of Rotterdam. *Parabolae, Sive Similia . . . Cum vocabularum liquot non ita vulgarium explicatione.* Basle, 1565.

8° Folger H.a.i

'X$^{li}$ 4 Gabrielis Harveij.' Below title: 'Gabriel Harvejus. mense Januario: 1566'. Also inscribed on title-page: 'Quas Ipse profitetur esse *exquisitae Gemmae*'. 'A quibus nihil boni spero, quia nolunt: ab ijs nihil mali metuo, quia non possunt'. 'Vel Arte, vel Marte'. Harvey perhaps used this as a school text. In 1566 he would have been about sixteen. The little volume is copiously annotated. (See pp. 138–9 above.)

★ESTIENNE, Henri. *L'Introduction au traité de la conformité des merveilles anciennes avec les modernes. Ou traité preparatif à l'Apologie pour Hérodote.* [Geneva], 1566.

8°

On fol. 209$^r$ of Guicciardini under the rubric 'Propria Eulogi bibliotheca

---

4 Harvey also uses this symbol on the title pages of Castiglione's *Courtier*, Freigius's *Mosaicus*, Guazzo's *La Civil Conversatione*, Hugkel's *De Semeiotice Medicinae*, and Oldendorf's *Loci communes iuris civilis*. Perhaps it signified 'ex libro' and referred to the use to which he had put the marginalia therein (?).

. . . 1590' Harvey lists 'Apologiam Herodoti', on fol. 8 of Euripides: 'Stephanum in Apologia Herodoti', and on sig. Ff8 (fol. 72) of Porcacchi: 'L'Apologie pour Hérodote; ou Traité de la Conformité des Merveilles anciennes avec modernes. Par Henri Estienne'. 'L'Apologie pour Hérodote' is the running title of the original 1566 edition listed above. This famous satire which catalogued the various vices of monks, friars, and nuns was suppressed immediately after publication and later editions were much altered.

EUCLID, *Liber Primus Geometrie.* Translated by Boethius. Paris, 1527.
    Folio     Brit. Lib. 533.K.1
Published in one volume with Sacrobosco and Bonetus (see listings of these). At end of the Euclid tract Harvey comments: 'No Euclide more profitable, then that lately published in Inglish; with a large and notable preface of M. Dee, one of the profoundest mathematicians now living'.

*EUNAPIUS Sardianus. *De vitis philosophorum et sophistarum.* Greek and Latin (translation by H. Junius). Antwerp, Chr. Plantin, 1568.
    4°
Under heading of 'Propria Eulogi bibliotheca . . . 1590' in Guicciardini (fol. 209ʳ) Harvey lists 'delicata omnium medulla in Eunapii vitis'. There was also a 1564 Latin edition and a 1579 English translation.

EURIPIDES. *Hecuba, & Iphigenia in Aulide . . . in latinam tralatae, Erasmo Roterdamo interprete. Eiusdem ode de laudibus Britanniae, Regisque Henrici septimi, ac regiorum liberorum eius, Eiusdem ode de senectutis incommodis.* Venice, Aldus, 1507.
    8°    Houghton *E.C.H263.ZZ507e
'Gabrielis Harvey'. in ornamental script at top of title-page and 'Gabriel Harvey'. in secretary hand to right of anchor–dolphin device. Harvey notes: 'Huc Sophoclis Antigone: novissima edita a Th. Watsono'. [Watson's edition was published in 1581.] On fol. 6 Harvey remarks: 'Erasmi fere judicium acre, et serium: nec dubium est, quin delectum ad libuerit in sapientissimis tragoediis cligendis exquisitum'. Volume contains numerous marginalia relating to classical and contemporary literature.

FABRICIUS MARCODURANUS, Franc. *M. Tullii Ciceronis Historia, per Consules descripta, & in annos LXIV distincta.* Editio Secunda. Cologne, Maternus Cholinus, 1570.
    8°    Cambridge Univ. Library F157.d.l.1
'Gabrielis Harveij. 1572'. No annotations, but red chalk-like mark crosses the title-page horizontally. At the end the volume has a number of handy indexes which may have been one of Harvey's reasons for acquiring it.

[FILLS, Robert]. *The Lawes and Statutes of Geneva, as well concerning ecclesiastical Discipline, as civill regiment, with certeine Proclamations duly executed, whereby Gods religion is most purelie mainteined, and their common wealth quietli governed: Translated out of Frenche into Englishe by Robert Fills.* London, Rouland Hall, 1562.

8° STC 11725 [whereabouts unknown]

Harvey's autograph and MS. notes (according to W. Carew Hazlitt in the 1899 Quaritch catalogue). A manual on canon law.

FIRMINUS [Firmin de Beauval]. *Firmini reportorium de mutatione Aeris; Tam Via Astrologica, quam metheorologica; pristino nitori restitutum, per Phillippum Iollainum Blereium, cum scholiis eiusdem.* Paris, Jacob Kerver, 1539.

Folio. Brit. Lib. C.60.o.9

'GH.' 'gabrielharvejus.' There is also a combined monogram 'GIRH' of the three Harvey brothers (Gabriel, John, and Richard). 'iii$^s$vi$^d$'. '*Regulae generales de tempore pluvioso*: de tonitruis, et coruscationibus: de ventis: de aspectibus planetarum, pro mutatione aeris: in Tansteteri Canonibus Astronomicis: Judicia tempestatum brevissima'.

[FITZHERBERT, A.]. *In this booke is contained the office of Shiriffes, Bayliffes of liberties, Escheatours, Constables, and Coroners, and sheweth what everyone of them may do by vertue of theyr offices, drawen out of bookes of the common lawe and of the Statutes.* London, Thomas Marshe? 1560.

8° STC 10991 [whereabouts unknown]

'G.H.'s autograph and notes', according to W. Carew Hazlitt in Quaritch catalogue (1899) and Moore Smith (1913).

FLORIO, John. *Florio his first fruites: a perfect induction to the Italian and English tongues.* London, Thomas Dawson for Thomas Woodcocke, 1578.

4° STC 11096 Houghton *70–81

Title-page missing. Copious notes in Harvey's hand within the volume. On sig. ***ii$^v$ he writes: 'G.H. of Messer Florios first fruytes. Tantus amor Florum, et generandi gloria mollis'. Marginalia on a number of consecutive pages are devoted to 'The politique historie of Doctor Stephen Gardiner, bishop of Winchester, & afterward Chancelour of England'. This volume was from the library of the antiquary Maurice Johnson who inserted a title-page in his own hand on which he noted that the book 'heretofore belonged to the learned and Ingenious Gabrael Harvey Esquire sometime Poet Laureat to her Majesty Queen Elizabeth'. (See pp. 154–6 above.)

FOORTH, Joannes. *Synopsis Politica.* London, H. Bynneman, 1582.

8° STC 11128 County Branch Library, Saffron Walden

'GabrielisHarveij.' 'Canones. Aphorismi. Apophthegmata. Axiomata'. 'Use Legges, & have Legges. Use Law, and have Law. Use nether, & have nether'. 'Animus, optimus medicus'. At end: 'Haec synopsis vix trium horarum pensum: ut ter sum expertus: Mense Augusto: 1582. Gabriel Harvejus. J.C. [Jurisconsultus]'. Very copious MS. notes, many interlinearly.

FOXE, John. *De Christo crucifixo concio*. London, John Day, 1571.
    4°     STC 11247     Folger 11247, copy I
    'GH. 1572'. 'Gabriel Harvejus'. No annotations but some underlinings and a few simple marginal symbols. This is the first Latin edition of Foxe's famous sermon at St. Paul's Cross after the Papal Bull excommunicating Queen Elizabeth.

FREIGIUS, J. T. *Ciceronianus. Joan. Thomae Freigii . . . Libris Decem*. Basle, Sebastian Henricpetrus [1575].
    8°     Worcester College, Oxford c.m.3(2)
    '1575. Gabriel Harvejus. GH.' 'X$^{li}$' 'prec./2$^s$'. 'Specimen maioris atque locupletioris operis; ex quibusque optimis, ac selectissimis auctoribus descripti; ut semper facile circumferietur, et Bibliothecae virem suppleret'. At the end of the epistle: 'Unicus Commentarius, totius fere instar humanitatis'. Manuscript notes within and at end of the volume. It is bound, perhaps by Harvey, between Ramus, *Ciceronianus*, and Livy, *Conciones*.

—— *Mosaicus; Continens Historiam Ecclesiasticam. 2494 annorum, ab orbe condito usque ad Mosis mortem*. Basle, 1583.
    Folio in sixes     Brit. Lib. c.60.f.4
    'GabrielisHarveij, et amicorum'. 'Gabriel Harvejus'. 'X$^{li}$' On sig. α3$^r$: 'Spiritus Melanchthonis, non Lutheri, Rami, non Bezae'. On flyleaf at the beginning of the book: 'Certe Mosaica historia, liber librorum est, et fons fontium.' At the end of the volume: 'gabrielharvejus. 1584'. Red mark across lower part of the title-page.

—— *Paratitla seu synopsis pandectarum juris civilis*. Basle, Sebastian Henricpetrus, 1583.
    8°     Private Collection, United States
    'Gabrielis Harveij. 1583'. 'GH.' 'Arte, et virtute'. Opposite title-page: 'Aurcum Alphabetum Mnemonicon'. Many red markings in volume.

FRONTINUS, Sextus Julius. *The strategemes, sleyghtes, and policies of warre gathered togyther, by S. Julius Frontinus, and translated into Englyshe, by Richard Morysine*. London, Thomas Berthelet, 1539.
    8°     STC 11402     Houghton Lf.18.54.8*
    'Gabrielis Harveij'. 'Precium-xx$^d$. 1578'. 'vel Arte, vel Marte'. 'Sero, sed serio. Nulla dies sine Linea'. '1588. Revolutio meae Reformationis, seu Annus Assuetudinis'. Copious MS. notes written at different periods. Many of the annotations refer to historical events and to British heroes.

*—— *Stratagemi Militari Di Sesto Giulio Frontino; tradotti in lingua Italiana* . . . *da Marc'Antonio Gandino; Con una aggiunta dell' istesso* . . . *Tratta da moderni historici. Con due Tavole, &c.* Venice, Bolognino Zaltiero, 1574.

4°

On fol. 209ʳ of Guicciardini under the heading 'Propria Eulogi bibliotheca . . . 1590', Harvey writes: 'cum utilissimis dulcissimisque stratagematibus Marcantonij Gandini'. On fol. 37ᵇ (sig. Bb5ᵛ) of Domenichi he writes 'Meus Gandinus'. Harvey seems to have valued this translation of Frontinus especially for its historical 'aggiunta'.

FULKE, William. *OYPANOMAXIA. Hoc est Astrologorum Ludus, Ad bonarum artium, & Astrologiae in primis studiosorum relaxationem comparatus, nunc primum illustratus.* . . . London, Thomas East & Henry Middleton for William Jones, 1572.

    4°    1571 ed. is STC 11445.    Houghton 24232.6.25*

'Gabrielis Harveij'. '[pret.] iiiᵈ'. No MS. notes. An astrological game with diagrams (played somewhat like chess) by Dr. William Fulke (1538–89), Master of Pembroke College, Cambridge, and later Margaret Professor.

*GALEN, Claudius [a suppositious work]. *Liber Galeni de urinis* . . . *una cum commentariis* . . . *Ferdinandi a Mena* . . . *eodem autore interprete*, 1553. (Gr. & Lat. compluti)

    4°

On verso of final leaf of Hugkel's *De Semeiotice Medicinae*, Harvey writes: 'at habeo Galeni Librum περὶ οὔρων [about urines]'.

*GANDINI, Marco Antonio. *See* Frontinus, *Stratagemi*.

G., R. *Beginning* Salutem in Christo. Good men and evill delite in contraryes. London [J. Day], 1571.

    8°    STC 11505 differs in spelling of 'delyght' & 'contraries'. Brit. Lib. G5443(2)

Title-page missing. No signature, but a few notes of Harvey's in margins of the text and at bottom of the final page where he writes: 'The Lord Treasurers hed, supposed to be on the conveyance of this Letter missive. I heard it reported in the Court, & affirmed in London, this november'. This is a letter concerning the second commitment of the Duke of Norfolk to the Tower, dated 13 October, 1571, and signed 'R.G.' [Richard Grafton?]. *British Museum Catalogue of Printed Books* attributes its authorship to William Cecil, Lord Burghley.

GASCOIGNE, George. *The Posies* . . . *Corrected, perfected, and augmented by the Authour.* London, Richard Smith, 1575.

    8°    STC 11637 is 4°.    Bodleian Malone 792(1)

'Gabriel Harvey. Londini, Cal. Sept. 1577'. 'Aftermeales'. Published with *Hearbes, Supposes, Jocasta, Weedes, The pleasant Fable of Ferdinando Jeronimi and Leonora de Valasco,* and *Certayne notes of Instruction.* All have copious MS. notes. At the end of *Weedes* Harvey makes the following comment on Gascoigne: 'Sum vanity; & more levity; his special faultes, & the continual causes of his misfortunes. Many other have maintained themselves gallantly upon sum one of his qualities: nothing fadgeth with him, for want of Resolution, & Constancy in any one kind'. Bound with *The Steele Glas.*

—— *The Steele Glas. A Satyre compiled by George Gascoigne Esquire. Togither with the Complaint of Phylomene. An Elegie devised by the same Author.* London, Henrie Binneman, for Richard Smith, 1576.
　　4°　　STC 11645　　Bodleian Malone 792(2)
'Gabriel Harvey.' 'GH.' 'Speculum mundi'. Copiously annotated. Bound with *The Posies.*

[GASSER, Achilles Pirminus]. *Historiarum, et Chronicorum Totius Mundi Epitome.* . . . [Basle?), 1538.
　　8°　　Brit. Lib. C.60.e.13
'ijˢ'. 'Gabriel Harvejus. 1576'. 'GH.' 'Fasciculus Temporum'. 'Hora est jam nos e somno surgere. Byshop Gardiners Text'. End of the book is dated by Harvey: 'Valdini Mense Februario 1577'. Copious MS. notes.

GAURICO, Luca. *Lucae Gaurici Geophonensis, Episcopi Civitatensis, Tractatus Astrologicus. In quo agitur de praeteritis multorum hominum accidentibus, per proprias eorum genituras ad unguem examinatis.* Venice, Curtius Troianus Navo, 1552.
　　4°　　Bodleian 4°Rawl.61
'GH.' 'gabrielharvejus. 1580'. 'Arte, et virtute'. On sig. F1ᵛ Harvey characterizes Gauricus as 'Veridicus vates' and observes: 'Certe non contemnenda tanta experientia, et autoritas, tantae scientiae coniuncta: idque in Aula prudentis pontificis'. At end: 'gabrielis harveij, et amicorum. 1580'. Copious MS. notes. Included in the volume are horoscope charts for various eminent individuals.

[GRAFTON, Richard]. *A brief treatise conteinyng many proper Tables, and easie rules, verye necessarye and nedefull, for the use and commoditie of al people, collected out of certaine learned mens workes.* London, Jhon Waley, 1576.
　　8°　　STC 12156　　Rosenbach Foundation
'GH. Gabrielis Harveij'. 'Emptus Eboraci, 1576 mense Augusto'. Below title Harvey notes: 'bie Richard Grafton'. On verso of title-page: 'I. hand dial. 2. prognostication perpetual. 3. Regulae generales, ad calendarum Ad paedium popularem'. At end (sig. H4ᵛ): 'gabrielisharveij, et amicorum. One of mie Yorke pamflets. 1576. then fitt for mie natural & mathematical

studies, & exercises in Pembrooke hall'. Bound with Dionysius Periegetes, Rowlands, Turler, and Lhuyd; the Grafton tract is third. Harvey refers to them as bound together, e.g. he alludes on sig. H4$^v$ of Grafton to 'the breviarie of Britaine/ j$^a$ [i.e. *infra*]' and 'the post of the world/ S$^a$ [i.e. *supra*].' On the title-page of each of the five pamphlets Harvey has inscribed a circle with dot in centre, evidently symbolizing a 'natural & mathematical' (or geographical) tract.

GREVERUS, Jodocus. *Secretum, et alani Philosophi Dicta de Lapide Philosophico, item alia nonnulla eiusdem materiae pleraque jam primum edito a Justo a Balbian.* Leyden, 1599.

    8°    [whereabouts unknown]

'With MS. notes in the autograph of Gabriel Harvey', according to sale catalogue of the Revd. Philip Bliss, Principal of St. Mary's Hall, Oxford (Sotheby's, 5 July 1858). Moore Smith refers to listing of this book in G. H. Puttick sale catalogue of 14 December 1893, no. 349, as noted by W. Carew Hazlitt. This brief tract is an alchemical text.

GUARINI, Battista. *Il Pastor Fido. Tragicomedia Pastorale.* Per J. Volfeo [John Wolfe] a spese di G. Castelvetri, 1591.

    12°    STC 12414    Columbia Univ. Spec. Coll. B851G93 U5(1)

No signature of Harvey's but a number of notes throughout in Harvey's tiny Italic hand. Notes in text are mainly cross-references, typographical corrections, and translations of certain words. Title-page has a later signature by 'N. Moore'. The little volume also contains Tasso's *Aminta*, 1591, which was printed with the Guarini. Pagination of the two is consecutive. (See p. 146 n. 30 above.)

GUAZZO, S. Stefano. *La Civil Conversatione. . . . Divisa in IIII. Libri.* Venice, Gratioso Percachino, 1581.

    16°   Brit. Lib. C.60.a.1(1)

'1582. gh. Gabrielis Harveij'. 'X$^{li}$' On final page: 'Supplemento delli dialogi d'Aretino'. 'plus Artis et virtutis'. On sig. ††6: 'Thesoro della Lingua, discorso, e conversatione Italiana'. Also in *Tavola* section: 'Play with me & hurt me not: Jest with me & shame me not. A notable rule of Civilitie'. Many MS. notes. Some red chalk-like markings.

—— *The Civile Conversation of M. Steeven Guazzo written first in Italian, and nowe translated out of French by George Pettie, devided into foure books. In the first is conteined in generall, the fruites that may be reaped by conversation and teaching howe to knowe good companie from yll. In the second, the manner of conversation, meete for all persons . . . betweene young men and olde, Gentlemen and Yeomen, Princes and private persons, learned and unlearned, Citizens and Strangers, Religious and Secular, men and women. In the third . . . betweene the husband and the wife, the father and the sonne, brother and brother, the Maister and the servant. In the fourth, the report of a Banquet.* London, Richard Watkins, 1581.

    4°    STC 12422    Newberry Case Y712.G939

    '1582. Gabriel Harvey'. A few brief notes and underlinings.    [tpx]

GUICCIARDINI, Lodovico. *Detti et Fatti Piacevoli, et Gravi; Di Diversi Principi, Filosofi, Et Cortigiani. Raccolti Dal Guicciardini; Et Ridotti A Moralita.* Venice, Christoforo de Zanetti, 1571.

    8°    Folger MS. H.a.2(3)

    'GH.' 'gabriel harvejo. Ratione, et diligentia. 1580'. 'Non alius: sed aliud. Vade mecum Discursum vix ullus finis: effectuum vix ullum principium'. On fol. 209 (a flyleaf near the end) is a list of books headed 'Propria Eulogi bibliotheca . . . 1590'. Copious notes in this volume. Guicciardini is bound following the remaining portion of Domenichi and of the Porcacchi tract. (See pp. 186–90 above.)

H., J., Chester Heralt [John Hart, Chester Herald]. *An Orthographie, conteyning the due order and reason, howe to write or paint thimage of mannes voice, most like to the life or nature.* London, W. Seres, 1569.

    8°    STC 12890    Nat'l. Lib. of Australia, Canberra

    No signature of Harvey's but fairly copious marginalia in several hands (apparently annotated from the 1570s onwards). On sig. Aii^r Harvey writes: 'Scriptura est pictura vocis humanae', and on sig. !!2^r: 'Orthographia: recte scribendi ars'. In the margins of the text Harvey comments on spellings and pronunciations in English and other languages. On sig. Kiii^v dealing with the pronunciation of 'sh', he observes: 'The tongue is all in all for the sound of this letter'. 'I like not this biting of the lip. It helps not mee to sound this letter'.

    In 1969 the Scolar Press reproduced this book in facsimile but were unable to identify the marginalia. The handwriting and manner of annotating make evident that the marginalia are Harvey's.    [facsimile seen]

*HARVEY, Gabriel. Bound volume of printed works containing manuscript letters and notes. London, 1577–93.

    4°    STC 12899–904    [whereabouts unknown]

    Described by Thomas Baker, the antiquary (1656–1740), in his volume of some of Harvey's printed works (Bodleian Bliss A.110) and in the Baker manuscripts at Cambridge University Library (Baker MS. XXXVI, 107–14). The Bodleian volume contains the separately published editions of: *Ciceronianus* (London, H. Binneman, 1577) (STC 12899); *Rhetor* (London, H. Binneman, 1577) (STC 12904); *Foure Letters* (London, John Wolfe, 1592) (STC 12900; *A New Letter* (London, John Wolfe, 1593) (STC 12902); *Pierces Supererogation* (London, John Wolfe, 1593) (STC 12903). In this volume

Baker entered transcriptions of Harvey's marginalia from his collection of his own works. On title-page of the *Ciceronianus* now at the Bodleian, Baker copied the following notes: 'secunda editio, paulo quam prima, emendatior. The next Title—of my Rhetorique orations put legistae Gabrielis Harveij Rhetoricarum Orationum liber. In Academia—cantabrigiens: publice habitarum.' Perhaps Harvey was then contemplating bringing out an edition of his combined works. Various other memoranda of Harvey's, plus some biographical data relating to him, are entered by Baker in the margins of this and other of the Harvey texts here bound together. In Baker's manuscripts he also transcribes a number of these annotations together with letters to and from Harvey and an epitaph for Nicholas Bacon written by Harvey. Presumably all were originally found in the volumes of his works seen by Baker.

—— *Gabrielis Harveij Ciceronianus, Vel Oratio post reditum, habita Cantabrigiae ad suos Auditores.* London, Henry Bynneman, 1577.
> 4°     STC 12899     Hatfield House
> Annotations in Greek, apparently in Harvey's hand. Bound with his *Gratulationes Valdinenses.*

[——] *Ode Natalitia, Vel Opus Eius Feriae, quae S. Stephani protomartyris nomine celebrata est anno 1574. In memoriam P. Rami, optimi, et clarissimi viri.* London, Thomas Vautrollier, 1575.
> 8°     Brit. Lib. Bagford 5990
> The title-page, all that remains of Harvey's own copy of his work, bears the monogram 'GH'. This ode to Ramus concluded on sig. Avii$^v$ with the printed initials 'A.P.S.' (Aulus Pembrochianus Socius). Harvey was a fellow of Pembroke College, Cambridge, from 1570 to 1578. The work is referred to by E.K. in the *glosse* to the September Eclogue of Spenser's *Shepheards Calender,* 1579.

—— *Gabrielis Harveij Valdinatis; Smithus; Vel Musarum Lachrymae: Pro obitu Honoratissimi viri, atque hominis multis nominibus clarissimi, Thomae Smithi, Equitis Britanni, Majestatisque Regiae Secretarii. Ad Gualterum Mildmaium, Equitem Britannum, & Consiliarum Regium.* London, Henry Bynneman, 1578.
> 4°     STC 12905     Newberry ICN Case Y682.H25
> 'GH.' monogram. Underlinings and minor textual and punctuation corrections.                                                                      [tpx]

—— *Gratulationum Valdinensium Libri quatuor.* London, H. Bynneman, 1578.
> 4°     STC 12901     Brit. Lib. C.60.h.17(2)
> No signatures but various corrections of the text in Harvey's hand.

—— Same title.
> 4°     STC 12901     Hatfield House

Similar corrections in Harvey's hand. Bound with *Smithus*; *vel Musarum Lachrymae*.

H., J. Physition [Harvey, John]. *A discoursive Probleme concerning Prophesies, How far they are to be valued, or credited, according to the surest rules, and directions in Divinitie, Philosophie, Astrologie, and other learning: Devised especially in abatement of the terrible threatenings, and menaces, peremptorily denounced against the kingdoms, and states of the world, this present famous yeere, 1588, supposed the Greatwoonderfull, and Fatall yeere of our Age.* London, John Jackson for Richard Watkins, 1588.
    4°     STC 12908      Folger STC 12908, copy 2
'Gabriel Harveij. Ex dono Jo. fratris'. The inscription has been scored through with double lines, presumably by a later owner of the little book.

*\*HESE, Joannes de. *Itinerarius Joannis de Hese presbiteri a Hierusalem describens dispositiones terrarum, insularum, montium et aquarum . . . Tractatus de decem nationibus et sectis Christianorum. Epistola Joannis Soldani ad Pium papam secundum. Epistola responsoria eiusdem Pij Pape ad Soldanum . . . Epistola ad Emanuelem Rhome gubernatorem, de ritu et moribus Indorum. . . . Tractatus . . . de situ . . . regionum et insularum totius Indie necnon de rerum mirabilium ac gentium diversitate.* Cologne, 1500 (or possibly other edition).
    4°
On title-page of Billerbege, Harvey comments that this work and the tract of Georgirios Peregrinus, 'De Turcarum Moribus', are 'Two *necessary Treatises*, to be annexid unto my other Discourses of Aestates, & Governements. *Itinerarius Joannis de Hese*, de Indorum Moribus: de Imperio presbyteri Joannis apud Indos, et Aethiopas Christianos: de Decem Sectis Christianorum: item de Soldano Babiloniorum. Sultano Memphitico'.

*\*HEYWOOD, John. *John Heywoodes workes. A dialogue conteyning the number of the effectuall proverbes in the English tongue. . . . With one hundred of epigrammes: and three hundred of epigrammes upon three hundred proverbes: and a fifth hundred of epigrams. Whereunto are now newly added a sixte hundred of epigrams by the sayde J.H.* London, T. Marsh, 1576.
    4°     STC 13287
On fol. 210ᵛ of Guicciardini, after referring to books in 'singulares bibliothecas', Harvey writes: 'ecce Eutrapeli . . . propriae dactilothecae . . .' (Eutrapelus signifying Harvey in the *persona* of the urbane man). This is followed on fol. 211 by a list of books, presumably Harvey's, which includes 'Heywoods proverbs, & epigrams'.

HILL, Thomas. *The Schoole of Skil: Containing two Bookes: the first, of the Sphere, of heaven, of the Starres, of their Orbes, and of the Earth, &c. The second, of the Sphericall Elements, of the celestiall Circles, and of their uses, &c. Orderly set forth according to Art, with apt Figures and proportions in their proper places.* London, T. Judson for W. Jaggard, 1599.

4° STC 13502 University of Wisconsin

'gabrielis harveij. 1599'. On verso of page 267: 'Scholia astronomica, ad methodum Scribonii sphaericam facile redigenda: aut ad Freigii quaestiones physicas de Sphaera. Qualis etiam nostri Blundevili usus in perspicuo tractatu de Sphaera, nuper edito'. A few other MS. notes and underlinings.

[desc and tpx]

HOLYBAND, Claudius [Desainliens, Claude]. *The pretie and wittie historie o, Arnalt & Lucenda with certen rules and dialogues for the learner of the Italian tong.* 2 pts. London, T. Purfoote, 1575.

16° STC 6758 Brit. Lib. C.60.a.1(2)

No title-page; there remains only sig. V2 (p. 305) to end. On sig. V2, which is the final page of *D'Arnalte et Lucenda*, is Harvey's signature and date '1582'. On sig. V2ᵛ begins the second part of the book, *Familiar talks, etc.*, containing Italian conversation on the left-hand page and an English translation on the adjacent right-hand page. On page 336 where the Italian text gives proper speech to propose to a young lady, Harvey notes: 'piu per dolcezza che per forza'. The book has many annotations.

*HOMER. *Homericae Illiados Liber VI, VIII, & X–XXIV interprete H.E.H.* [*Helius Eobanus Hessus*], n.p., 1573.

8°

In Gauricus on sig. T4ᵛ Harvey writes: 'Habeo Eobani Hessi Iliada Homericam, sane perelegantem'.

HOOD, Thomas. *The marriners guide, set forth in forme of a dialogue, wherein the use of the plaine sea card is briefly and plainely delivered.* London, Thomas Est, for Thomas Wight, 1592.

4° 1596 ed. is STC 13696. Brit. Lib. C.60.f.8(2)

'gabriel harvey. The most sensible & familiar Analysis of the Sea-Card, that ever yet cam in print'. Dated 'this 1594' on verso of final page (sig. f2ᵛ). Some MS. notes on this page. Bound with Borne, *Regiment for the Sea.*

HOPPERUS, Ioachim. *In veram Iurisprudentiam Isagoges ad filium Libri octo.* . . . Cologne, Cholinus, 1580.

8° Brit. Lib. C.60.e.14

'Gabrielis Harveij, 1580'. 'GH.' On sig. A4: 'Synopsis totius juris civilis'. Copious MS. notes, some (e.g. on sig. XX3ᵛ) as late as 1605. Wide red ochre markings on the title-page and on the outside edge of the last forty pages. (See p. 164 above.)

*HOPPERUS, Ioachim. *Seduardus, sive de vera Iurisprudentia, ad Regem, libri XII . . . sive de Iuris & Legum Codendarum Scientia, libri IIII. Rerum Divinarum*

*et Humanarum, sive de Iure civili publico, libri IIII. Ad Pandectas, sive de Iure civili privato, libri IIII.* Antwerp, Plantin, 1590.
Folio
In Hopperus, *In veram Iurisprudentiam Isagoges* at the top of sig. R4ʳ, Harvey writes: 'Hopperi *Vera Jurisprudentia Ad Regem.* Extat tandem: et habeo'.

*[HOTMAN, François (E. Varamundus, pscud.)]. *A true and plaine report of the furious outrages of Fraunce, & the horrible and shameful slaughter of Chastillion the Admirall, and divers other Noble and excellent men.* Striveling, 1573.
  8° STC 13847
  In Harvey's 'Letter-Book' (Sloane MS. 93) on fol. 90ᵛ he writes of 'the furius outragies of Fraunc in Inglish' as one of the books of his which he will be glad to lend Arthur Capel. The text recounts the massacre of St. Bartholomew's Day.

HOWARDE, Henry (Earl of Northampton). *A Defensative against the Poyson of supposed Prophesies.* London, J. Charlewood, 1583.
  4° STC 13858  Houghton STC 13858.2
  '1583. Gabriel Harvey'. 'Quae supra nos, nihil ad nos'. 'GH.' Two leaves with MS. notes and a few pages briefly annotated. This book is an answer to Richard Harvey's 1583 prognostication, *An Astrological Discourse.*

HUGGELIUS, Joannes Jacobus [Hugkel]. *De Semeiotice Medicinae Parte, Tractatus: Ex probatis collectus authoribus, & in tabulae formam redactus.* Basle, Nicolaus Brylingerus, 1560.
Folio  Brit. Lib. C.60.0.10(2)
  'gabrielisharveij'. 'Idiota est, cui Huggelii, et Weccheri Semeiotica non est ad unguem: ad praesentem analys in urinae, egestionis, sputi, pulsus'. Another signature within this volume and the date '1584'. At the right top corner of the title-page is the monogram 'JH.' of Gabriel's physician brother, John Harvey. This book apparently belonged to him at some time before his death in 1592. At the bottom of the title-page Gabriel writes: 'Xˡⁱ Nondum lectum, quod nondum tuum ad unguem'. There are copious MS. notes both of John's and Gabriel's including many medical prescriptions and some red chalk-like markings. A tract on medical diagnosis with tables at the end on diseases and cures.

HUMFREDUS, Laurentius [Humphrey, Laurence]. *Interpretatio Linguarum: seu de ratione convertendi & explicandi autores tam sacros quam profanos, Libri tres.* Basle, Hieronymus Frobenius & Nicolaus Episcopus, 1559.
  8° Trinity College, Cambridge III.18.74
  'Gabrielis Harveij'. Elaborate monogram 'GH'. At top of sig. R7ᵛ (Book 3): 'Gabriel Harvejus, 1570'. Copious annotations in Latin and Greek in the

three books. Humphrey (1527?–90), Queen's Professor of Divinity at Oxford, was a writer on linguistics as well as theological controversy.

[HURAULT, Michel]. *An Excellent Discourse upon the Now Present Estate of France. Faithfully translated out of French, by E.A. [Edward Aggas].* London, John Wolfe, 1592.

    4°    STC 14005     Huntington 49490

    'gabrielharvey. 1592'. 'given mee bie Mr Woolfe, for a special rare Discourse'. A long folded sheet pasted in between fols. 3 and 4 contains lists in Harvey's hand of: (1) duchies, with their dependent counties; (2) counties dependent on the crown; (3) peers of France; (4) archbishoprics; (5) academies. On reverse of the sheet in Harvey's hand: 'A compendious description of France. A proffitable Table'. On the final leaf: 'Gabrielharvey: this August: 1592. Il legere nutrica lo ingegno'.          [desc]

ISOCRATES. *The Doctrinal of Princes made by that noble Oratour Isocrates, and translated out of Greke by Syr Thomas Eliot.* London, Thomas Berthelet [1534].

    16°    STC 14277     Newberry     Case Y 642185

    No signature (title-page missing). Tract has underlinings throughout and a page of Harvey's notes at the end.[5]

JAMES VI, of Scotland. *The essays of a prentise in the divine art of poesie.* Edinburgh, T. Vautrollier, 1585.

    4°    STC 14374     Magdalene College, Cambridge     Lect.26(1)

    'GH.' 'Gabriel Harvejus. le pris, iisiiiid'. 'Ex dono praestantissimi Doctoris, Bartholomaei Clarci, Arcuum Decani'. Copious MS. notes and Latin and English poems of Harvey's written in his hand and signed 'Axiophilus'. At the end of the preface of 'A Short Treatise' on sig. Kiii, Harvey writes: 'The excellentist rules, & finest Art, that a King could learne, or teach, in his Kingdom. The more remarkable, how worthie the pen, & industrie of a King. How mutch better, then owr Gascoigne's notes of instruction for Inglish Verse, & Ryme'. At conclusion of book (sig. Piiiiv): 'Legi xxiiii. februarii. 1585. Gabriel Harvei'. Bound with *poeticall exercises* and two works of Saluste du Bartas.

—— *His majesties poeticall exercises at vacant houres.* 3 pts. ('The Furies', 'the Lepanto', 'la Lepanthe'). Edinburgh, Robert Waldegrave [1591].

    4°    STC 14379     Magdalene College, Cambridge     Lect. 26 (2–4)

    'gabriel harvey. le pris, xiid'. At end (sig. P1v): 'gabrielisharveij, et amicorum'. Manuscript notes throughout. At end of line 915 of 'The Lepanto' (sig. L2): 'A gallant & notable poem, both for matter, & forme'.

JONSON, Ben. *The Workes of Benjamin Jonson.* London, William Stansby, 1616.

    Folio    STC 14751     Houghton 14426.4F*

[5] I am grateful to Eugene Sheehy for suggesting the probable location of this volume.

At top of the title-page: 'GH. pretium 19ˢ'. Top of the page has been cut off but there are traces of an additional inscription. No other markings by Harvey except in *Sejanus* (which has a number of underlinings and some lines scored through in brown ink) and in the *Epigrams*, most of the titles of which are checked at left with a short slanted ink stroke.

JOVIUS, Paulus. *Novocomensis Libellus de Legatione Basilii Magni Principis Moschoviae ad Clementem VII. Pontifex Max. in qua situs Regionis antiquis incognitus, Religio gentis, mores, & causae legationis fidelissime referentur. Caeterum ostenditur error Strabonis, Ptolomaei, aliorumque Geographiae scriptorum, ubi de Rypheis motibus meminere; quos hac aetate nusquam esse, plane compertum est.* Basle, 1527.
    4°    Houghton *70–84
'Gabrielis Harveij'. 'At hodie agnoscunt Baro ab Herberstein in sua Moscovia: et Neander in sua Samartia'. Harvey apparently read and annotated this book a number of times, for there are marginalia written in early and later hands of his. The volume subsequently belonged to the antiquary Maurice Johnson (1688–1755) who had it bound with Harvey's copies of Thomas and Billerbege.

JUSTINIAN. *D. Iustiniani Imp. Institutionum Libri IIII. Francisci Accursii glossis illustrati.* Lyons, Bartholomeus Vincentius, 1577.
    8°    Emmanuel College, Cambridge 324.862
'Gabrielis Harveij. 1579'. 'G.H.' 'mense, Martjo. CIↃ IↃ LXXIX'. 'fide, et taciturnitate'. Harvey's annotations include an analytical survey of the contents.                                      [tpx]

LENTULO, Scipio. *An Italian Grammer; written in Latin by Scipio Lentulo a Neopolitane: and Turned in Englishe: by H. G. [Henry Grantham].* London, T. Vautrollier, 1575.
    8°    STC 15469    Huntington 62184
'Gabrielis Harveij. 1579. mense Aprili'. 'Poco, y bueno'. Some MS. notes, but cropping of the volume has mutilated them. A group of Italian plays were evidently originally bound with the grammar, for Harvey writes on page 155: 'No finer, or pithier Examples, then in the Excellent Comedies, & Tragedies following: full of sweet, & wise Discourse. A notable Dictionarie, for the Grammer'.                       [desc]

LHUYD, Humfrey [Llwyd, Humphrey]. *The Breviary of Britayne . . . Contayning a learned discourse of the variable state, & alteration thereof; under divers, as wel natural; as forren princes, & conquerors. . . . Writen in Latin by Humfrey Lhuyd of Denbigh . . . and lately Englished by Thomas Twyne.* London, Richard Johnes, 1573.
    8°    STC 16636    Rosenbach Foundation

'Ex dono Mʳⁱ Browghton, Christensis. Gabriel Harvei'. Vertically at the right margin of the title page: 'Tractatus, cuique Anglo necessarius; non ignoranti, rudique suae patriae'. Also some notes within the book. Bound, probably by Harvey, at end of the volume containing Dionysius Periegetes, Rowlands, Grafton, and Turler. All have Harvey's same marginal symbol ⊙ and there are references from one work to another in the volume.

LITTLETON, Sir Thomas. *Littletons Tenures in Englishe*. London, R. Tottel, 1581.

8º    STC 15771    [whereabouts unknown]

According to W. C. Hazlitt in the 1899 Quaritch catalogue, this volume has 'Harvey's autograph and MSS. Notes'. It was a standard work on English law used as a textbook for law students.

LIVY, Titus. *T. Livii Patavini Conciones cum argumentis et annotationibus Joachimi Perionii*. Paris, S. Colinaeus, 1532.

8º    Worcester College, Oxford c.m.3(3)

'Gabriel Harveijus'. 'Thomas Smyth' (in Smyth's hand). At right (partially trimmed off): 'Gabrielis H . . .' /'ex dono C . . .'/'Smithi/ 1578'. This evidently was a gift to Harvey by one of the Smyth family after Sir Thomas's death in 1577. On the final page (sig. L8ᵛ) Harvey has written: 'Adoptatio Arrogatio. J.C. Linea Eunapii adamantina. Poco, y bueno. 1578'. Some red chalk or minium markings on the title-page. Many underlinings and a few notes within the text. Bound (possibly by Harvey) with Ramus, *Ciceronianus*, and Freigius, *Ciceronianus*. Moore Smith believed these books were used by Harvey to prepare for his lectures on rhetoric. The *Conciones* would not have been in his possession early enough for this but it may, of course, have been borrowed from Sir Thomas at some time before the spring of 1576.

—— *T. Livii Patavini, Romanae Historiae Principis, Decades Tres, cum dimidia; partim caelii secundi curionis industria, partim collatione meliorum codicum iterum diligenter emendatae. . . . Simonis Grynaei de utilitate legendae historiae . . . Doctorum virorum in hunc auctorem Annotationes, Glareani annotationibus, suis locis commode & diligenter infertae. Chronologia Henrici Glareani, ab ipso recognita & aucta. Eiusdem Chronologia, in alphabeticum ordinem, ut longius quaerendi abesset labor, a Iodoco Badio Ascensio redacta*. Basle, Jonnes Hervagios, 1555.

Folio in sixes    Private collection, United States

'Gabriel Harvejus. Mense Julio. Anno. 1568'. 'maturo consilio et strenua actione.' 'Laurentii Vallae adversus Livium argutissima Disputatio post Glareani annotationes'. 'arte, et virtute'. 'ex dono Dʳⁱˢ Henrici Harveij. A. 1568'. 'Gabrielis Harveij'. On sig. a3: 'Livius, Romanae historiae facile princeps. Singuli Libri, in suo genere praestantissimo'. At the end of the index is a long note signed by Harvey and dated '1590', and on the final page the inscription: 'gabrielis Harveij, et amicorum'. Copious MS. notes made at various times (at least three readings). A few red markings.

**\*LUCIANUS.** *Opera quae quidem extant omnia, graece & latine in quatuor tomos divisa quorum elenchos post aliquot pagines reperies, una cum Gilberti Cognati & Johannis Sambucci annotationibus utilissimis.* Basle, Henricus Petrus, 1563.

This must have been the 'four-volume Lucian' (Greek and Latin) which Harvey was to forfeit to Edmund Spenser if Harvey did not fulfil the conditions (stipulated in his hand) in a copy of Murner's *Howleglas.* For terms of the wager, see entry under [Murner], below.

**MACHIAVELLI,** Niccolò. *The Arte of Warre; written in Italian by Nicholas Machiavel; and set foorth in English by Peter Withorne, student at Graies Inne, with other like Martial feates and experiments; as in a Table in the ende of the booke may appeare. Newly imprinted with other additions.* n.p., 1573

4° STC 17165 Private collection, United States

'1580. Gabriel Harvey. precium iij^d'. On verso of the title-page: 'The Art of Defence, newly published in Italian. A necessary preamble to the Art of Warr.' On sig. A1^v: 'Machiavels intention to inform militar Discipline: & to restore the auncient orders of warly services, & martial vertu'. Numerous annotations in English and Latin, and Latin poems, presumably composed by Harvey.

—— *The Arte of Warre written first in Italian by Nicholas Machiavell, and set forth in Englishe by Peter Whitehorne, studtent at Graies Inne . . . Newly imprinted with other additions. (Certaine ways for the ordering of Souldiours in battelray . . . Gathered and set foorth by Peter Whitehorne. Most Briefe Tables to knowe redily howe many ranckes of footemen . . . [by] Girolamo Cataneo Novarese, Tourned out of Italian into English by H.G.)* 3 Parts: 1573, 1574, [1574].

4° [whereabouts unknown]

According to W. Carew Hazlitt (in Bernard Quaritch, *Dictionary of Book Collectors,* Part XIII, London, 1899, chapter on Gabriel Harvey), these titles are bound together in one volume with the autograph on the first title 'Gabriel Harvey, 1572', and the motto 'Unus quisque Fortunae faber' in Harvey's hand. Hazlitt further describes the book: 'The autograph is repeated in the course of the volume, which also has several signatures of his brother Richard, "Ricardi Harveij", and is filled with Gabriel's MSS. notes of various kinds, including mottoes, Latin verses, and bibliographical references'. The date '1572' may refer to the reading of the manuscript before publication or perhaps to the reading of an earlier edition of *The Art of War* (e.g. the Italian one of 1537). [desc]

**MAGNUS,** Olaus. *See* OLAUS MAGNUS.

**\*MANLIUS,** Joannes. *Locorum communium collectanea: a J. Manlio per multos annos; pleraque tum lectionibus, D.P. Melanchthonis, tum ex aliorum doctissimorum virorum relationibus excerpta, & nuper in ordinem ab eodem redacta.* 2 pts. Basle [Joannis Oporinus], 1563.

8°

In Guicciardini (fol. 209ʳ) under the heading of 'Propria Eulogi bibliotheca
. . . 1590', Harvey lists 'locos communes Manlij'. This Renaissance Latin
quotation book is also referred to by Harvey on sig. Dd7ᵛ of his Domenichi.

*MARTIALIS, Marcus Valerius. *M.V. Martialis Epigrammaton lectoris castimonia
dignorum liber: ubi omnia veneris illius despuendae quasi irritamenta quibus passim
sordidatus, lectorem nares corrugabat, accurata F. Sylvii Ambianatis diligentia,
deletili spongia detersa sunt, et eluta. In fine libri habes . . . Martialis vitam ex
Crinito.* Paris, Jacob Kerver, 1535.
4°
Harvey refers to his Martial on fol. 37ᵛ of Domenichi where he comments:
'Meus Rabelaisius: meus Martialis: Domenicus meissimus. Ad unguem
totus'. Although it is not known what edition of Martial, Harvey owned, he
seems generally to have preferred those editions of a man's works which
contained a biography.

[MEIER, Georg, M.D., of Würzburg]. *In Judaeorum Medicastrorum calumnias,
& homicidia; pro Christianis pia exhortatio. Ex Theologorum, & Juresconsultorum
Decretis.* [Speyer], 1570.
4°       Brit. Lib. C.60.h.18
'gabrielharvejus'. 'This very book sent to the famous Doctor Erastus, by
Doctor Struppius. Given me by Mistris Castelvestra, late wife of Erastus'.
On sig. A4ᵛ: 'Si tantum miraculorum posset *Cabala* Judaeorum, et illa
egregia naturalis Magia, quam *Cosmologiam* vocant; quantum ipsi gloriantur:
certe Judaei, soli essent Medici Mundi: nec ulli Physici, illis similes'. Annotations throughout.

MELANCHTHON, Philip. *Selectarum Declamationum Philippi Melanthonis, quas
conscripsit; & partim ipse in schola witebergensi recitavit, partim aliis recitandas
exhibuit.* Tomus primus. Strasbourg, 1564.
8°       Private collection, United States
'Gabrielis Harveij. 1583'. 'Emi a D. Pinello Gallo'. 'GH.' monogram. On
sig. Ziʳ: 'Astrologia et Geografia coniunctos Artes'. On sig. aiiᵛ: 'Nihil
facilius Arithmetica'. On inside of back cover Harvey comments on various
astronomical beliefs. Some underlinings and red markings but marginalia
are scanty.

MOHAMMED II, Sultan [from papers collected by Laudivius]. *The Turkes
Secretorie. Conteining his Sundrie Letters Sent to divers Emperours, Kings, Princes,
and States; full of proud bragges, and bloody threatnings: With severall Answers
to the Same, both pithie and peremptorie. Translated truly out of the Latine tongue.*
London. M.B[radwood]., 1607.
4°       STC 17996       Houghton Ott.251.1.20
'Gabrielis Harveij, mense Maio 1607'. 'Rerum in mundo maximarum

epistolae'. Text, which lacks a few pages, comprises a group of letters from the Turkish conqueror of Constantinople. These letters, originally written in Syrian, Greek, Scythian, and Slavonian, were intercepted by Laudivius, Knight of Jerusalem, and translated by him into Latin. As explained on sig. B3, M.B. subsequently published them in English 'for the good of those who desire to know the affaires and proceedings of the Turke'. On sig. Bi^r, re Mohammed II, Harvey writes: 'An Arch-imitatour of great Alexander, & greater Cesar combined. Full of pollicie, & valour in extent. But more brave, like Alexander, then fine, like Cesar. A worthie king if not a barbar[ous][6] Tirant'. Below letter of 'The Turke to the Venetians' (C2^r) Harvey writes: 'Veni, vidi, vici. Dictum, et ictum'. On C3^r, 'The Turke to the Genuensians': 'Menelorus contra Turca quasi Demosthenes contra Philippum'. On D2^r after 'The Turke to the Pannonians': 'Vana est sine viribus ira'. On Fi^r next to 'The Turke to King Ferdinand': 'Courting termes on both parts. Pragmatice, et sophistice'.

[MURNER, T.]. *A merye jeste of a man called Howleglas*. London, William Copland [dated *c.* 1528 by Bodleian Library].
    4°    Bodleian 4° Z.3. Art. Seld.
    All but the last few pages of the text are missing. On verso of the final page (sig. M4^v) is a faded inscription by Harvey (effaced portions indicated by square brackets): 'This Howletglasse, with Skoggin, Skelton, and L[a]zarill, given me at London of M^r Spensar XX. Decembris 1[5]78, on condition [I] shoold bestowe the reading of them over, before the first of January [imme]diately ensuing: otherwise to forfeit unto him my Lucian in fower volumes. Whereupon I was the rather induced to trifle away so many howers, as were idely overpassed in running thorowgh the [foresaid] foolish Bookes: wherein methowgh[t] not all fower togither seemed comparable for s[u]tle & crafty feates with Jon Miller whose witty shiftes, & practises ar reported amongst Skeltons Tales'. Till Eulenspiegel ('Howleglas'), a German peasant's son born about 1300 (according to tradition), was the subject of a collection of satirical tales first published in 1519.

NAPIER, John. *A description of the admirable table of logarithmes*. London, S. Waterson, 1618.
    12°    STC 18352    [whereabouts unknown]
    According to G. C. Moore Smith in *MLR*, xxviii (1933), 81, this book has G. H.'s autograph on the title-page but no marginalia. It was listed in Pickering & Chatto's Catalogue 281 (1933), Lot 71 but its present location is unknown.                                                                    [desc]

*NIFO, Agostino. *Augustini Niphi Medices Philosophi Suessani Collectanea ac Commentaria in libros de Anima.* 1522.
    Folio

---

[6] Outside edge of book has been trimmed.

This is a tract defending the Catholic doctrine of immortality against the attacks of Pomponazzi and others. On sig. B4ᵛ of Harvey's Simlerus bibliography is printed 'Petrus Pomponacius Mantuanus philosophus vide in Augustino *Nipho de immortalitate animae.*\*' (Asterisk and underlinings are Harvey's). Below this Harvey comments: 'habeo, legi, nec inficior, nec tamen assentior'.

NORTH, George. *The Description of Swedland, Gotland, and Finland; the auncient estate of theyr kynges; the most horrible and incredible tiranny of the second Christiern, kyng of Denmarke, agaynst the Swecians; the poleticke attayning to the Crowne of Gostave, wyth hys prudent providying for the same. Collected and gathered out of sundry laten Aucthors, but chieflye out of Sebastian Mounster.* London, John Awdely, 1561.

    4°     STC 18662     Folger STC 18662

'GabrielHarvey. 1574'. 'Gabrielis Harvey'. On sig. G2 after text description of languages of this area Harvey notes: 'The same radical of owre Inglish, & Scottish: notwithstanding sundrie dialects, or idioms, even amongst owrselves'. Other MS. notes include comments on law and customs, and praise of Thomas Stukeley to whom this book is dedicated by the author.

OLAUS MAGNUS. *Historia de gentibus Septentrionalibus, earumque diversis statibus, conditionibus, moribus, ritibus, superstitionibus, disciplinis, exercitiis, regamine, victu bellis, stricturis, instrumentis, ac mineris metallicis, et rebus mirabilibus; nec non universis pene Animalibus in Septentrione de gentibus, eorumque nativa. . . . Auctore Olao Magno Gotho, Archiepiscopo Upsalense, Suetiae, et Gothae Primate.* Rome, Viotti, 1555.

    Folio     Private collection, United States

Original title-page is missing but Harvey has inscribed the full title on the recto of the first flyleaf. On this page he has also written: 'Gabrielis Harveij. 1578. Vincenti Gloria Victis'. Many MS. notes. This is a history of the peoples of the north, dealing with their sagas, lore, and science, with many illustrative woodcuts.

(OLDENDORPIUS, Joan.) *Loci Communes Juris Civilis. Ex mendis tandem, & barbarie, in gratiam studiosorum utiliter restituti. Addita sunt Praesumptionum fere omnium, quae in foro frequentantur. Exempla; cum Joan. Oldendorphii Epistola nuncupatoria.* Lyons, Seb. Gryphius, 1551.

    18 cm.     Brit. Lib. Bagford 5991

'GH.' 'Juris regulae: pluribus locupletatae pragmaticis Sententijs. J.C. Gabriel Harvejus. Gnomae, et Aphorismi Pragmatici'. 'Xˡⁱ Huc etiam *Regulae Brocardicae Juris*: a Doctore Kaufero collectae: *cum pragmaticis clausulis*, solitis inseri *Instrumentis Publicis*'. At the end of the index: 'gabrielis-harvey, et amicorum. 1579. Disce: doce: age'. Of this book there now remains only the title-page, end of index, page with colophon emblem, and a number of blank pages filled with Harvey's closely written MS. notes, some evidently relating to the text and others comprising a group of astrological notes.

OVIDIUS, Publius. *Metamorphoses . . . argumentis quidem soluta oratione; enna-rationibus autem & allegoriis elegiaco versu accuratissime expositae; summaque; diligentia ac studio illustratae, per M. Johan. Sprengium Augustan.* Frankfurt, George Covinus, etc., 1563.
    12°     Houghton A1447.5.10
'gh.' 'gabrielisharveij'. According to H. S. Wilson (*HLB*, ii. [1948], 351 ff.), the volume must have been intended as a school text, designed to accompany the Ovidian text, which is not printed in this edition. The order followed throughout for each of the Ovidian poems is: (1) illustration by Vergilius Solis; (2) prose summary; (3) elegiac version; (4) allegorical explanation of poem's meaning, deriving a Christian moral from it.

PERCYVALL [Perceval], Richard. *Bibliotheca Hispanica. Containing a Grammar; with a Dictionarie in Spanish, English, and Latine . . . the Dictionarie being in-larged with the Latine by the advise of Master Thomas Doyley, Doctor in Physicke.* 2 pts.    London, John Jackson for Richard Watkins, 1591.
    4°    STC 19619    Huntington 56972
'gabrielharvejus' on title-page of the grammar. 'GH. Huc meum Dic-tionarium Homogeneum, proprie, et mere Hispanicam', on title-page of the dictionary. Marginalia in Latin and Spanish, and various passages marked.
                                                               [desc]

PINDAR. *Olympia, Pythia, Nemea, Isthmia. Caeterorum Octo Lyricorum carmina, Alcaei, Sapphus, Stesichori, Ibyci, Anacreontis, Bacchylidis, Simonidis, Alcmanis. Nonnulla etiam aliorum. Editio IIII Graecolatina, H. Steph. quorundam inter-pretationis locorum; & accessione lyricorum carminum locupletata.* Paulus Stephanus, 1600.
    8°    Folger, PA4274 A2 1600 Cage
'gabrielharvejus. 1580'. 'GH.' Since this was not published until 1600, Harvey's '1580' probably signifies the date when he first read these works of Pindar, possibly in an earlier edition of Henri Estienne's. Most of Harvey's annotations in this volume are in Latin on the Latin side of the dual Greek–Latin text. Some underlinings in red and green coloured markings.

PORCACCHI, Thomaso. *Motti Diversi Raccolti per Thomaso Porcacchi; Et aggiuntovi alle Facetie di M. Lodovico Domenichi.* [Venice, Andrea Muschio?], 1574.
    4°    Folger MS. H.a.2(2)
The above title occurs on sig. Dd8ᵛ of the Domenichi volume. This initial page of the Porcacchi pamphlet is covered with Harvey's marginalia but has no signature, monogram, or date. At bottom of the page Harvey writes:

'*His jest more serious, than anie mans earnest*'. On Ee1ʳ vertically at right margin: '*Make most* of such Examples, as may serve for *Mines* of Invention; *Mirrours* of Elocution; & *fountains* of pleasant devises. The finest *platforms*'. These pages like the rest of the volume are copiously annotated.

PTOLEMY. *La Geografia di Claudio Ptolomeo Alessandrino, Con alcuni comenti & aggiunte fattevi da Sebastiano Munstero Alamanno, Con le tavole . . . aggiuntevi di Messer Jacopo Gastaldo Piamontese cosmographo, ridotta in volgare Italiano da M. Pietro Andrea Mattiolo Senese medico Eccelentissimo. Con l'Aggiunta D'Infiniti nomi moderni, di Citta, Provincie, Castella, et altre luoghi, fatta con grandissima diligenza da esso Meser Jacopo Gastaldo. . . .* Venice, Gioan. Baptista Pedrazano, 1548. 2 pts., each with separate pagination.

8°     Collection of Charles Tanenbaum, New York.

No signatures but some annotations unmistakably in Harvey's hand, as well as underlinings and coloured markings. The brief inscription in the middle of sig. G6ᵛ may have been made by a later owner. Text of the first part is that of the second century A.D. geographer (and astronomer) still venerated as an authority well into the sixteenth century. In the pages that list towns (especially in the section entitled 'Sito Dell L'Hispagna Tarra-conense') Harvey apparently has placed a few annotations noting modern spellings or changed place-names. For example, on sig. G7ʳ next to the text listing of 'Libisocca' is the MS. annotation 'Hodie Lezuza'. The second part of the volume contains engraved double-page maps of various parts of the world: first the map made by Ptolemy and following it a contemporary one of the same area by Gastaldo. On the blank verso of a number of modern maps are annotations in Harvey's Italian hand referring to the work of other cartographers, e.g. on sig. 2iiᵛ (verso of 'Anglia e Hibernia Nova') are the following MS. notes: 'plinio: lib. 4, C. 16 Solino. c. 34: habraham ortelio. 9, 10, 11, 12, 13, 14, girava pag. 89; millac. 1720, leguac 430:' On the map captioned 'Germania Nova Tabula MDXXXXII' (sigs. 9iᵛ and 9ii) are some forty or fifty small red-brown paint or ink splotches picking out various points of interest; the map of 'Polonia et Hungaria' is treated in a similar way but with fewer coloured markings. In Harvey's marginalia one can find allusions to Ptolemy's work, as for example those on fol. 6 of Dionysius Periegetes. On fols. 4ᵛ–5ʳ of the Simlerus bibliography 'Claudius Ptolomaeus . . . de geografia' is underlined and also 'Venetiis . . . Geographia Ptolomaei Italice excusa est in 8°. Petro Mathiolo Senensi interprete, cum tabulis elegantissimis et novis quibusdam, in aere sculptis'.

QUINTILIAN, M. Fabius. *M. Fabii Quintiliani Oratoris eloquentissimi, Institutionum oratoriarum Libri XII.* Paris, Rob. Stephanus, 1542.

8°     Brit. Lib. C.60.l.11

'Gabrielis Harvey. mense Martio. 1567. precium—iiiˢviᵈ'. Elaborate 'GH' monogram. Below printer's device: 'gabrielharvejus. 1579'. The latter date evidently a rereading of the volume. On sig. a3ᵛ Harvey summarizes: 'The first two bookes, preparative. The next five, Logique for Invention, and Disposition. The fower following, Rhetorique for Elocution, & pronunciation: Logique for memory: an accessary, and shaddow of disposition. The last, a supplement, and discourse of such appurtenaunces, as may otherwyse concerne an Orator to knowe, and practise. As necessary furniture, and of no lesse use, or importaunce in Oratory Pleas, then the Praemisses'. Copious MS. notes including notes for Harvey's disputation at Audley End in 1578.

★RABELAIS, François. *Les Œuvres . . . contenans la vie, faicts & dicts Heroiques de Gargantua, & de son filz Panurge. Avec le Prognostication Pantagrueline.* [Paris], 1553.

8ᵒ

On fol. 209ʳ of Guicciardini under the rubric 'Propria Eulogi bibliotheca . . . 1590', Harvey includes 'dicta factaque heroica Gargantua'. On fol. 37ᵛ of Domenichi he writes of 'meus Rabelaisius'. Harvey's copy of Rabelais could have been either the 1553 Paris edition or the small thick 12ᵒ 1588 volume published in Lyons by Jean Martin. Harvey evidently read Rabelais in French as there was no early Italian or Latin translation and no English translation until Urquhart's in 1653, although 'Gargantua his Prophesie' was entered to John Wolfe on 6 April 1592, and apparently never published.

RAMUS, Joannes. *Oikonomia seu Dispositio Regularum utriusque Juris in Locos Communes brevi interpretatione subiecta: quae commentarii & locorum communium Joannis Rami Jureconsulti ad easdem Regulas, instar sit Enchiridij.* Cologne, ad intersignum Monocerotis, 1570.

8ᵒ    County Branch Library, Saffron Walden

'1574.' 'Gabrielis Harveij'. 'Ratione, et Diligentia'. On sig. A2ʳ: 'This whole booke, written & printed, of continual & perpetual use: & therefore continually, and perpetually to be meditated, practised, and incorporated into my boddy, & sowle'. Volume contains later datings of '1580' and '1582'. Copious MS. notes and some red and green chalk-like markings.

RAMUS, Petrus [Ramée, Pierre de la]. *P. Rami, Regii Eloquentiae, et Philosophiae Professoris, Ciceronianus; ad Carolum Lotharingum Cardinalem.* Paris. Andreas Wechelus, 1557.

8ᵒ    Worcester College, Oxford C.m.3(1)

'Gabrielis Harveij. GH.' 'cum Freigii etiam Ciceroniano: et Perionii Liviano: meis corculis. Pretium xviiiᵈ'. 'Labor improbus omnia vincit'. At top of the title-page in a later hand is the signature of 'W. Gower' and to right of printer's device in an early hand 'Sam: Edgley'. At end (sig. R7ᵛ) in Harvey's hand: 'I redd over this Ciceronianus twice in twoo dayes, Being then Sophister in Christes College. [1568–69] Gabriel Harvey'. Bound in contemporary binding, presumably Harvey's own, with Freigius, *Ciceronianus,* and Livy, *Conciones*. The Ramus volume is described as follows by Walter Ong (*Ramus and Talon Inventory*, Harvard University Press, 1958, p. 296): 'Ramus here gives his observations on the aims and methods of education, based on an account of Cicero's training and achievements. The imitation of Cicero's style here favoured by Ramus is moderate, like Erasmus's rather than Bembo's'. Some red chalk or minium markings.

[ROWLANDS, Richard]. *The Post of the World. Wherein is contayned the antiquities and originall of the most famous cities in Europe. With their trade & traficke, with their wayes and distance of myles, from country to country. With the true and perfect knowledge of their coynes, the places of their Mynts: with al their Martes and Fayres. And the Raignes of all the kings of England.* . . . 2 pts. London, Thomas East, 1576.

    4°     STC 21360     Rosenbach Foundation

    'Gabriel Harvey. 1580'. This first part contains seven leaves; then follows new title-page of second part:

*The Post For divers partes of the world: to travaile from one notable Citie unto an other; with a descripcion of the antiquitie of divers famous cities in Europe* . . . *very necessary & profitable for Gentlemen, Marchants, Factors or any other persons disposed to travaile. The like not heretofore in English.* London, Richard Rowlands for Thomas East, 1576.

    'Poste per diversi parti del mondo: con il viaggio di S. Jacomo di Galitia'. Pamphlet contains MS. notes, a number of which are comments on the characteristics of various cities, also a six-line English verse entitled 'Conceptiones verborum, ad matrimonium contrahendum aptae'. This two-part tract is bound with Dionysius Periegetes, Grafton, Turler, and Lhuyd.

SACCHI DE PLATINA, Bartholomacus. *Platinae hystoria de Vitis pontificum periucundae, diligenter recognita: & nunc tantum integro impressa.* . . . 2 pts. Paris, in vico sancti jacobi sub intersignio divi claudii [*c.* 1505].

    8°     University Library, Cambridge Rel. d.50.2

    On front flyleaf in Harvey's hand: 'Gabrielis Harveij liber emptus a Joanne Hutchinsono, Pembrokiano'. Above this inscription in another hand: 'Guilelmi Hutchinsoni liber, emptus ab. H.A.L.' On title-page: 'Gabrielis Harveij'. On flyleaf at end also in Harvey's hand is an index of popes and page references. This is completed on verso of flyleaf and has the signature 'Gabriel Harvejus'. The volume is a pejorative history of the popes. Part two has no title-page but has separate pagination beginning on sig. A1 and has the heading 'Liber primus *Divo Sixto, iiii. Pon. Max. Platynae dialogus de falso & vero bono.* Incipit. . . .'

SACROBOSCO, Joannis de. *Textus de Sphaera* . . . *Introductoria Additione* . . . *commentarioque, ad utilitatem studentium philosophiae Parisiensis Academiae illustratus, Cum compositione Annuli astronomici Boneti Latensis: Et Geometria Euclidis Megarensis*. Paris, Simon Colinaeus, 1527. 3 pts.

    Folio     Brit. Lib. 533.k.1

    'gabrielharvejus'. 'Plus in recessu, quam in fronte'. 'Arte, et virtute. 1580' Diagrams and copious marginalia. On sig. aii: 'Elenchus insignium materiarum'. 'Sacrobosco, & Valerius, Sir Philip Sidneis two bookes for the Spheare. Bie him specially commended to the Earle of Essex, Sir Edward Dennie, & divers gentlemen of the Court. To be read with diligent studie, but sportingly, as he termed it'. First flyleaf has intricate alchemical diagram (probably in the hand of an earlier owner of the volume), to left of which Harvey writes: 'Gemma salutaris, quae nescitur orbicularis: Est Lapis occultus, secreto fonte sepultus'. For descriptions of subtitles following Sacrobosco's astronomical text, see Bonetus and Euclid.

SALUSTE DU BARTAS, Guillaume de. *A Canticle of the victorie obteined by the French King, Henrie the Fourth, at Yvry*. Translated by J. Silvester. London, R. Yardley, 1590.

    4°    STC 21669    Magdalene College, Cambridge Lect. 26(6)

    No title-page. First page contains red diagonal mark and inscription by Harvey of two lines from Ovid's *Amores* (I. XV. 35–6): 'Vilia miretur vulgus: mihi flavus Apollo Pocula Castalia plena ministret aqua'. At bottom of sig. A2 Harvey writes: 'Nine leaves: or an howers reading: sed rosa solis'. Bound with *Triumph of Faith* and four works of James VI.

—— *The Triumph of Faith*. Translated by J. Silvester. London, R. Yardley & P. Short, 1592.

    4°    STC 21672    Magdalene College, Cambridge Lect. 26(5)

    'gabrielharvey. gh.' 'written in French by W. Salustius lord of Bartas; and translated by Josuah Silvester, Marchaunt Adventurer'. On sig. A4ᵛ: 'Francis the French King karried theise Triumphs abowt him, for his jewell: as Alexander used Homer'. 'Excellent Commonplaces, & of singular much use in the world for both matter, & forme'. On C4ᵛ: 'Bionis epitaphium: but sublimed. Henrici Epinicion: but quintessenced. Ilias Iliados: but multi-plied. Gargantuisme: but refined. Aretinisme: but disciplin'd. Singularitie: but civilis'd'. Copious MS. notes. Red marks on title-page.

*SCHORUS, Henricus. *Specimen et forma legitima tradendi sermones et rationis disciplinas ex P. Rami scriptis collecta*. Strassburg, 1572.

    8°

    In a letter to [Humphrey?] Hales (Sloane MS. 93, fol. 99ᵛ) Harvey writes that he gave his copy of this book 'to a friend of mine above a munth ago'.

*SIDNEY, Sir Philip. *The Countesse of Pembrokes Arcadia*. London, J. Windet for W. Ponsonby, 1590.

    4°    STC 22539

On fol. 209ʳ of Guicciardini under the rubric 'Propria Eulogi bibliotheca
. . . 1590' Harvey lists 'Arcadia Comitissa'. It was probably the 1590 edition
that he owned since he refers in his marginalia on sig. 7ᵛ of Euripides to
'Sidneium in novissima Arcadia' and on sigs. G3–G4 of *Pierces Supererogation*,
1593 (in a passage where he discusses and praises Sidney's *Arcadia*), Harvey
refers to Zelmane, Amphialus, and a number of other characters found only
in the 'New Arcadia'. Moore Smith lists as Harvey's an annotated copy of
the 1613 *Arcadia* (now at the Houghton Library) but I rather doubt that
Moore Smith ever had the opportunity of personally inspecting this book,
for I find the handwriting and character of the notes unlike Harvey's.

SIMLERUS, Josias. *Epitome Bibliothecae Conradi Gesneri, conscripta primum a
Conrado Lycothene Rubeaquensi: nunc denuo recognita & plus quam bis mille
authorum accessione (qui omnes asterisco signati sunt) locupletata: per Josiam
Simlerum Tigurinum.* Zurich, Chr. Froschoverus, 1555.
    Folio     Houghton A1447.3.100F
'Gabrielis Harveij'. Title-page also has partial signature and notes of a
1666 owner at Clare Hall, Cambridge. On sig. *5ᵛ Harvey describes this
important early bibliography: 'Magna adhuc opus est Gesneri bibliotheca:
praesertim ad argumenta variorum auctorum, et censuras. Quae magni sunt
momenti in classicis, multisque aliis scriptoribus considerate, ut refert, et
utiliter perlegendis. . . .' At the bottom of these notes is the signature 'gabriel
harvejus. 1584'. Although Harvey values this bibliography, he realizes that
it is far from complete, for on sig. H4ᵛ he adds: 'At multos ego libros legi,
et manuscriptos, et typis editos, variosque scriptores pervolvi, nunnullos
etiam memorabiles, de quibus hic ne gry quidem'. He evidently went over
the listing carefully, underlining a number of titles (probably those that he
had read, for many correspond with books owned or mentioned in his
marginalia) and marking a great number of the entries with symbols denoting
their subject matter, e.g. † (language, grammar, rhetoric), J.C. (legal matter),
LL. (the laws themselves), ⊕ (natural history), ⊖ (specific references to
Pliny), ☐ (theology), ∞ (controversy), ◯→ (military matters), ʃʃ (enlight-
ening, figuratively or literally shedding light). The listing is alphabetically by
authors' first names. At various points Harvey has made MS. additions of
other works by these authors or has inserted personal comments.

SLEIDANUS, Johannes. *De Statu Religionis Et Reipublicae, Carolo Quinto Caesare,
Commentarii Varia Ac Multiplici Rerum Utilissimarum Cognitione Referti. Nunc
Recens Accurata Diligentia, Summaque Fide Recogniti, Et Novis Summariis Singu-
lorum Librorum, Pro Faciliori Rerum Cognitione, Et Inventione, Aucti, Et Illustrati.
Adiecta Est Etiam Appendix, Seu Continuatio Eorum Commentariorum, Ab Anno
Christi M.D.LVI. quo Autor è Vita excessit, usque ad praesentem M.D.LXVIII.
Annum. . . . Auctore . . . D. Iustino Goblero, Goarino, V.I.Doctore. Cum Indice*

*Rerum omnium locupletissimo.* Frankfurt-on-Main, Petrus Fabricius for Hieronymus Feyrabend, 1568.

 Folio     Collection of Virginia F. Stern, New York

 On first title-page Harvey signs 'GabrielisHarveii' in his very early hand. The book was probably acquired shortly after publication, when he was about eighteen. In the title itself he inserts a comma after 'Commentarii' and below the title inscribes a decorative monogram 'GH'. On verso of title-page he lists Popes of this period as well as rulers of France and of England and at bottom of page notes other useful histories of the Reformation. Brief Latin notes throughout text and some underlinings. On title-page of Goblerus's 1556–68 continuation Harvey again inscribes his name and on verso of final page writes: 'gabrielis harveii, et amicorum'.

SMITH, Sir Thomas. *De recta & emendata Linguae Anglicae Scriptione, Dialogus, Thoma Smitho Equestris ordinis Anglo authore.* 2 pts. Paris, Robertus Stephanus, 1567. *De Recta & Emendata Linguae Graecae Pronuntiatone, Thomae Smithi Angli, tunc in Academia Cantabrigiensi publici praelectoris, ad Vintoniensen Episcopum Epistola.* Paris, Robertus Stephanus, 1568.

 4°     Private collection, United States

 'gh.' 'ex dono Joannis Woddi, aulici, et amici mei singularis. Anno 1569'. 'Acceptissima semper munera sunt, Auctor quae preciosa facit'. 'G. Harvejus'. 'Joannis Woddi liber, ex ipso Auctoris dono'. Some annotations, underlinings, and red markings in volume. Separate titles, pagination, and Harvey signatures in each of two parts.

STRAPAROLA da Caravaggio, Gio. Francesco. *Le notti . . . nelle quali si contengono le Favole, con i loro Enimmi da dieci donne, & da duo giovani raccontate.* 2 pts. Venice, Francesco Lorenzini da Turino, 1560.

 8°     Folger PQ 4634 S7 P53 1560 Cage

 No signatures on first title-page but some annotations and marginal symbols of Harvey's and red horizontal mark. Volume has been closely cropped at top. On title-page of part two: 'gabrielis harveij. secundae mensae'. On final page are faded and almost illegible marginalia of Harvey's which place this book in the genre of 'Canterburie tales. Boccatij Decameron'.

TALON, Omer. *Audomari Talaei Academia. Eiusdem in Academicum Ciceronis fragmentum explicatio. Item in Lucullum Commentarii; cum indice copiosissimo eorum, quae in his continentur, ad Carolum Lotharingum Cardinalem Guisianam.* Paris, Matthew David, 1550

 4°     All Souls, Oxford     a-11-4 (1)&(2)

 'GH.' 'Gabrielis Harveij'. Also signature of a previous owner, Thomas Hatcher (M.A. 1563 from King's College, Cambridge). Harvey's annotations consist of brief comments and some cataloguing of subject matter. Red horizontal mark on first title-page and greenish-brown markings within text. Although there are two title-pages, the tracts were published together.

TASSO, Torquato. *Aminta Favola Boschereccia*. Per J. Volfeo a spese di G. Castelvetri, 1591.

    12°     STC 12414     Columbia Univ. Lib. Spec. Coll. B851G93 U5(2)

Italian text. Published with Guarini's *Il Pastor Fido*, 1591. No signatures but some annotations of Harvey's and a manuscript index at the end.

TERENCE, P. *Le Comedie di Terentio Volgari; di nuovo ricorette, et a miglior tradottione ridotte*. Venice, Aldus, 1546.

    8°     Houghton *EC.2623 Zz546t

'Gabrielisharveij'. 'Quasi Synopsis omnium Mundi Comoediarum'. 'Speculum mundi $^{\text{vulgaris}}_{\text{nobilis}}$'. Some MS. notes and a few underlinings and red markings within this volume.

THOMAS, William. *The Historie of Italie*. London, Thomas Marshe, 1561.

    4°     STC 24019     Houghton *70–82

Title-page is missing but transcription of it is supplied by Maurice Johnson, later owner of the volume, who bound it with Billerbege and Jovius. Volume is filled with Harvey's marginal comments. At beginning of 'The Table' he writes: 'Theise Descriptions, & Guicciardin' of the Low Countries, full of most pregnant Instructions for the affaires of the World: & a little Glas of the great World'. and: 'Excellent Histories, & notable Discourses for everie politician, pragmatician, negotiatour, or anie skillfull man. A necessarie Introduction to Machiavel, Guicciardin, Jovius'.

TURLER, Jerome. *The Traveiler . . . devided into two Bookes. The first conteyning a notable discourse of the maner and order of traveiling oversea, or into straunge and forein Countreys. The second comprehending an excellent description of the most delicious Realme of Naples in Italy*. London, William How for Abraham Veale [1575].

    8°     STC 24336     Rosenbach Foundation

'Gabrielis Harveij.' 'ex dono Edmundi Spenserii, Episcopi Roffensis Secretarii. 1578'. 'Cosmopolita'. 'Methodus apodemica Zwingeri'. At bottom of final page: 'Legi pridie Cal. Decembrus. 1578. Gabriel Harvey'.

TUSSER, Thomas. *Five Hundred Pointes of Good Husbandrie, as well for the Champion, or open countrie, as also for the woodland, or Severall, mixed in everie Month with Huswiferie. . . .* London Henrie Denham . . . being the assigne of William Seres, 1580.

    4°     STC 24380     Rathbone (1966)

'Gabriel Harvey. xviii August. 1580'. 'gratum opus agricolis'. 'Heywoods proverbs, Floyds Discourses of the nine worthyes'. At end of the text: 'Now sum, and then sum: to lack is of paine'. 'Tusser, & Heywood poets for common life, and vulgar discourse. Full of matters to the purpose, in everie howse. The new art of Conie-catching in three, or fower parts'. Filled with

MS. notes. Tusser's verse gives practical advice of a general nature as well as specifically for each month. The above volume is listed (together with a reproduction of the title-page) in the Sotheby Sales Catalogue of March 21, 1966. It was sold to 'Rathbone' but its present whereabouts are unknown.

[desc]

VALERIUS MAXIMUS. *Valerii Maximi Dictorum factorumque memorabilium exempla*. Paris, Robertus Stephanus, 1544.

8°    University Library, Cambridge Adv.d.8.1

'Gabrielis Harveij' and a very large monogram 'GH'. 'Recentium exemplorum novem Libris, a Baptista Egnatio amplificata'. Red markings and a few annotations in the text. Valerius Maximus, who lived in the reign of Tiberius (first century A.D.), has here compiled a collection of historical anecdotes and curious events.

[VON HUTTEN, Ulrich (and others)]. *Duo Volumina Epistolarum Obscurorum Virorum, ad. D.M. Ortui Gratium, Attico lepore referta; denuo excusa & a mendis repurgata*. [Rome], 1570.

12°    Balliol College, Oxford 700.a.20 (1)

These are mock-serious letters written about 1515 to defend the scholar Reuchlin against his opponents' stupid accusations. On the title-page is the signature '1572. Gabrielis Harveij', and on the recto of the preceding flyleaf is the following comment in Harvey's hand: 'Scholarismus. Eadem fere adhuc vena nonnullorum Academicorum scholastica: tam nostrorum aliquando, quam alienorum frequentius'. The tract is bound with *De generibus ebriosorum*.

[tpx]

WHITEHORNE, Peter. *Certeine wayes for the orderin𝔰 [sic] of Soldiours in battelray, and setting of battayles, after divers fashions with their maner of marching: and also Fugures of certayne new plattes for fortification of Townes: And moreover, howe to make Saltpeter, Gunpowder, and divers sortes of Fireworkes. . . .* London, W. Williamson: for Jhon Wight, 1573.

4°    STC 17165 (Part 2)    Ministry of Defence Library, Old War Office, Whitehall: 211a

'Gabriel Harvey'. 'Agrippae pyrographia, sive pyromachia, non est in praedicamentis. Vanocii Pyrotechnia exstat: et nonnulla *Cardani Experimenta pyria*, De Subtilitate. Qualia etiam plura suggerit *naturalis magia*'. In the title Harvey has corrected 'Fugures' to 'Figures'. Volume has copious annotations and Latin verses presumably composed by Harvey. This work may be part of the Machiavelli *Arte of Warre* listed on lower half of p. 226.

WILSON, Sir Thomas. *The Art of Rhetorike, for the use of all suche as are studious of Eloquence*. London, Jhon Kingston, 1567.

4°    STC 25803    Rosenbach Foundation

'Gabrielis Harveij.' Extensively annotated. On fol. 69: 'One of my best for the art of jesting: next Tullie, Quintilian, the Courtier in Italian, the fourth of mensa philosoph. Of all, the shortest, & most familiar, owr Wilson'.

\*—— *A discourse uppon usurye.* London, R. Tottell, 1572.
    8º    STC 25807
Included in listing headed 'Propria Eulogi bibliotheca . . . 1590' on fol. 209ʳ of Guicciardini. On fol. 178ᵛ: 'cunning contractes, & bargaines in . . . Wilson's discourse of usurie'. On final page of Quintilian text: 'Wilson's Rhetorique & Logique, the dailie bread of owr common pleaders, & discourses. With his dialogue of usurie, fine & pleasant'.

[——]. *The Rule of Reason, conteinyng the Arte of Logike.* 2 pts. London, Jhon Kingston, 1567.
    4º    STC 25813    Rosenbach Foundation
'Gabriel Harvejus.' On last page of 'The Table': 'gabrielis harveij, et amicorum. 1570. Great varietie of rhetorique, logique, & much other learning'. Many annotations, a number alluding to Thomas More. Bound with *Art of Rhetorike*, possibly in Harvey's time.

XENOPHON. *Xenophontis Philosophi et Historici Clarissimi Opera, Quae Quidem Graece extant, omnia; partim jam olim, partim nunc primum, hominum doctissimorum diligentia, in latinam linguam conversa; ac multi quam ante accuratius recognita. Quorum elenchum versa pagella reperies.* Basle, Mich. Isingrinius, 1545.
    8º    Bodleian 90.C.19
'1570. GabrielisHarveij'. 'Scientia, et virtute'. Elaborate monogram 'GH'. On final page (sig. F3ᵛ): 'Ante et post Plutarcham, meorum alterum radiantum oculorum'. 'Gabriel Harvejus: Valdini: 1576 faventibus Etesijs. In scientia, et virtute omnis spes. Caesaris ipsius axioma'. Manuscript notes and underlinings in the text.

### WORKS OF UNCERTAIN AUTHORSHIP
### (LISTED BY PUBLICATION DATE)

1549. *The Images of the Old Testament; Lately Expressed, Set Forthe in Ynglishe and French, with a Playn and Brief Exposition.* Lyons, Johan Frellon. Prose translations by H. Corrozet. With ninety-four woodcuts after Holbein.
    4º    STC 3045    Huntington 56974
'GH.' 'Gabrielis Harveij, 1580'. 'Sallust, du Bartas, the only brave Poet in this sacred vein'. Manuscript notes and underscorings in the text. This is a French–English edition of the popular *Historiarum Veteris Testamenti*.

[desc]

1565. *De generibus ebriosorum, et ebrietate vitanda.*
  12°      Balliol College, Oxford      700 a.20(2)
  On title-page 'Gabriel Harvejus'. Some underlinings and a few marginal symbols within this tract about drink-loving men. Bound with *Duo Volumina Epistolarum Obscurorum Virorum.*

[1567]. *Institutions, or Principal Grounds of the Laws and Statutes of England, newly corrected and amended.* London, R. Tottel.
  8°      STC 9293b      [whereabouts unknown]
  W. Carew Hazlitt (in Quaritch catalogue) lists this title as having 'Harvey's autograph and MSS. notes'. It is also listed in Moore Smith's *Marginalia* (1913).

*[1567]. *Merie tales newly imprinted and made by Master Skelton.* London, T. Colwell.
  8°      STC 22618
  Referred to in *Howleglas* marginalia as 1578 gift to Harvey from Spenser. For Harvey's inscription, see [Murner].

*1573? *The pleasaunt historie of Lzarillo de Tormes.* Tr. by D. Rouland. Entered to T. Colwell 1568-9; sold to H. Bynneman 19 June 1573.
  'L[a]zarillo' is referred to by Harvey in his marginalia in *Howleglas* as having been received from Spenser in 1578 (see Murner). Earliest extant English edition is the 1586 octavo of A. Jeffes but date of entry in the Stationers' Register suggests that there was an earlier edition. Harvey seems to have been studying Spanish about 1590 and at that time may have read 'Lazarillo' in its original tongue as he refers to the Spanish title on sig. S4ᵛ of Corro's *Spanish Grammer* (1590).

1582. *An excellent, perfect, and an approoved medicine and waie to helpe and cure the stone in the raines.* n.p.
  Folio sheet      Brit. Lib. C.60.0.10(3b)
  Attached to the end of the Hieronymus von Braunschweig volume but evidently printed separately and considerably later. It has MS. notes and underlinings of Harvey's. At end of the sheet in small type is printed: 'This was written at the request of maister Simon Boyer, one of the gentlemen ushers to hir most excellent Majestie, by a friend of his, that tried the same upon his owne bodie, and hath found great helpe therein; and hath thereby holpen manie others thereof. G.G.'

1583. *Calendarium Gregorianum Perpetuum.* Antwerp, Chr. Plantin.
  16°      National Library of Wales, Aberystwyth Peniarth MS. 526.
  'precium ijᵈ iijˢ. gh.' 'Gabrielis Harveij. 1583'. Ornamental monogram 'GH'. On verso of flyleaf preceding title-page: 'Confer Sepulvedam de

correctione Anni, Mensiumque Romanorum: Ad Gasparem Contarenum Cardinalem. Ante Calendarium Gregorianum. Post hoc Calendarium, scrupulosas observationes Maestlini: quo nemo fere curiosior astronomus'.

[xerox copy seen]

[1583?]. *An Abstract, of Certaine Acts of Parlement: of certaine her Majesties Injunctions: of certaine Canons, Constitutions, and Synodals provinciall, established & in force, for the peaceable government of the Church, within her Majesties Dominions and Counries, for the most part heretofore unknowen and unpractized.* (2 Pt.). [R. Waldegrave?]

4° STC 10394 Durham Univ. Library G.111.33(2)

'GH.' 'gabrielharvey'. A few annotations within the tract. Bound with the two R. Cosin works.

[1585?]. *These Oiles, Waters, Extractions, or Essence Saltes, and other Compositions; are at Paules wharfe ready made to be solde, by John Hester, practisioner in the arte of Distillation; who will also be ready for a reasonable stipend, to instruct any that are desirous to learne the secrets of the same in few dayes. . . .*

Broadsheet Brit. Lib. C.60.0.6

Underlinings and brief notes by Harvey. At end: 'Now M. Keymis, the great Alchymist of London'. 'gabrielharvey. 1588'.

*1593. *Articles accorded for the Truce-generall in France.* London, John Wolfe for A. White.

4° STC 13117

Mentioned on sig. A2ʳ of *New Letter of Notable Contents* (1593) as one of the books Harvey is grateful for having received from Wolfe.

*1593. *A True Discourse . . . the wonderful mercy of God shewed toward the Christians against the Turke before Sysek 22 June 1592.* Tr. from High Dutch, John Wolfe, 1593.

4° STC 5202

Referred to on sig. A2ʳ of *New Letter of Notable Contents* (1593) as read with pamphlet about the Duke of Mayne, both of which had just been received from John Wolfe.

*1593. *Remonstrance to the Duke de Mayne.* London, John Wolfe.

4° STC 5012

Mentioned on sig. A2ʳ of *New Letter* (1593) as received from Wolfe. It refers to Charles de Lorraine, Duke of Mayenne.

1626. At end of his listing of Harvey's books in the Quaritch catalogue, W. Carew Hazlitt notes: 'A tract, published in 1626, occurred many years ago at an auction, bearing Harvey's autograph; it attested to his survivorship to that date; but I unfortunately neglected to register the title'.

# 3. Manuscripts belonging to Gabriel Harvey

Brit. Lib. Add. MS. 36,674. The first four articles in this group of treatises relating to magic have annotations in Harvey's hand, which probably date from about 1577.

36,674(1). The Book of King Solomon called the Key of Knowledge. Harvey notes on fol. 5: 'Clavicula Salomonis. extat latine: et legi'. 'Cabalistica: sed sophisitica'. This MS. runs from fol. 5 to 22.

36,674(2) is described in Harvey's annotations (fol. 23) as follows: 'This torne booke was found amongst the paperbookes, & secret writings of Doctor Caius: master & founder of Caius colledg. Doctor Legg gave it to M^r Fletcher, fellowe of the same colledg, & a learned artist for his time'. On fol. 45 at the end of this MS. Harvey notes: 'The best skill, that M^r Butler physician had in nigromancie, with Agrippas Occulta philosophia: as his coosen Ponder upon his Oathe often repeated, seriously intimated unto mee'. This MS. runs from fol. 23 to 45.[1]

36,674(3). Simon Forman's 'An excellent booke of the arte of Magicke, first begoone the xxii^th of Marche Anno Domini 1567'. Believed written in Forman's hand but evidently later owned by Harvey since it has a few brief annotations by him, e.g. on fol. 47: 'Nova methodus, et praxis magica Academici philotechni. . . . (Ad artem notoriam inspiratam.) Speculum omniscium'. This MS. runs from fol. 47 to 57^v.

36,674(4). Harvey captions this MS. 'Certaine straung visions, or apparitions of memorable note. Anno 1567. Lately imparted unto mee for secrets of mutch importance. A notable Journal of an experimental Magitian' (fol. 58). At the bottom of fol. 59 Harvey comments: 'The visions of S^r Th. S. himself: as is credibly supposed. Thowgh M^r Jon Wood imagins one G.H. Tempus demonstrativum revelabit'. This article, which runs from fol. 58 to 62^v relates to 'visions' seen by 'H.G.' [Humphrey Gilbert?] and his 'skrier' John Davis.

Brit. Lib. Add. MS. 32,494. A miscellaneous commonplace book of fifty-two closely written folios. Here Harvey has recorded some of his studies and opinions and many random observations and precepts. About half of this is transcribed by Moore Smith in his *Gabriel Harvey's Marginalia* (1913). The little volume is about 3⅛ by 5½ inches. On fol. 2 is the signature 'Gabriel Harvejus'. and an inscription beginning, 'Natura, Ars, Industria; perpetua

[1] Thomas Legge, Master of Caius College from 1573–1607; John Fletcher, famous astrologer and seer; William Butler of Clare Hall, physician to Prince Henry in 1612, reputed an eccentric; John Davys, navigator and explorer.

Inductiones; colligere debent Bona Animi, Corporis, Fortunae'. The notes in the volume seem to span a number of years.

?? Small octavo-sized commonplace book, *c.* 1584, noted by W. Carew Hazlitt and Moore Smith. It is said to contain a poem entitled 'A View, or Spectacle of Vanity', at the end of which Harvey appends 'Incerti Authoris Anno 1584', an English fragment relating to enclosures, and a few other fragments in Latin and Italian. Only five leaves of this book are extant.

[whereabouts unknown]

Brit. Lib. Lansdowne MS. 120, fol. 12. Harvey's MS. of his *Xaipe vel Gratulatio Valdinensis ad . . . Dom . . . Burgleium. . . .* This work presumably was written during July 1578 (time of Harvey's disputation at Audley End) and presented to Lord Burghley then.

Brit. Lib. Sloane MS. 93. Harvey's so-called 'Letter-Book', transcribed and printed by E. J. L. Scott in 1884 (Camden Society). Scott dates its contents between 1573 and 1580. It includes an assortment of letters to and from Harvey (chiefly the latter) together with some MS. verses and an account of his sister 'Marcie' Harvey's attempted seduction by Philip Howard, Lord Surrey, and of how Gabriel put a stop to it. It is followed several pages later by a letter from Gabriel to Lady Smith (wife of Sir Thomas) requesting a place in her employ for his 'pore sister'. At the end of the notebook are five pages (fols. 100$^r$–101$^r$ and 103$^v$–104$^r$) of memoranda written upside down (in a hand unlike any of Harvey's hands) listing various moral admonitions and wise precepts to observe, with Biblical references given for each. At the top of the last of these pages Harvey comments: 'M. Wilkes *fooleryes.*' I cannot identify 'M. Wilkes' whose notes I assume these are. Perhaps he was a member of the family of Sir Thomas Wilkes (1545?–98), secretary to Valentine Dale and ambassador to France in 1573.

Bodleian Rawlinson MS. Poet. 82. Manuscript (in an unidentified hand) of a poem printed in Surrey's *Songs and Sonnettes* (1587). It is titled in Harvey's holograph, 'Totus mundus in maligno positus'. On the recto following the poem is the signature 'Gabriel Harvey' and written above this, apparently in Harvey's hand, are a few lines headed 'Sir John Cheek', and beginning, 'Who can persuade, when treson is above reson; and Might ruleth right. . . .'

*Harvey, Gabriel. *Anticosmopolita, or Britanniae Apologia.* Entered S.R. on 30 June, 1579, to Richard Day, but apparently never printed. In *Three Proper and wittie familiar Letters* (1580) sig. Dii$^r$, Harvey writes to Edmund Spenser: 'Some newe occasion, or other, ever carrieth me from one matter to another, & will never suffer me to finishe eyther one or other . . . which my *Anticosmopolita*, though it greeve him, can both testifye, remayning still as we saye, *in statu, quo,* and neither an inche more forward nor backwarde, than he was fully twelvemonth since in the Courte, at his last attendaunce uppon my Lorde [Leicester?] there'. Richard Harvey also refers to this work in the dedication of his *Astrological Discourse* (1583), where he alludes to 'my brothers Anticosmopolita'.

# 4. Short-title listing by publication date

Date in square brackets is date of acquisition by Harvey or of his first reading of the work

1570? Ascham, *Affaires and State of Germany* [by 1590]
1570 Fabricius, *Ciceronis Historia* [1572]
1570 Meier, *In Judaeorum Medicastrorum calumnias*
1570 Ramus, J., *Oikonomia, seu Dispositio Regularum* [1574]
1570 Von Hutten, etc., *Epistolarum Obscurorum Virorum*
1571 Buchanan, *Ane Admonition*
1571 Buchanan, *Ane Detection*
1571 Buchanan, *De Maria Scotorum Regina* [1571]
1571 Castiglione, *Comitis de Curiali sive Aulico* [by 1572]
1571 Domenichi, *Facetie, motti, et burle*
1571 Foxe, *De Christo crucifixo Concio* [1572]
1571 G., R., *Salutem in Christo.* . . .
1571 Guicciardini, L., *Detti et fatti* [1580]
1572 Dionysius Periegetes, *Surveye of the world* [1574]
1572 Fulke, *Ouranomachia. Hoc est Astrologorum Ludus*
1572 Schorus, *Specimen et forma . . . ex Rami scriptis collecta* [by 1580]
1572 Wilson, *Discourse uppon Usurye* [by 1590]
1573 Barnaud, *Dialogus . . . quae Lutheranis et Hugonotis acciderunt*
1573 Du Faur, *De rebus gallicis*
1573 Homer, *Iliados* [by 1580]
1573 Hotman, *Report of the furious outrages of Fraunce* [by 1580]
1573? *Lazarillo de Tormes* [1578]
1573 Lhuyd, *Breviary of Britayne*
1573 Machiavelli, *Arte of Warre* [1580]
1573 Whitehorne, *Certaine wayes for the Ordering of Soldiours*
1574 Frontinus, *Strategemi Militari*, Gandino commentary [by 1590]
1574 Machiavelli, *Arte of Warre* [1572]
1574 Porcacchi, *Motti Diversi*
1575 Freigius, *Ciceronianus* [1575]
1575 Gascoigne, *Posies* [1577]
1575 Harvey, G., *Ode Natalitia*
1575 Holyband, *Arnalt & Lucenda* [1582]
1575 Lentulo, *Italian Grammer* [1579]
1575 Turler, *The Traveiler* [1578]
1576 Cheke, *Hurt of Sedition* [by 1580]
1576 Gascoigne, *Steele Glas & Phylomene* [1577?]
1576 Grafton, *Brief treatise conteyning many . . . Tables* [1576]
1576 Heywood, J., *Woorkes*
1576 Rowlands, *The Post of the World* [1580]
1577 Davies, *Funerall sermon . . . Walter Earle of Essex*
1577 Harvey, G., *Ciceronianus* (2 copies)
1577 Harvey, G., *Rhetor*
1577 Justinian, *Institutiones* [1579]
1578 Du Ploiche, *Treatise in Englishe and Frenche* [1580]

1578    Florio, *First Fruites* [1580]
1578    Harvey, G., *Gratulationes Valdinenses* (2 copies)
1578    Harvey, G., *Smithus, vel Musarum Lachrymae*
1580    Blundevill, *Fower chiefyst offices belonging to Horsemanship*
1580    Hopperus, *In veram Jurisprudentiam Isagoge* [1580]
1580    Tusser, *Five Hundred Pointes of Good Husbandrie* [1580]
1581    Guazzo, *La Civil Conversatione* [1582]
1581    Guazzo, *The Civile Conversation* [1582]
1581    Littleton, *Tenures in Englishe*
1582    Foorth, *Synopsis Politica* [1582]
1582    'To helpe and cure the stone in the raines' (folio sheet)
1583    *Calendarium Gregorianum Perpetuum* [1583]
1583    Chytraeus, *De tribus . . . Caesaribus*
1583    Freigius, *Mosaicus* [1584]
1583    Freigius, *Paratitla* [1583]
1583    Howarde, *Defensative against . . . Prophesies* [1583]
1583?   *An Abstract, of certaine acts of Parlement*
1584    Cosin, *An answer to . . . a certain factious libell*
1584    Astley, *Art of Riding*
1584?   Billerbege, *Straunge Discourses of Amurathe*
1585    Blagrave, *Mathematical Jewel* [1585]
1585    Bruele, *Praxis medicinae theorica* [1589]
1585    James VI, *Essayes of a prentise* [1585]
1585    'These Oiles, Waters, Extractions . . . to be solde' (broadsheet) [1588]
1587    Heywood, J., *Woorkes* [by 1590]
1588    Harvey, J., *Discoursive Probleme concerning Prophesies*
1588    Aretino, *Quattro comedie*
1590    Corro, *Spanish Grammer* [1590]
1590    Hopperus, *Seduardus, sive de vera Iurisprudentia*
1590    Saluste du Bartas, *Canticle of . . . victorie*
1590    Sidney, *Arcadia* [1590]
1591    Guarini, *Il Pastor Fido*
1591    James VI, *Poeticall exercises*
1591    Percyvall, *Bibliotheca Hispanica*
1591    Tasso, *Aminta*
1592    Borne, *Regiment for the Sea* [1594]
1592    Eliot, J.? *Description of France* [1592]
1592    Harvey, G., *Foure Letters*
1592    Hood, *Marriners Guide* [1594]
1592    Hurault, *Discourse upon the . . . Estate of France* [1592]
1592    Saluste du Bartas, *Triumph of Faith*
1593    *Articles accorded for the Truce-generall in France* [1593]
1593    Barnes, *Parthenophil and Parthenophe* [1593]
1593    Chute, *Beawtie dishonoured* [1593]

1593    Cosin, *An Apologie for Sundrie Proceedings* [1593]
1593    *A True Discourse* [1593]
1593    Eliot, J., *Ortho-epia Gallica* [1593]
1593    Harvey, G., *New Letter of Notable Contents*
1593    Harvey, G., *Pierces Supererogation*
1593    *Remonstrance to the Duke de Mayne* [1593]
1598    Chaucer, *Woorkes* [1598]
1599    Greverus, *Secretum . . . de Lapide Philosophico*
1599    Hill, *Schoole of Skil* [1599]
1600    Pindar, *Olympia, Pythia* [1580]
1607    Mohammed II, *The Turkes Secretorie* [1607]
1616    Jonson, *Workes*
1618    Napier, *Table of logarithmes*
1626    Volume with Harvey signature seen by Hazlitt (title unknown)

# 5. Size of the library estimated

How extensive was Harvey's complete library? No contemporary listing has survived to aid us, but we do have a report by Thomas Nashe in *Have with you to Saffron-walden*[1] that Harvey told friends of having 'a Library worth 200. pound'. If Nashe is reporting accurately, this statement was made by Harvey at some time before 1596 (the publication date of Nashe's work) when the Saffron Walden bibliophile was about forty-six and still had ahead of him some thirty-four years of collecting. I believe that one may give considerable credence to Nashe's report. Harvey was not prone to exaggeration, and although Nashe's veracity as a reporter may be questioned, one finds that most of his factual statements are indeed accurate; it is only their interpretation which he bends askew.

Extrapolating from a record of William Drummond's that he paid £17. 14s. 9d. for seventy-six English and three hundred and twenty-three French books, R. H. MacDonald has estimated that Drummond's library of some 1,600 books would have been worth £90.[2] In line with this relatively low-price-per-book estimate (of one shilling, one and a half pence per book), Harvey's £200 library (of 1596 or earlier) could have comprised over 3,500 titles.

The average price that Harvey paid per volume would have been influenced by several factors. He had a great number of books from continental presses (Paris, Basle, Lyons, Rome, Antwerp, etc.) and these cost far less than those of English printers. The usual price for a printed book in England (unbound) was one halfpenny a sheet; illustrated or legal books were about eight-tenths of a penny per sheet. By contrast, a book in Paris would have cost less than one-half its price in London.[3] Secondly, although a number of Harvey's books were bought in the year of publication, many were purchased second-hand some years later,[4] and these would therefore have been of reduced price. A third factor influencing cost would have been the type of binding. Bindings ranged in price from one or two pence for a simple wrapper to a top price of fifteen shillings for an elaborate metal clasped folio. If Harvey's extant books are a fair sample, he seems to have preferred a sturdy simple leather binding which would have cost not more than two or three shillings. Many of his volumes are in a so-called 'Cambridge binding' and include three titles. Among his identified volumes I have noted only five particularly handsome bindings; that of the Thomas

---

[1] *Works*, iii. 89.

[2] MacDonald, op. cit., p. 42 and *passim*.

[3] MacDonald, p. 38.

[4] Harvey frequently enters the date of acquisition on the title-page.

Smith tracts and those of the Livy, Chaucer, Sleidanus, and Jonson folios. Of these, the Smith and Livy and possibly the Sleidanus were gifts.

Harvey has left evidence of the price paid for some of his books, for occasionally he makes a note of this amount next to his signature on the title-page.[5] These prices range from a few pence to a few shillings for a folio volume (except in the unique case of the large Ben Jonson which was purchased at least twenty years after Nashe's statement for the extravagant sum of nineteen shillings). While there are among the extant volumes about thirteen folios (all those pre-1596 except the Blagrave were purchased considerably after the year of publication), most are less expensive books of quarto or octavo size and a good many are smaller still. If one makes an estimate of three shillings per volume, including binding, as an average cost per title (perhaps an overly generous estimate when one considers Harvey's tendency toward thrift), the sum of £200 would represent a library of at least 1,300 books. By 1596 Harvey's collection thus could have included anywhere between this minimum and a maximum of over 3,500 volumes, the figure arrived at on the basis of Mac-Donald's approximations. Probably because of Harvey's various economies it was nearer the latter figure. By the time of Harvey's death in 1631 the total would have certainly increased considerably.

Harvey was justifiably proud of this library which he housed in his Walden home. In *Ciceronianus*, the lecture delivered in the spring of 1576 to Cambridge students after Harvey's return from a period spent at home (see p. 30 above), he introduces the topic of Ciceronianism with the following description:

In my Tusculan villa (for it pleases me to appropriate this name) . . . I have so spent my leisure that I did not seem altogether without occupation in my idleness nor without leisure amid my occupation. Often there lay at my hand . . . your friend Cicero; sometimes the champions among the historians, Caesar and Sallust (what great men were those!); or the most illustrious [Latin] poets . . . Virgil, Horace, and Ovid, writers whom I had long neglected. To tell the truth, I also had by me some of the newer writers, men who are not only lamps of the present age but ornaments of all posterity, Sturmius, Manutius, Osorius, Sigonius, and Buchanan, all of them thoroughly polished from the very school of eloquence and on so many accounts most dear to me. The rest of the more cultivated and humane authors, including all the Greeks, I left at home, caged up, as it were, in the library, as books reserved for more serious studies. At least I should have had nothing to do with them, had I not by chance run across a certain little oration of Isocrates, wherein he inveighed against the sophists with no less wisdom than urbanity and elegance.

---

[5] The following are examples of prices inscribed by Harvey in his volumes: Fulke, *Ouranomachia* (London, 1571), 4° (probably purchased new), 3 pence; Frontinus, *Stratagemes* (London, 1539), 8° (purchased 1578, probably bound), 20 pence; Gasser, *Historiarum . . . totius Mundi Epitome* (Basle? 1538), 8° (purchased 1576, probably bound), 2 shillings; Duarenus, three legal tracts (Paris, 1564), 4° (probably purchased bound in the late 1570s), 2 shillings, 6 pence; Quintilian, *Institutionum . . .* (Paris, 1542), 8° (purchased 1567, probably bound), 3 shillings, 6 pence; Alciato, three large legal works (Lyons, 1530–2) in folio (probably purchased bound, late 1570s), 10 shillings; Bruele, *Praxis medicinae theorica* (Antwerp, 1585), folio (probably purchased 1590s, bound), 10 shillings.

But I dropped Isocrates immediately, and returned daily to my habitual and marvelously exhilarating association with the Latin authors. And with them, indeed, I lived so familiarly and assiduously (I would like you to imagine this, my dear listeners) that now your friend Tully invited me to breakfast, now Julius Caesar himself to lunch, now Virgil to dinner, now the others to their desserts, which were very delightful for a change. I shall not relate what dishes, what courses, what banquets, what delicacies, sweetmeats each one served. . . . But I shall speak of the words of my hosts, which were the equivalent of the sweetest condiment.

All the others uttered conversation that I thought delightful, and indeed nothing less than pure delight; but Cicero alone (my heart leaps at the recollection) had the sweetest voice to invite me, the clearest to entertain me, and far the pleasantest to dismiss me. On the tongues of the others I thought there dwelt the Muses, and the Graces, and that 'Marrow of Persuasion' so celebrated by the ancients, but in his utterance there was an indescribable distinction, more perfect and divine than Apollo himself and Minerva.[6]

In a letter written about 1573 to Arthur Capel, who had matriculated at Trinity College, Cambridge, in 1571,[7] Harvey tells us more about his books and his habit of circulating them among favoured friends and students in whom he has especial interest, so that they, too, may derive benefit:

M. Capel, I dout not, but you have ere this sufficiently perusid, or rather thurroughly red over thos tragical pamflets of the Quen of Scots: as you did not long ago that pretti elegant treatis of M. Chek against sedition: and verry lately good part of the Mirrur for Magistrates: thre books iwis in mi judgment wurth the reading over and over, both for the stile and the matter. Now, if your leisure wil serv you, (for truly I praesume of your good will) to run thurrough ani part of M. Ascham, (for I suppose you have canvassid him reasnably wel alreddi) or to hear the report of the furius outragies of Fraunc in Inglish, or to read over the Courtier in lattin (whitch I would wish, and wil you to do for sundri causis) or to peruse ani pes of Osorius, Sturmius, or Ramus, or to se ani other book, ether Inglish, or lattin, that I have, and mai stand you in stead, do but cum your self, or send on for it, and make your ful account not to fail of it. Perhaps you wil marvel at the sudden proffer: In good sooth mi purpose is nothing els, but this: I wuld have gentlemen to be conversant and occupied in thos books esspecially; whereof thai mai have most use, and practis, ether for writing, or speaking, eloquently, or wittely, now, or hereafter. Fare wel, good M. Arthure, and account of lerning, as it is, to be on of the fairist, and goodliest ornaments that a gentleman can bewtifi, and commend him sclf with al. This morning. In hast.

There is a frend of mine, that spake unto me yesterniht, for mi book of the Quen of Scots. If you have dun withal, I prai you send me it praesently, otherwise he shal for me tarri your leisure. Or if you send it now, assure your self to have it again at your pleasure. Iterum vale.[8]

[6] Gabriel Harvey, *Ciceronianus*, ed. Harold S. Wilson, tr. by Clarence A. Forbes (Univ. of Nebraska Press, Nov. 1945), pp. 45, 47.

[7] Venn and Venn, *Alumni Cantabrigiensis*, Part i (Cambridge, 1922) records that Arthur Capyll, son of Sir Henry, matriculated Fellow-Commoner from Trinity, Michaelmas 1571. Harvey may have been Arthur Capel's tutor.

[8] Sloane MS. 93, fols. 90ᵛ–91. This has been transcribed by Edward John Long Scott in *The Letter-Book of Gabriel Harvey*, pp. 167–8.

Harvey's large stock of valuable books and his habit of embellishing them with annotations is referred to in the Cambridge satiric comedy *Pedantius* of which he is the protagonist.[9] When financial straits, it is suggested, made it necessary for him to sell his books:

Pedantius now asks all men whoever and of whatever sort they may be whether they wish to buy today the most accurate authors of every kind, Greek, Latin, ancient, modern: 'since I have now embellished them enough and beyond for contemplative use in reading, writing, commenting, and have gilded them over with marginal annotations as if they were gems or stars, it now pleases me to return them to a practical purpose'. (IV, iv, 2194–2201)[10]

Although Harvey was frequently in financial straits, there is no reason to believe that he ever tried to sell his books. However, the satire's allusion to Harvey's pragmatism rings very true. His books were for a 'practical purpose' and were esteemed according to their pragmatic value to him in acquiring skills or deepening his understanding and thereby equipping him to function in the world of his day.

Some of his more important volumes of marginalia have been mentioned on pages 148 to 190 above, but these do not cover the full range of his library content. For instance, he kept himself informed about current events on the European continent with three 1571 tracts by Buchanan on the intrigues of Mary, Queen of Scots, with Ascham's discourse on Germany and Charles V (1570?), Chytraeus on Charles V, Ferdinand I, and Maximilian II (1583), Billerbege on Amurathe and the methods of Turkish diplomacy (1584?) ,and *The Turkes Secretorie* (1607), the two last books clearly defining the Turkish threat. In addition, Harvey owned a number of news pamphlets some of which were published by his printer friend John Wolfe.[11]

Harvey's library contained books on the topography of France and of the Scandinavian countries, as well as Turler's guide for the traveller. Thomas's *Historie of Italie* (1561) was so closely read that subsequent marginalia in Florio which appear to indicate an actual visit of Harvey's to the Academy at Florence may be only the result of his study of Thomas's and other writers' descriptions.[12]

---

[9] See p. 69, above.

[10] 'Homines omnes quicunque qualescunque sint, interrogat nunc Pedantius, numquid authores omnis generis exactissimos, Graecos, Latinos, veteres, neotericos coemere velint hodie. Hos cum satis jam superque ad contemplativum usum legendo, scribendo, commentando ornaverim, & annotationibus marginalibus tanquam gemmis aut stellis deauraverim, placet nunc ad activum finem referre'.

[11] For example, *Remonstrance to the Duke of Mayne* (J. Wolf, 1593) and *Articles accorded for the Truce-generall in France* (J. Wolfe for A. White, 1593).

[12] In Florio's *First Fruites* (1578) at the bottom of sigs. P1ᵛ and P2ʳ are the following marginalia below the inscribed symbol for eloquence and the word 'Toscanismo': '[Words missing where page has been trimmed] assemble at three o'clock in the afternoon, in a Hall, purposely appoyntid; where on of them mounteth into a place, callid the Harange, a lyttle hygher, then the rest; and in *his owne Moother* tongue maketh an *Oration of an hower*

Throughout his life Harvey kept abreast of new developments in scientific, medical, and mathematical fields. In 1618 at the age of sixty-five, he purchased John Napier's important new treatise on logarithms. However, neither this nor the 1616 Jonson folio show signs of the studious perusal he gave to earlier acquisitions. But although Harvey's annotating probably diminished in later years, his practice of expanding his library apparently did not cease. His signature is said to appear in a book published as late as 1626[13] (when he was about seventy-six years old).

*longe,* of what matter, and argument himself thynketh best. . . . [This paragraph clearly derives from Thomas; underlinings are Harvey's.]

    I never saw there fare so slender, but any honest gentlemen woold have bene right well contentid therewithall: and if men generally in other places could follow it, the rich should live more healthfully, and the poore fynde more plenty. Besydes that theyr fyne and neate service, the sweetnes of theyr houses, the good and coomly ordre of all thynges, and the familiar conversation of those men every way so curtous, fyne, and clenly; were [page trimmed here].'

Although Harvey uses a few Thomas phrases, most of the content of this paragraph is new and expressed differently, but one cannot be completely certain that Harvey's comments come from his first-hand impressions.

13 Hazlitt mentions this fact at the end of his listing of Harvey's books in the Quaritch, *Dictionary of Book Collectors.* See p. 241, above.

# 6. Conclusion

ONE may wonder what became of Harvey's valuable library after 7 February 1630/1, for he left no will. All of his books may have been sold and dispersed shortly after his death. Certainly no evidence to the contrary exists, and he made no provision for preserving his collection intact. Perhaps he felt that its pragmatic value could best be appreciated by allowing it to be scattered among many individuals.

Shortly after his demise, at least two volumes were acquired by men who would have cherished them for sentimental reasons. His handsome folio Chaucer is inscribed with the signature 'Henry Capell' (of the Hadham Hall family), perhaps a son of Arthur Capel whom Harvey had known since Cambridge days.[14] It was at Hadham Hall in Hertfordshire that Harvey in 1578 presented his *Gratulationes Valdinenses* to the Queen. The other volume is Harvey's copy of Ovid's *Metamorphoses* which bears the signature 'Stephanus Jones' and the notation that it was purchased on 14 January 1632 for the price of three shillings eight pence, a large sum for such a little book, but perhaps it was bought for old time's sake by a Cambridge crony, the Praelector of Philosophy Jones to whom Harvey alludes in *Ciceronianus*.[15]

Today Harvey's books are found widely scattered in more than twenty-five libraries in England and America; nor can one trace the provenance of more than three or four books to a single owner. The largest early groupings are those from the Royal Library and from the collection of the early eighteenth-century barrister and antiquary, Maurice Johnson.[16] Those libraries which

[14] See Harvey's letters to Arthur Capel in Sloane MS. 93, fols. 90ᵛ and 102ʳ.

[15] See p. 31, above. Harvey usually dated the year from 1 January (not from Lady Day). If Jones did the same, this would make his purchase within the year following Harvey's death.

[16] Maurice Johnson (1688–1755) owned Harvey's copy of Florio's *First Fruites* and a volume containing Thomas's *Historie of Italie*, Billerbege, and Jovius, all having copious marginalia. To replace a missing title-page of the Florio, Johnson wrote: 'Florio's Italian Grammar in Italian and English compiled by Seign: John Florio Teacher of the Italian Tongue in London where this book was printed by Thomas Dawson Anno Dni. 1578. It heretofore belonged to the learned and Ingenious Gabrael [*sic*] Harvey Esq. sometime Poet Laureate to her Majesty Queen Elizabeth & most of the old MSS Notes therin & at the end are of his hand writeing'. On the hand-written 'title-page' of the Thomas–Billerbege–Jovius volume, Johnson refers to the 'MSS Notes written by the learned Gabriel Harvey Esquire Author of the Aulicus, Gratulationes ad R[egi]nam Elizabetham and her Majesties Poet Laureat a Gentleman who well understood most languages particularly Greek Latin and Italian and had traveld thro Italy &c.' Was the last statement a surmise of Johnson's based on the Florio marginalia or did he have definite knowledge, unknown today, that Harvey had travelled through Italy? The possibility of such travel is suggested on p. 132, above.

currently have the greatest holdings of Harvey marginalia have acquired them at various times and from various sources.

A portrayal of Gabriel Harvey's library is in effect the truest portrait of the man himself. He is here revealed not by the acid pen of a contemporary but through his treasured books and voluminous records—the man of erudition who aspired to make his personal imprint upon the sands of history, left little mark, but retained indomitable spirit and health well beyond the limits of most men of his day. In addition, Harvey's marginalia constitute a unique and intimate memorial of his age. His notes tell us what he thought and felt and dreamed of. If Harvey was right in believing that each of us is a microcosm of all mankind.[17] then this most prolific of annotators has left us a rich account of the intellectual life and interests of his times.

I have in this study reversed the usual procedure of the biographer: to select a person who has in some respect become outstanding and then to examine his training and the forces which have shaped him towards this achievement. Instead I have chosen as subject a man who systematically and assiduously shaped and trained himself for greatness but somehow failed to achieve it— whether through force of circumstances or from faults in his personality. I prefer to believe that the former were primary and the latter a consequent reaction. May Harvey's too long accepted caricature soon be replaced by a more rounded portrait fashioned by the goddess Mnemosyne, wise preserver of so much of his library and marginalia!

[17] On sig. G2ᵛ of Guicciardini, Harvey writes: 'Unus instar omnium: praeclarus quisque in seipso microcosmus'.

# APPENDIX A

## D/ABW 18/180   Will of Christian Harvye of Walden, 3 December 1556   Copie proper

In the name of God Amen the third daye of December in the y[ea]re of our lord God 1556 I Christian Harvye of Walden in the countie of Essex wydowe lieying sycke in body but in good mynde & memorye thanckes be unto almightye God/ do make & ordeyne this my p[rese]nt Testament & laste will in manner & for[m]e followyng. Firste I gyve & bequeth my soule in to the handes of Almighty God & my bodye to be buryed in the church yarde of Walden aforesaid by my late husband Richard Harvye/ And as touchynge my landes both Free and copye situate lying and beyng in Walden aforsaid this is my devise   I will & bequethe them unto Elizabeth Stockbredge my doughter the wife of Richard Stockbredge of Stansted Mountefichet to have & to hold the said landes bothe free & copye with all & singular thapperteninge/ after the deceasse of me the said Christian & unto the saide Elizabeth my doughter hir heres & Assignes for ever with the profites & comoydeties ther comyng & growyng upon the said ground   And as consyrnyng my moveables implementes & household stuffe this is my mynde I will & bequethe to Gabriell Harvey the sonne of John Harvye my sonne/ a great Brassepotte with vjˢ viijᵈ of good lefull mony of Inglond/ Item I bequethe to Alice Harvye the doughter of the saide John Harvye/ my folse sable with xxˢ of good & lefull monye Item I will to Marye Harvye hir Sister/ a Brasse pan with ij eares & vjˢ viijᵈ/ Item bequeth to Marye Harvye the doughter also of John Harvye/ a lytell Brasse potte & vjˢ viijᵈ Item I will also to Elizabeth Ewkelye the doughter late of John Ewkelye a vjˢ viijᵈ   Item I bequeth to Maric Stockbredge a Brasse pott with iii longe fete which was my Grandfathers   And I will to the same Marie Stockbredge my Bullocke/   Item I will & bequethe to my said Sonne John Harvye my Fetherbed & bolster with a Matteres/ And I will & bequeth to my said doughter Elizabeth my bed which I lye on with ij mattereses & ij lytell bolsters with a litell Coverlett & a great Kettell with a pan to cole in worte¹ And I will my cubbord in the hall to go with my house Item I will to Ales Harvye the wife of my said Sonne John Harvye/ my beste Gowne & my beste petycot² with ij railes of the best   And the reste of all my weryng gere bothe lynen & wollen I will & bequeth them unto the said Elizabeth my doughter   Item I will &

---

¹ 'Wort' was a malt liquor formed in brewing and used occasionally for medicinal purposes as in the treatment of an ulcer. The wort pan was a shallow pan in which the infusion of malt was placed to cool.

² 'kyrtell' has been scored through in the original and 'petycot' written above it.

bequethe to my Sister Jone Cockoroe a xiij$^s$ iiij$^d$ [i.e. 1 mark]   Item I bequethe
to John Korke a iij$^s$ iij$^d$ Item I will to John Jennynge vij$^s$   And I will that one
combe [4 bushels] of Barley or Malte to be divided amonges the pore householders in Golstrete in Walden aforesaide   Item to the pore in the almessehouse
toward ther care xij$^d$ Item to every one  of my god child[r]en   iiij$^d$  a pece
The rest of all my goods chatells bothe reall & personall moveable & immoveable
whatsoever  not  before  gyven  nor  bequethed  I  will  them  to  be  equally  &
indifferently   devided bytwene my said Sonne John Harvye & Elizabeth my
doughter parte & parte like of the same   And I ordeigne and make the saide
John Harvye & Richard Stockbredge my Sonnelaw myne Executores/ these
beyng witnes  John Brampton  Richard lyon  John White & Nicholas Pratt
And I have Surrendered my copy hold land in to the handes of John Brampton
in pre[senc]e of Richard lyon & John White to the use of this my said will
& testament.

# APPENDIX B

The following are three poems of Harvey's in the *ACADEMIAE CANTABRIGIENSIS LACHRYMAE TUMULO Nobilissimi Equitis, D. Philippi Sidneij Sacratae PER Alexandrum Nevillum*, London, 1587 (16 Feb.). Two are addressed to Sidney, the third to Leicester (see pp. 78–9, above):

Academiae Cantabrigiensis lachrimae, in obitum clarissimi Equitis, Domini
Philippi Sidneii.

O Fili, dilecte Deo, dilecte parenti,
(Heu, nimium, nimiumque orbae, viduaeque parenti.)
Et *Musis* dilecte bonis, *Genijsque* beatis,
Et Virtuti ipsi dilecte, & Amoribus ipsis,
Et Pietati ipsi, quam pura mente colebas:
Chara orbis soboles, matris grande incrementum,
Divinum ingenium, suadaeque, artisque medulla,
Tantarumque sperum Haeres, tantorumque bonorum:
Quis dolor, ah quantus, quantus dolor, optime fili,
Dulce tuum amisisse caput, nimiumque suavi
Deprivari anima, charitum quam gratia tota
Mirifice insignivit, & omni ornavit honore:
Sive honor armatus peteretur, sive togatus,
Sine alius quicunque illustri stemmate dignus?
Siccine te cunctae Virtutes, totaque *Pallas*,
Omnisciusque *Hermes*, & quicquid nostra *Lycaea*
Ornamentorum decantant, paxque, duellumque,
Et validus *Mavors*, infoelix bestia *Mavors*,
Dotibus instruxere suis, et protinus esses
Ille ingens Marcellus, ad omnia summa creatus,
Solis sed castris summi fabrifactus olympi?
Ecce meus nuper quam fortunatus Alumnus,
Sive aula, sive urbe incederet inclytus, unum,
Unum omnes ut spectarent, blandoque viritim
Ore salutarent; addentes saepius, hic est:
Aut hic, aut nemo generosae gloria stirpis:
Gemma iuventutis: flos aulae: sidus honoris:
Laurea virtutis: magnorum splendor avorum.[1]

At nunc: sed praestat lachrymarum flumina tanta
Littoribus cohibere suis, luctuque refertum

---

[1] The printed original is not divided into stanzas.

Obvoluisse caput, linguamque dicare stupori;
Unica qui veri facundia vivida luctus,
Tantorum solet Heroum deflere sepulchra.
Quid crinem incomptum; aut sparsos sine lege capillos;
Turgentesve genas: aut pullas denique sordes
Conquerar? aut fundam gemibunda voce querelas,
Qualia vix tragicum sensit lamenta Theatrum?
In cassum furor: Ah perijsti splendida proles,
Quam solis poterunt lachrymis decorare Camenae;
Artibus haud ullis animare, aut fletibus ullis:
Hippolitum ut quondam, Tuberonemque, Aviolamque,
Totque alios, subito redivivos fabula finxit.
Dulce decus tamen affari quaeruloque tremore
Compellare iuvat, planctumque abstergere planctu.
Dum tu illustrem animam generoso in corde fovebas,
Ac validum in pulchro gestabas corpore pectus;
Ecquaenam mater celebratarum undique matrum,
Totius augustas Europae habitantium Athenas,
Tale orbi potuit tantumque ostendere pignus?
Pignus honoratae plenum virtutis, & artis,
Seu *Mavors*, seu *Mercurius* certare iuberet.
Heu virtus, nulli virtuti funera parcunt.
Heu lachrymae, desunt tanto, tantoque dolori:
Et stupeo magis, atque magis venerabile bustum,
Inclyta quo sese, quasi mortua, fama recondit,
Ac monumento uno centum monumenta reponit,
Queis *Sidneianum* decus inclaresceret orbi,
Socraticisque alis cunctas volitaret in oras.
Quid faciam? Divine pater qui condere solus
*Sidneiana* potes miracula rerum, hominumque,
Redde novos Musis, immortalesque *Philippos*,
Qui nostrum illustrent fulgenti nomine nomen,
Nec squallere sinant literatae gentis Alumnos.
Interea horribili mactatum strage Philippum,
Heu iterumque, iterumque, etiamque, etiamque dolendum;
Aeterno Elogio, quod possumus, ac debemus,
Luctisonisque himnis, suadaque ornabimus omni:
Dum Musae calamos, aut linguas gratiae habebunt,
Aut virtuti aderit comes inseperabilis Hermes,
Magnanimos alto decorans Heroas honore.
Ah, magis, atque magis iuvat incubuisse sepulchro,
Et Thamesi, & Nilo, lachrymarum, nobile saxum
Alluere, impositumque cadaver plangere versu:
Hic situs utriusque Lycei filius ingens,

Et coelo, & terris, hominique, Deoque, chorisque
Angelicis, supra reliquos optatus alumnos:
Et vitae, & morti charus, cunctisque cupitus:
Coelicolis absens primum, praesensque Britannis,
Coelicolis subito praesens, absensque Britannis:
Hic, illic, absens praesens, vivusque, iacensque,
Aequales praeter, multum, multumque petitus.
Tanta erat in valido divina, humanaque virtus
Pectore, & omnigenos sic delibavit honores.
O dulce ingenium, nimium nisi dulce fuisses,
Dignius & coelo, quam tetro carcere mundi,
Indigni tanta virtute, ac numine tanto.
Sors tua splendidior: nostra est iactura, scholarum
Extinctum quibus est lumen praelustre duarum.
Sancte Deus, miserere mei, miserere sororis,
Et *Sidneiani* splendoris lumina plura
Coelitus in nostris quamprimum accende Theatris.
                                                    G.H.


De subito & praematuro interitu Nobilis viri, Philippi Sydneij, utriusque
militiae, tam Armatae, quam Togatae, clarissimi Equitis: officiosi amici
                                    Elegia.

Post alios, aliosque sepultos ordine amicos,
    Tu quoque amicorum prime, *Philippe*, iaces?
Aut quorsum hinc *Hermes*, hinc *Mars*, hinc patria virtus,
    Hinc omnes charites, spesque, laresque dolent?
Nimirum occubuisti, animae pars aurea nostrae:
    Et merito charites, spesque, laresque dolent.
Angliades, Belgaeque dolent, longumque dolebunt,
    Magnanimus quibus est fusus honore cruor.
Et socer, & coniux, & avunculus: & quotus eheu
    Non dolet affinis, sanguineusve tuus?
Et Cantabrigia, Oxonium, Musaeque remotae:
    Utraque castra dolent: Arma, togaeque dolent.
Nobilium ecce cohors: en ipsa Hera, magna suorum
    Curatrix procerum, terque, quaterque dolet.
Patria tota, Britannia tota, Hollandia tota,
    Utriusque Arete tota corona dolet.
Prae cunctis dolet, eternumque dolebit amicus,
    Unum qui funus, funera mille putat.
Ah, quantus, quantusque fuit, potuitque fuisse,
    Tam praeclari usus, magnificique viri?

Ornavere libri iuvenem: maturior etas
    Quas non praeconi multiplicasset opes?
Area larga nimis, nimium nimiumque dolendi,
    Seu commune agito, seu proprium ipse bonum.
Seu Martis, seu Musarum seu civica castra
    Cogito, seu iusta consociata fide.
Crescite ab hoc Cedri, Lauri, Palmaeque sepulchro:
    Magnarum hic situs est unicus Alpha sperum.
Immortalem animam venerabitur inclyta virtus,
    Belgarum, aut Britonum dum recoletur honor.
Hoc validae, dulcique animae, lachrymabile carmen,
    Devoto zelo, squallida Musa dicat.[2]


Ad illustrissimum Dominum Comitem Leicestrensem protheoreticon.

Dicitur audaces fortuna iuvare labores,
Ceptaque magnanimi decorare ingentia cordis.
Quis neget, interdum generoso pectore raptos
Ausibus immensis cumulasse trophea stupenda?
Maxima terrarum ostenta, ac miracula quondam
Sic incredibiles coacervavere triumphos.
Magnus *Alexander* validum sectatur *Achillem*:
*Caesar Alexandri* duplicat, triplicatque vigorem:
*Caesaris* exemplo, quot speravere monarchae
Se quoque praerapido absorpturos, cuncta furore?
Successum est non nunquam: & forte audacia sola
Nobile hiperbolici transcendit culmen honoris.
Ast certum mortale nihil: cumque omnia mundi
Fluxa ruunt, tum bellorum impense alea ludit,
Que contingenti innumeros decepit elencho.
Respice tot validas acies, turmasque potentes,
Invictosque duces, heroasque ad summa creatos,
Nec vi, nec virtute, nec ullo Marte, nec arte,
Nec proprio, nec communi molimine tutos.
    Proh dolor: heu virtus: eheu fiducia cordis
Mascula, & intrepidi mens imperterrita Martis:
Aspice magnanimi *Sidneij* exsangue cadaver,
Vivida quem dudum subvexit ad aethera virtus.
Sic volvit fortuna rotam sic *Caesaris* ausus
Cum vult, confundit, cum vult extollit ad astra.
Non animus, non consilium, non techna *Philippo*,
Non ars defuerat, non ulla in pectore virtus:

---

[2] In some copies this poem is unsigned as above; in other copies it is followed by the inscription 'Cantab. 10 Novemb. 1586'.

Solius abfuerat Martis fortuna faventis,
Que modo dat montes, modo celsum sternit Olympum,
Nec quicquam, aut quenquam solida ratione tuetur.
Mercurium Base quadrata statuere vetusti:
Fortunamque pila: quam raro fallitur ille?
Sed quoties fallit? quoties sed fallitur illa?
Ah quantum, quantumque fefellit nobile pectus
Magnanimi iuvenis, summos satagentis honores,
Ac valido gladio decus immortale operantis:
Dum sclopa terribilis, violento intorta furore,
Horrifico foderet generosum vulnere corpus,
Atque uno tot virtutes contunderet ictu?

O te non fallat, cura infinita bonorum.

                                                    G.H.

# APPENDIX C

Additional epigrams from Thomas Bastard's *Chrestoleros*, 1598 (as referred to on p. 121 above):

### Liber Secundus: Epigr. 1 (sig. C6ʳ)

Thou which deluding raisest up a fame,
And having shewd the man concealst his name:[1]
Which canst play earnest as it pleaseth thee,
And earnest turne to jest as neede shall be,
Whose good we praise, as being likt of all.
Whose ill we beare, as being naturall,
Thou which art made of vinegar and gall,
Wormewood, and *Aqua fortis* mixt with all.
The worldes spie, all ages observer,
All mens feare, fewe mens flatterer.
Cease, write no more to aggravate thy sinne:
Or if thou wilt not leave, now ile beginne.

### Liber Quartus: Epigr. 11 (sig. G2ᵛ)

*Publius* hath two brothers fowle and cleane:
The fowle is honest, and the cleane a foole:
He in the middest maketh up the meane,
Sitting in vertues place: so saith our schoole.
Of his extremes neither alowe he can,
The cleane foole, nor the filthy honest man.[2]

[1] I take this line to mean that Nashe has shown Harvey's human foibles but has concealed his reputation (or worth).

[2] If this epigram refers to the Harvey brothers, which seems likely, the 'cleane foole' would be the younger brother Thomas who never attended the University nor achieved anything noteworthy so far as is known. The filthy honest man would be Richard who although respectable may have been filthy in appearance or in the use of abusive language. John, of course, had died in 1592, six years before this volume was published.

# APPENDIX D

## Additional books probably owned by Harvey

*AESOP. *Aesopi Phrygis Fabulae* . . . (*Aesopi* . . . *vita a M. Planude conscripta*)
. . . . [*The Fables and Life of* Aesop translated by A. P. Manutius]. Lyons, I.
Tornaesius, 1582 (or earlier edn.).
16°

On sig. a3ᵛ of Simlerus, *Epitome Bibliothecae Conradi Gesneri*, Harvey
places a + next to 'Aesopi Phrygis vitam & fabulas, Maximus Planudes
conscripsit', probably indicating a book that he either owns or contemplates
acquiring. On sig. y4ᵛ of Simlerus next to the entry for 'Maximus Planudes'
Harvey underlines *Aesopi fabulatoris vita & apologi*. Underlinings in this
volume evidently indicate books Harvey has read, for he often notes next to
them that he has also read other works by the same man. In the marginalia
in sig. F3ᵛ of Xenophon, Harvey refers to 'Meo Aesopo' and on sig. L8ᵛ of
Erasmus to 'Aesopici Apolog[ia]' and on sig. D5 of Livy, *Romanae Historiae*
to 'fabulis . . . Aesopi'.

*ALTHUSIUS, Joannes. *Iuris Romani libri duo ad leges Methodi Rameae conformati:
& tabula illustrati*. Basle, 1586.
8°

On fol. 37ᵛ of Domenichi, *Facetie et Motti*, Harvey alludes to 'Meus
Althusius'. On fol. 173ʳ of Guicciardini, *Detti et Fatti*, among the law volumes
('pro privata professione') he would least like to do without is 'Althusii
Ars iuris'. It is referred to again on fol. 159ᵛ.

*AUGUSTINE, Saint, Bishop of Hippo. *Divi Aurelii Augustini de civitate Dei
libri XXII*. Basle, Hier. Frobenius & Nic. Episcopius, 1555.
Folio

Mentioned frequently in various marginalia together with chapter and
line references. See especially Livy, *Romanae Historiae*, sig. B2ᵛ, C4, G5, and
M5. About 1570 Harvey's references to this work are so frequent and so
specific that it is likely he owned it, although he may have had a 1520 or
earlier edition rather than the one listed above.

*BION, of Smyrna. Βίωνος Σμυρναίου Εἰδύλλια, *Idyllia*. [Geneva], Henri
Estienne, 1566.
Folio

Harvey makes frequent allusion to Bion's epitaph on Adonis (i.e. his
third idyll), for example, on sig. diiᵛ of Quintilian, C4ᵛ of Du Bartas, *Triumph*

*of Faith*, nn4 of Pindar, and on fol. 211$^r$ of Guicciardini. In the latter passage he lists 'Bionis Epitaphium' together with a number of other books, most of which we know he owned.

*BLUNDEVILL, Thomas. *M. Blundevile His Exercises, containing sixe Treatises . . . verie necessarie to be read of all yoong Gentlemen that . . . are desirous to have knowledge as well in Cosmographie, Astronomie, and Geographie, as also in the Arte of Navigation. . . . To the furtherance of which Arte of Navigation, the said M. Blundevie speciallie wrote the said Treatises. . . .* London, John Windet, 1594.

4°

In Hood's *The marriners guide* (sig. f2$^v$) Harvey alludes to 'M$^r$ Blundevilles newist art of navigation. this 1594', and on sig. 6$^v$ of Grafton to 'Blundeviles new Kalendar for ever in his newest Art of Navigation. this 1594. A notable general Almanack; & a pregnant Analysis of most points in these Astronomical tables.' Harvey prided himself on his knowledge of navigation and thought highly of Blundeville so it is likely that he would have acquired the above.

*BODIN, Jean. *Jo. Bodini Andegavensis, De Republica Libri Sex, Latine Ab Autore Redditi, Multo Quam Antea Locupletiores. Cum indice copiosissimo.* Lyons and Paris, Jacob DuPuy$^s$, 1586.

Folio

Harvey refers to 'Bodinus de Republica' on sig. xiiij$^v$ of Duarenus, *De Sacris Ecclesiae*, to 'Bodino L[inea].4 de Republica C[aput].2' on sig. Aiiij$^v$ of Gauricus, and elsewhere in his marginalia both frequently and specifically.

*CARDANUS, Hieronymus. *Liber de libris propriis.* Lyons, Gryphius, 1545.

8°

In Simlerus's edition of the Gesner bibliography on sig. n4$^r$ Harvey underlines this title. In Gauricus on sig. Rii$^r$ Harvey writes: 'Cardani . . . Item alia brevissima de Libris propriis', and on sig. GG4$^v$: 'Cardani autem brevissima Apologia, de libris propriis'. Harvey frequently compares other men's writings to those by Cardano in the same field and almost invariably finds the latter sounder or more knowledgeable. There are allusions to so many of Cardano's writings that it seems likely Harvey had read most of his works and undoubtedly owned a number of them.

—— *Hieronymi Castellionei Cardani Medici Mediolanensis de malo recentiorum medicorum medendi usu libellus.* Venice, 1536.

8°     Brit. Lib. 1038.e.4

No signatures, but MS. notes in hand that strongly resembles Harvey's. Content of notes suggests Harvey's habits of annotations. In Simlerus's bibliography Harvey has underlined this title.

*—— *De subtilitate lib. 21. Primus de principiis. 2. De elementis 3. De Coelo 4.*

*De luce & lumine 5. De mixtione & mixtis imperfectis seu metallicis 6. De metallis
7. De lapidibus 8. De plantis 9. De animalibus quae ex putredine generantus 10.
De perfectis animalibus 11. De hominis necessitate & forma 12. De hominis natura
&c.* . . . Nuremberg, Petreius, 1550.

Folio

Harvey underlines this title in his Simlerus bibliography and refers to
many of the above books in his marginalia. In Ramus, *Dispositio Regularum*
(sig. C3^r) Harvey writes: 'Inutiles Cardani subtilitates negligenda: sola
pragmatica. et cosmopolitica curanda.' In his commonplace book (Add. MS.
32,494) on fol. 3 he refers to 'Cardani L.12 subtilitatum: De hominis natura:
838' and on the title-page of the Machiavelli tract at the Ministry of Defence
Library he also refers to Cardano's 'De subtilitate'.

*CRESPIN, Jean. *Lexicon Graeco-Latinum, repurgatum studio E. G[rant]*. London,
H. Bynneman, 1581.

8°      STC 6037

Near the end of this thick Greek–Latin dictionary are printed Harvey's
two orations on the study of Greek, entitled *De Discenda Graeca Lingua*
(9 pp.) identified by T. W. Baldwin in *Shakespere's Small Latine & Lesse
Greeke* (Univ. of Illinois, 1944), i, 436–7. It is likely that Harvey owned this
book since he probably owned at least one copy of each of his printed works.
His copy of each published work except the above and *Three Proper, and
wittie familiar Letters* has already been identified.

*DIGGES, Leonard. *A Prognostication everlastinge of righte good effecte, fruictfully
augmented by the auctour, contayning* . . . *rules to judge the Weather by the Sunne,
Moone, Starres, Comets, Rainebow, Thunder, Cloudes, with other extraordinarye
tokens, not omitting the Aspects of Planets, with a briefe judgement for ever, o,
Plenty, Lacke, Sickenes, Dearth, Warres &c.* . . . *To these* . . . *are joyned divers
General pleasant Tables, with many Compendious Rules.* . . . *Published by Leonard
Digges Gentleman. Lately corrected and augmented by Thomas Digges his Sonne.*
London, Thomas Marsh, 1578.

4°      STC 6865

Thomas Digges may have been a personal friend of Harvey's (see Blagrave
title-page and Dionysius Periegetes, fol. 3) and there are many references in
Harvey's marginalia to his esteem for his works. On sig. Ai^v of Grafton,
Harvey refers to 'Digges general prognostication' and on sig. Aii to 'Digges
perpetual prognostication'. There are also several references to Digges'
*Stratioticos* (1579) e.g. on sig. N7^r of Frontinus. Harvey may also have owned
this work.

*DURANDUS, Gulielmus ('Speculator'). *Speculum Juris.*

8°

A thirteenth-century synthesis of Roman and ecclesiastical law by the
Bishop of Mende. Contains a general explanation of civil and canonical

procedure. There were many editions of this highly esteemed work. Harvey refers to 'Speculator meissimus' on fol. 37ᵛ of Domenichi and in Duarenus on sig. A6ʳ writes: 'Duarenus: Vigelius: Speculator: all three in 8°; & worthy to be fayerly bound together in one volume'. In Guicciardini on fol. 173ʳ he mentions 'Durandi speculum pragmaticum' for use in his law practice but one cannot be certain which octavo edition he owned.

GEMMA FRISIUS. *Gemmae Frisii Medici Ac Mathematici De Astrolabo Catholico Liber quo latissime patentis Instrumenti multiplex usus explicatur, & quicquid uspiam rerum Mathematicarum tradi possit continetur.* Antwerp, Joan. Seelsius, 1556.
    8°    Brit. Lib. 531. f. 8
    No signatures but copious notes, charts, and diagrams. The manner of note making seems remarkably like Harvey's and the handwriting resembles his secretary hand (as found in the page of closely written notes preceding the title of Hopperus). In the marginalia of Blagrave, Harvey refers to Gemma and possibly specifically to the above book on sigs. d5ᵛ, d6, and d7ᵛ of the Sacrobosco volume.

*GODEFROY, Denis (Gothofredus). *Codicus Dn. Justiniani . . . repetitae prae-lectionis libri XII. Commentarius D. Gothofredi illustrati.* 1585, etc.
    8°
    On sigs. Ii2ᵛ and Ii3ʳ of Hopperus, Harvey praises 'Gothofredi Institutiones' and indicates that it is an important source for him. In Guicciardini on fol. 159ᵛ he writes that 'Gothofredi Institutiones' is one of the books that he has at his fingertips ('ad unguem').

*[HARVEY, Gabriel and SPENSER, Edmund]. *Three Proper, and wittie familiar Letters: lately passed betweene two Universitie men: touching the Earth-quake in Aprill last, and our English refourmed Versifying. With the Preface of a wellwiller to them both.* London, H. Bynneman, 1580.
    4°    STC 23095
    Second title (on sig. G2ʳ): *Two Other very commendable Letters, of the same mens writing: both touching the foresaid Artificiall Versifying, and certain other Particulars: More lately delivered unto the Printer.* It is probable that this book was in Harvey's library as he seems to have owned a copy of each of his printed works. At least one of each has been so identified except for the above and *De Discenda Graeca Lingua* (see Crespin, above).

*JULIAN THE APOSTATE. *Juliani Imperatoris Misopogon et epistolae, graece latineque nunc primum edita & illustrata a Petro Martinio . . . addita est praefatio de Vita Juliani eodem authore.* Paris, A. Wechelus, 1566.
    8°
    Harvey frequently refers to the *Misopogon* (e.g. on sig. B4ᵛ of *The Turkes Secretorie* and C2ʳ of Wilson's *Art of Rhetorike*, and on fol. 178ʳ of Guicciardini

to this edition). Harvey was a great admirer of the Emperor Julian and of his ironical use of the mock encomium.

\*LANCELOTTI, Joannes Paulus. *Corpus Juris Canonici emendatum et notis illustratum: Gregorii XIII jussu editum. . . . Appendice P. Lanceloti . . . adauctum . . . Accesserunt novissime Loci Communes . . . ex ipsis Canonibus collecti, etc.* 4 Pt. Lyons, 1591.

4°

Harvey refers frequently to Lancelot's *Institutions* and on sig. A6 of Duarenus's *De Sacris Ecclesiae* writes: 'Lancelots Institutions, & the Alphabetical, or memorative, compendium of Petrus Ravennas in 4°, woold hansomly be combined in on fayr Book'.

\*MACHIAVELLI, Niccolo. *I discorsi di N. Machiavelli sopra la prima deca di T. Livio.* Palermo, gli heredi d'Antoniello degli Antonielli [London, John Wolfe], 1584.

8° STC 17159

Harvey makes many specific references to Machiavelli's *Discorsi*, e.g. to 'Machiavel. Discorsi lib. 2, cap. I' on fol. 20$^r$ of the Commonplace Book (Add. MS. 32,494) and to 'Discorsi cap. 43 lib. 3' on fol. 70$^r$ of Florio. It is likely that Harvey acquired the above surreptitiously printed John Wolfe edition, although, since the reading of Livy and the *Discorsi* with Thomas Preston at Trinity Hall (see p. 150, above) probably antedated 1584, Harvey may have owned an earlier Italian text.

—— *I Sette Libri Dell'Arte Della Guerra Di Niccolo Machiavelli Cittadini, et Secretario Fiorentino.* [London, John Wolfe], 1587.

8° STC 17163.2 Folger 17163.2

Annotations in at least three hands. Those in Italian on front flyleaf and in the text margins look very much like Harvey's hand. Accompanying the latter are several instances of the use of a rudimentary flower symbol (also found in other volumes which seem to be Harvey's). Notes on title-page apparently were inscribed by other owners. On sig. N6$^v$ of Frontinus, Harvey alludes to 'Machiavels Rules in the end della guerra'. And on N7$^v$ he refers to 'Martiall Praecepts these of order, omitted in Machiavel della guerra'. In other marginalia Harvey makes reference to Machiavelli's *Life of Castruccio*, '*Il Prencipe*', '*Lasino doro* [the spellings of John Wolfe's 1584 and 1588 edd.]', *Florentine History*, *Mandragola*, and *Clitia*. Possibly Harvey owned some of these as well.

\*MACROBIUS, Ambrosius Theodosius. *Hoc volumine continentur Macrobij interpretatio in somnium Scipionis a Cicerone confictum. Eiusdem Saturnaliorum libri septem. Haec omnia Nicolaus Angelius . . . summa diligentia correxit, imprimique; curavit.* Florence, P. Iuntae, 1515.

8°

Sig. B4ᵛ of *Ciceronianus* refers to Harvey's reading of Macrobius' *Dialogues of the Saturnalia*, presumably in his library at Walden.

★OSORIO DA FONSECA, Jeronimo. *The five Bookes of . . . H. Osorius, contayninge a discourse of Civill and Christian Nobilitie . . . Translated out of Latine into Englishe by W. Blandie.* . . . London, T. Marsh, 1576.

4°

In a letter to Arthur Capel (Sloane MS. 93, fol. 90ᵛ), Harvey offers to lend Capel any of a number of books which he has including Osorius, Sturmius, and Ramus. On sig. B4ᵛ of *Ciceronianus*, Harvey refers to reading Osorius's oration *De Gloria*. Harvey could have owned the above translation or perhaps the Latin octavo edition of 1571.

PHILBERT DE VIENNE. *Le Philosophe De Court.* Paris, Estienne Groulleau, 1548.

16°       Brit. Lib. 231.k.40

No signatures but the book has been trimmed at the top. The inscription 'Utile nihil quod non honestum', is found near the top edge of the title-page in a hand that markedly resembles Harvey's Italian style. There are a few underlinings, especially in the text of the prologue. In one of Harvey's letters from Cambridge (Sloane MS. 93, fol. 42ᵛ–43) he refers to 'Philbertes Philosopher of the Courte' as one of the books then being avidly read at the university.[1]

[PUTTENHAM, George]. *Arte of English Poesie.* London, Richard Field, 1589.

4°       Bodleian 4° P.21 Art

No signatures but this contains a number of brief marginal notes which seem to be in Harvey's Italian hand. Notes are chiefly single word summaries of subject-matter, e.g. 'lirique', 'elegie', 'comick', 'tragical', 'eglogue', and 'spondeus—', 'trocheus—◡', 'iambus ◡—', etc. There are some underlinings in the text, e.g. on sig. Niiijʳ of '*dittie*', '*odes*', '*epigrammes*' next to which is written in the margin: 'not above 12 verses'.

The marginalia in this copy have (incorrectly I believe) been attributed to Ben Jonson apparently because of a misunderstood prefatory inscription by Edmund Malone which alludes to Jonson's comments in another volume of Puttenham. To confirm the unlikelihood of these marginalia being Jonson's, note that the page numbers alluded to by Malone do not correspond to those in this volume nor is the handwriting like Jonson's Italian hand.

★SOPHOCLES. *Sophoclis Antigone. Interprete Thoma Watsono. J.U. studioso* [*Juris Utriusque studioso*, i.e. student of both laws, canon and civil]. London, John Wolfe, 1581.

4°       STC 22929

[1] I am grateful to Daniel Javitch for suggesting to me that Harvey probably owned a copy of this work.

On the title-page of his Euripides, Harvey notes: 'Huc Sophoclis Antigone: novissima edita a Th. Watsono'. He refers to this book also on sig. F5$^v$ of Gascoigne's *Posies* and on the verso of the Dolce title page where he writes: 'eadem in Sophoclis Antigonem affectio, ab Episcopo Watsono tralatam'. He is evidently confusing (perhaps absent-mindedly) the poet Watson who translated the *Antigone* with Thomas Watson, Bishop of Lincoln, who at one time was chaplain to Stephen Gardiner. On recto of the flyleaf preceding the title-page of Folger 22929 is an inscription in a hand somewhat resembling Harvey's. No signatures or annotations are found in the text.

★STREBAEUS, Jacobus Lodovisus. *De Verborum Electione et Collocatione Oratoria, ad D. Joannem venatorem cardinalem Libri Duo*. Basle, 1539.

4°

In Sloane MS. 93, fols. 54$^r$ f., Harvey quotes from 'Strebaeus libro primo . . . verborum electione, et collocatione oratorja. Cap. 2', and heads the ensuing memoranda 'Fine Notes for mie rhetorique discourses'. On fols. 56$^v$–57$^r$ are notes which are headed 'ex praefatione libri secundi'. Harold S. Wilson in his MS. notes at Victoria University, Toronto, listed the Strebaeus volume as one containing Harvey's marginalia but Wilson did not specify its present location nor whether he himself had inspected it.

★STURMIUS, Joannes. *De amissa dicendi ratione, & quomodo ea recuperata sit libri duo*. Strasbourg, V. Ribelius, 1543.

8°

In a letter to Arthur Capel (Sloane MS. 93, fol. 90$^v$) Harvey mentions Sturmius as one of the writers whose work he owns and will gladly lend to Capel and in *Ciceronianus* (Forbes–Wilson, ed., p. 44) alludes to Sturmius as one of the writers whose work is in his library. On sig. C6 of Humphrey's *Interpretatio Linguarum* Harvey in an annotation of several lines refers to 'Sturmius in libro primo De amissa dicendi ratione'. Probably Harvey owned several of Sturm's works and it seems likely that this title was one of them.

★[TOMMAI] PETRUS, of Ravenna. *Foenix Artificiosa Memoria Clarissimi Juris Utriusque Doctoris & militis. . . .* Venice, Nicolino de Sabio, 1533.

4°

This popular mnemonic tract is mentioned several times in Harvey's marginalia, e.g. in Duarenus, *De Sacris Ecclesiae* on sig. A6 Harvey writes: 'Lancelots Institutions, & the Alphabetical, or memorative, compendium of Petrus Ravennas in 4°; woold hansomly be combined in on fayr Book'.

★VERGIL, Polydore. *Polydori Vergilii Urbinatus De Rerum Inventoribus libri octo*. Basle, Isingrinus, 1545.

8°

On the title-page of Jovius, Harvey refers to 'Polydori Virgilii, L[inea].4. C[aput].2.de Inventoribus rerum'. In Duarenus, *De Sacris Ecclesiae* (sig. Ciii$^v$), Harvey also refers to: 'Polydorus Vergilius l.4.C.2. de Inventoribus rerum.' There is a copy of this book in the British Library (1396.C.1) with notes in at least two hands, one of which may perhaps be Harvey's. See especially 6ᵛ, 8, and U3.

⋆VIGELIUS, Nicolaus. *Juris Feudalis totius absolutissima methodus . . . liber unus: nunc denuo ab auctore auctus. . . .* Basle, 1584.
8º

On sig. A6ʳ of Duarenus Harvey writes: 'Duarenus: Vigelius: Speculator: all three in 8º; & worthy to be fayerly bound together in one volume'. On fol. 187ᵛ of Guicciardini: 'Necessaria etiam ad unguem . . . Vigelii reportorium et ius controversam'. On fol. 173ʳ Harvey mentions: 'Vigelii methodo'.

⋆WECKER, Hans Jacob. *Practica Medicinae Generalis. . . .* 1585.
16º

On fol. 37ᵛ of Domenichi, Harvey refers to 'Weccherus meissimus'. On fol. 39ʳ of Domenichi he writes: 'Mei proprii auctores, Gandinus et Weccherus: Speculator et Domenicus. Spiritus meissimi'. Although the title listed above seems a likely one, Harvey does not specifically indicate which of Wecker's works he owns.

⋆ZWINGERUS, Theodor. *Methodus apodemica in eorum gratiam, qui cum fructu in quocunque; tandem vitae genere peregrinari cupiunt.* Basle, E. Episcopius, 1577.
4º

On sig. A5ʳ (at top of contents page) of Lhuyd, Harvey writes: 'Lege Zwingeri Methodum Apodemica: quatuor Athenarum exempla'. On the title-page of Turler is the annotation: 'Methodus apodemica Zwingeri'. On sig. B4 one finds the very specific reference: 'Zwingerus in Analyse Reip. Atheniensis L.3. Methodi Apodemicae.356. et 368'. On verso of the final page of Florio is a manuscript reference to 'Zwingerus, Methodi Apodemica Lib. 2, 71'.

# APPENDIX E

## Scholarly works dealing with Harvey's marginalia

BOURLAND, Caroline. 'Gabriel Harvey and the Modern Languages', *HLQ*, iv (1940), 85–106 [Corro, du Ploiche, Eliot, Hurault, Lentulo, Percyvall, and *Images of the Old Testament*].

CAMDEN, Carroll, Jr. 'Some Unnoted Harvey Marginalia' in *PQ*, xiii (Apr. 1934), 214–18 [Hill].

HOENIGER, F. D. (for the late Harold S. Wilson). 'New Harvey Marginalia on *Hamlet* and *Richard III*', *SQ*, xvii (Spring 1966), 151–5 [Domenichi].

JARDINE, L. 'Humanism and Dialectic in Sixteenth Century Cambridge: A Preliminary Investigation', *Classical Influences on European Culture:* A.D. *1500–1700* (Cambridge Univ. Press, 1976, edited by R. R. Bolgar, pp. 149–54 [Agricola, Quintilian].

LIEVSAY, John Leon. *Stefano Guazzo and the English Renaissance* (Univ. of N. Carolina Press, 1961), pp. 88–96 [Domenichi, Guazzo, Guicciardini].

MARCHAM, Frank. *'Lopez the Jew', An Opinion by Gabriel Harvey, with some notes* (Middlesex, 1927) [Meier, Euripides, Livy, North].

MARGOLIN, Jean-Claude. 'Gabriel Harvey, lecteur d'Erasme', *Arquivos do Centro Cultural Portugues*, iv (1972), 37–92 [Erasmus].[1]

MOORE, Hale. 'Gabriel Harvey's References to Marlowe', *SP*, xxiii, 3 (July 1926), 337–57 [Frontinus, Hurault].

MORLEY, Henry. 'Spenser's Hobbinol' in *Fortnightly Review*, v (1869), 274–83 [Quintilian].

RELLE, Eleanor. 'Some New Marginalia and Poems of Gabriel Harvey', *RES*, New Series, xxiii, 92 (Nov. 1972), 401–16 [Du Bartas, James VI].

RUUTZ-REES, Caroline. 'Some Notes of Gabriel Harvey's in Hoby's Translation of Castiglione's *Courtier* (1561)', *PMLA*, xxv (1910), 608–39 [Castiglione (Hoby trans.)].

SMITH, G. Gregory. *Elizabethan Critical Essays* (Clarendon Press, Oxford, 1904). i, 46–57 [Gascoigne, *Certayne Notes of Instruction*].

[1] The reader of this paper should be warned that Margolin's transcriptions are not always accurate.

Smith, G. C. Moore. *Gabriel Harvey's Marginalia* (Stratford-upon-Avon, 1913), contains selected transcriptions from twenty-three volumes.

Tannenbaum, Samuel. 'Some Unpublished Harvey Marginalia' in *MLR*, xxv, iii (July 1930), 327–31 [Wilson, *Arte of Rhetorike* and *Rule of Reason*].

Wilson, Harold S. 'Gabriel Harvey's Method of Annotating His Books' in *HLB*, ii (1948), 344–61, reproduces a few marginalia pages [Frontinus, Fulke, Guicciardini, Simlerus].

# THE FAMILY OF GABRIEL HARVEY

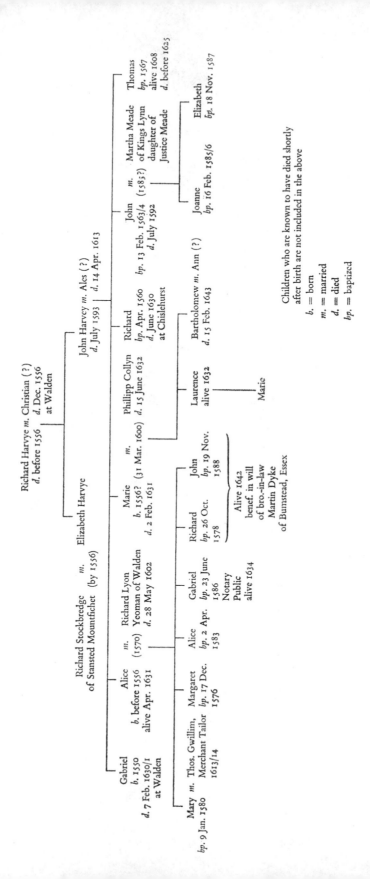

Richard Harvye *m.* Christian (?)  
*d.* before 1556  *d.* Dec. 1556 at Walden

Richard Stockbredge  *m.*  Elizabeth Harvye  
of Stansted Mountfichet  (by 1556)

Richard Lyon  *m.*  
Yeoman of Walden  (1570)  
*d.* 28 May 1602

John Harvey *m.* Ales (?)  
*d.* July 1593  *d.* 14 Apr. 1613

Elizabeth Harvye

Gabriel  
*b.* 1550  
*d.* 7 Feb. 1630/1 at Walden

Alice  
*b.* before 1556  
alive Apr. 1631

Marie  
*b.* 1556?  
*d.* 2 Feb. 1631

*m.* (31 Mar. 1600)  Phillipp Collyn  
*d.* 15 June 1632

Richard  
*bp.* Apr. 1560  
*d.* June 1630 at Chislehurst

John  
*bp.* 13 Feb. 1563/4  
*d.* July 1592

*m.* (1583?)  Martha Meade  
of Kings Lynn  
daughter of Justice Meade

Thomas  
*bp.* 1567  
alive 1608  
*d.* before 1625

Mary *m.* Thos. Gwillim,  
*bp.* 9 Jan. 1580  Merchant Tailor  
1613/14

Margaret  
*bp.* 17 Dec. 1576

Alice  
*bp.* 2 Apr. 1583

Gabriel  
*bp.* 23 June 1586  
Notary Public  
alive 1634

Richard  
*bp.* 26 Oct. 1578

John  
*bp.* 19 Nov. 1588

Alive 1642  
benef. in will  
of bro.-in-law  
Martin Dyke  
of Bumstead, Essex

Laurence  
alive 1632

Bartholomew *m.* Ann (?)  
*d.* 15 Feb. 1643

Joanne  
*bp.* 16 Feb. 1585/6

Elizabeth  
*bp.* 18 Nov. 1587

Marie

Children who are known to have died shortly  
after birth are not included in the above

*b.* = born  
*m.* = married  
*d.* = died  
*bp.* = baptized

# Index